Methods in Molecular Biology

Series Editor
John M. Walker
School of Life Sciences
University of Hertfordshire
Hatfield, Hertfordshire, AL10 9AB, UK

For further volumes:
http://www.springer.com/series/7651

RNA Mapping

Methods and Protocols

Edited by

M. Lucrecia Alvarez

Diabetes, Cardiovascular, and Metabolic Diseases,
Translational Genomics Research Institute, Phoenix, AZ, USA

Mahtab Nourbakhsh

Department of Pharmacy and Biotechnology, German University in Cairo, Berlin, Germany
Department of Plastic Surgery, University Hospital of the RWTH Aachen, Aachen, Germany

Humana Press

Editors
M. Lucrecia Alvarez
Diabetes, Cardiovascular, and Metabolic Diseases
Translational Genomics Research Institute
Phoenix, AZ, USA

Mahtab Nourbakhsh
Department of Pharmacy and Biotechnology
German University in Cairo
Berlin, Germany

Department of Plastic Surgery
University Hospital of the RWTH Aachen
Aachen, Germany

ISSN 1064-3745 ISSN 1940-6029 (electronic)
ISBN 978-1-4939-1061-8 ISBN 978-1-4939-1062-5 (eBook)
DOI 10.1007/978-1-4939-1062-5
Springer New York Heidelberg Dordrecht London

Library of Congress Control Number: 2014942861

Printed on acid-free paper

Humana Press is a brand of Springer
Springer is part of Springer Science+Business Media (www.springer.com)

Preface

About 40 years ago, scientists believed that only less than 3 % of human DNA encodes proteins. The remaining 97 % was labeled as "genetic junk"—DNA that is never transcribed and has no biological functions—and attributed to molecular trial-and-error that has accumulated over the course of evolution. However, it has recently become apparent that more than 85 % of the genome—including most of the non-protein-coding DNA—is indeed transcribed and performs essential biological functions. In fact, the Encyclopedia of DNA Elements (ENCODE) Project has recently claimed that at least 80 % of the human genome has a biochemical function. Therefore, the existence of "junk DNA" is gradually becoming a myth. Nowadays, the continuous discovery of a myriad of non-protein-coding RNAs with regulatory functions, such as microRNAs and long noncoding RNAs, contributes to the unraveling of the remarkable versatility of the RNA molecule. However, the discovery of new functional transcripts is yet to come.

The recent developments in RNA-based technologies in concert with the increasing importance of RNA molecules as biomarkers in diagnostics and therapeutics call for a summary of both modern and traditional strategies for characterization of cellular RNAs. "RNA Mapping: Methods and Protocols" intends to provide not only an update of many of the classic techniques but also an introduction, description, and summary of newer approaches that go beyond the pure biomedical applications. This book is particularly targeted to biochemists, molecular biologists, and any researcher in the life sciences interested in the molecular characterization of coding and noncoding RNAs. The purpose of "RNA Mapping: Methods and Protocols" is therefore to provide instruction and inspiration for all those scientists who are facing the challenges of the discovery and/or functional characterization of RNA molecules for a wide variety of applications ranging from novel biomedical diagnostics to therapeutics and biomaterials.

The book has been organized in two separate parts:

Part I contains 13 protocols for the structural mapping of cellular RNAs. These chapters encompass the definition of RNA boundaries, primary, secondary, and tertiary RNA structure as well as RNA identification and quantification protocols based on structural properties of target RNAs. For example, step-by-step descriptions of different methods for RNA profiling using quantitative real-time PCR or next-generation sequencing have been included in this section.

In *Part II*, 15 protocols focus on a variety of functional elements in RNA including the mapping of internal ribosome entry sites and regulatory mRNA elements as well as the identification of actively translated mRNAs. In addition, this section describes methods for the identification of novel RNA functions with a special focus on microRNA targets prediction and their experimental validation using state-of-the-art technology.

The collection of protocols in this volume is a consequence of the emerging interest in the characterization of cellular RNAs urged by their potential use as diagnostic biomarkers or therapeutic targets. In particular, the biological relevance of microRNAs in human physiology and disease development is highlighted in the 16 chapters focused on methods for their physical and functional mapping.

We would like to thank all the contributing authors for providing excellent chapters and sharing years of their experience in a specific technique, to John Walker for editorial guidance, and to the staff of Humana Press for professional production of this volume.

Phoenix, AZ, USA — *M. Lucrecia Alvarez*
Aachen, Germany — *Mahtab Nourbakhsh*

Contents

Preface .... *v*
Contributors .... *ix*

PART I STRUCTURAL RNA MAPPING

1 Full-Length Characterization of Transcribed Genomic Regions 3
Marc R. Reboll, M. Lucrecia Alvarez, and Mahtab Nourbakhsh

2 Rapid Mapping of RNA 3′ and 5′ Ends 19
Victoria Zismann and Mahtab Nourbakhsh

3 Single Nucleotide Mapping of RNA 5′ and 3′ Ends 27
Mahtab Nourbakhsh

4 Analysis of RNA Secondary Structure 35
Mahtab Nourbakhsh

5 Tertiary Structure Mapping of the Pri-miRNA miR-17~92 43
Steven G. Chaulk and Richard P. Fahlman

6 In Situ Hybridization Detection of miRNA Using LNA™ Oligonucleotides 57
Stefania Cotta Doné and Olga Beltcheva

7 Quantification of miRNAs by a Simple and Specific qPCR Method 73
Susanna Cirera and Peter K. Busk

8 RNA Isolation for Small RNA Next-Generation Sequencing from Acellular Biofluids 83
Kasandra L. Burgos and Kendall Van Keuren-Jensen

9 Sequencing Small RNA: Introduction and Data Analysis Fundamentals 93
Jai Prakash Mehta

10 Measuring Expression Levels of Small Regulatory RNA Molecules from Body Fluids and Formalin-Fixed, Paraffin-Embedded Samples 105
Adrienn Gyongyosi, Otto Docs, Zsolt Czimmerer, Laszlo Orosz, Attila Horvath, Olga Török, Gabor Mehes, Laszlo Nagy, and Balint L. Balint

11 MicroRNA Profiling in Plasma or Serum Using Quantitative RT-PCR 121
Marina C. Costa, Ana Lúcia Leitão, and Francisco J. Enguita

12 MicroRNA Profiling of Exosomes Isolated from Biofluids and Conditioned Media 131
Sweta Rani

13 Isolation of Urinary Exosomes for RNA Biomarker Discovery Using a Simple, Fast, and Highly Scalable Method 145
M. Lucrecia Alvarez

Part II Functional RNA Mapping

14 Identification of Actively Translated mRNAs 173
Marc R. Reboll and Mahtab Nourbakhsh

15 Mapping of Internal Ribosome Entry Sites (IRES) 179
Sarah Mehrtens and Marc R. Reboll

16 Mapping of Protein Binding RNA Elements 187
Marc R. Reboll

17 Purification of RNA-Binding Proteins 195
Birgit Ritter and Marc R. Reboll

18 De Novo Approach to Classify Protein-Coding and Noncoding Transcripts Based on Sequence Composition 203
Haitao Luo, Dechao Bu, Liang Sun, Runsheng Chen, and Yi Zhao

19 Computational Methods to Predict Long Noncoding RNA Functions Based on Co-expression Network 209
Yi Zhao, Haitao Luo, Xiaowei Chen, Yi Xiao, and Runsheng Chen

20 MicroRNA Biogenesis: Dicing Assay 219
Carlos A. Melo and Sonia A. Melo

21 Faster Experimental Validation of microRNA Targets Using Cold Fusion Cloning and a Dual Firefly-*Renilla* Luciferase Reporter Assay 227
M. Lucrecia Alvarez

22 Experimental Validation of Predicted Mammalian MicroRNAs of Mirtron Origin 245
Anita Schamberger and Tamás I. Orbán

23 A Guide for miRNA Target Prediction and Analysis Using Web-Based Applications 265
Ana Lúcia Leitão, Marina C. Costa, and Francisco J. Enguita

24 Tapping MicroRNA Regulation Networks Through Integrated Analysis of MicroRNA–mRNA High-Throughput Profiles 279
Anthony D. Saleh and Hui Cheng

25 miRWalk Database for miRNA–Target Interactions 289
Harsh Dweep, Norbert Gretz, and Carsten Sticht

26 A Schematic Workflow for Collecting Information About the Interaction Between Copy Number Variants and MicroRNAs Using Existing Resources 307
Harsh Dweep, Norbert Gretz, and Kyriakos Felekkis

27 SYBR® Green and TaqMan® Quantitative PCR Arrays: Expression Profile of Genes Relevant to a Pathway or a Disease State 321
M. Lucrecia Alvarez and Stefania Cotta Doné

28 Comprehensive Meta-analysis of MicroRNA Expression Using a Robust Rank Aggregation Approach. 361
Urmo Võsa, Raivo Kolde, Jaak Vilo, Andres Metspalu, and Tarmo Annilo

Index *375*

Contributors

M. LUCRECIA ALVAREZ • *Diabetes, Cardiovascular, and Metabolic Diseases, Translational Genomics Research Institute, Phoenix, AZ, USA*
TARMO ANNILO • *Estonian Genome Center, University of Tartu, Tartu, Estonia*
BALINT L. BALINT • *Department of Biochemistry and Molecular Biology, Center for Clinical Genomics and Personalized Medicine, University of Debrecen Medical and Health Science Center (UD MHSC), Debrecen, Hungary*
OLGA BELTCHEVA • *Department of Medical Chemistry and Biochemistry, Molecular Medicine Center, Medical University of Sofia, Sofia, Bulgaria*
DECHAO BU • *Key Laboratory of Intelligent Information Processing, Institute of Computing Technology, Chinese Academy of Sciences, Beijing, Republic of China*
KASANDRA L. BURGOS • *Neurogenomics Division, Translational Genomics Research Institute, Phoenix, AZ, USA*
PETER K. BUSK • *Department of Biotechnology, Chemistry and Environmental Engineering, Aalborg University Copenhagen, Copenhagen, Denmark*
STEVEN G. CHAULK • *Department of Biochemistry, University of Alberta, Edmonton, AB, Canada*
RUNSHENG CHEN • *Laboratory of Bioinformatics and Non-coding RNA, Institute of Biophysics, Chinese Academy of Sciences, Beijing, China*
XIAOWEI CHEN • *Laboratory of Bioinformatics and Non-coding RNA, Institute of Biophysics, Chinese Academy of Sciences, Beijing, China*
HUI CHENG • *Tumor Biology Section, Head and Neck Surgery Branch, National Institute on Deafness and Other Communication Disorders, National Institutes of Health, Bethesda, MD, USA*
SUSANNA CIRERA • *Faculty of Health and Medical Sciences, Department of Veterinary Clinical and Animal Sciences, University of Copenhagen, Frederiksberg, Denmark*
MARINA C. COSTA • *Faculdade de Medicina, Instituto de Medicina Molecular, Universidade de Lisboa, Lisbon, Portugal*
ZSOLT CZIMMERER • *Nuclear Receptor Research Group and MTA-DE Lendulet Immunogenomics Research Group, Department of Biochemistry and Molecular Biology, University of Debrecen Medical and Health Science Center, Debrecen, Hungary*
OTTO DOCS • *Department of Pathology, University of Debrecen Medical and Health Science Center, Debrecen, Hungary*
STEFANIA COTTA DONÉ • *Diabetes, Cardiovascular, and Metabolic Diseases, Translational Genomics Research Institute, Phoenix, AZ, USA*
HARSH DWEEP • *Medical Faculty Mannheim, Medical Research Center, University of Heidelberg, Mannheim, Germany*
FRANCISCO J. ENGUITA • *Faculdade de Medicina, Instituto de Medicina Molecular, Universidade de Lisboa, Lisbon, Portugal*
RICHARD P. FAHLMAN • *Department of Biochemistry, University of Alberta, Edmonton, AB, Canada*
KYRIAKOS FELEKKIS • *Department of Biological Sciences, University of Nicosia, Nicosia, Cyprus*

Norbert Gretz • *Medical Faculty Mannheim, Medical Research Center, University of Heidelberg, Mannheim, Germany*
Adrienn Gyongyosi • *Department of Biochemistry and Molecular Biology, Center for Clinical Genomics and Personalized Medicine, University of Debrecen Medical and Health Science Center, Debrecen, Hungary*
Attila Horvath • *Department of Biochemistry and Molecular Biology, Center for Clinical Genomics and Personalized Medicine, University of Debrecen Medical and Health Science Center, Debrecen, Hungary*
Kendall Van Keuren-Jensen • *Neurogenomics Division, Translational Genomics Research Institute, Phoenix, AZ, USA*
Raivo Kolde • *Institute of Computer Science, University of Tartu, Tartu, Estonia*
Ana Lúcia Leitão • *Faculdade de Ciências e Tecnologia, Departamento de Ciências e Tecnologia da Biomassa, Universidade Nova de Lisboa, Caparica, Portugal*
Haitao Luo • *Key Laboratory of Intelligent Information Processing, Institute of Computing Technology, Chinese Academy of Sciences, Beijing, Republic of China*
Gabor Mehes • *Department of Pathology, University of Debrecen Medical and Health Science Center, Debrecen, Hungary*
Sarah Mehrtens • *Robert Koch Hospital, Gehrden, Germany*
Jai Prakash Mehta • *Conway Institute of Biomolecular & Biomedical Research, University College Dublin, Dublin, Ireland*
Carlos A. Melo • *Division of Gene Regulation, The Netherlands Cancer Institute, Amsterdam, The Netherlands; Doctoral Program in Biomedicine and Experimental Biology, Centre for Neuroscience and Cell Biology, Coimbra, Portugal*
Sonia A. Melo • *Department of Cancer Biology, Metastasis Research Center, University of Texas MD Anderson Cancer Center, Houston, TX, USA; Genetic Dynamics of Cancer Cells, Institute of Molecular Pathology and Immunology of the University of Porto (IPATIMUP), Portugal*
Andres Metspalu • *Department of Biotechnology, Institute of Molecular and Cell Biology, and Estonian Genome Center, University of Tartu, Tartu, Estonia*
Laszlo Nagy • *Department of Biochemistry and Molecular Biology, Center for Clinical Genomics and Personalized Medicine, University of Debrecen Medical and Health Science Center, Debrecen, Hungary*
Mahtab Nourbakhsh • *Department of Pharmacy and Biotechnology, German University in Cairo, Berlin, Germany; Department of Plastic Surgery, University Hospital of the RWTH Aachen, Aachen, Germany*
Tamás I. Orbán • *Institute of Enzymology, Research Centre for Natural Sciences, Hungarian Academy of Sciences, Budapest, Hungary; Chemical Technology Transfer Ltd., Budapest, Hungary*
Laszlo Orosz • *Department of Obstetrics and Gynecology, University of Debrecen Medical and Health Science Center, Debrecen, Hungary*
Sweta Rani • *REMEDI, National Centre for Biomedical Engineering Science (NCBES), NUI Galway, Galway, Ireland*
Marc R. Reboll • *Molecular and Translational Cardiology, Medical School Hannover, Hannover, Germany*
Birgit Ritter • *Institute for Virology, Medical School Hannover, Hannover, Germany*
Anthony D. Saleh • *Tumor Biology Section, Head and Neck Surgery Branch, National Institute on Deafness and Other Communication Disorders, National Institutes of Health, Bethesda, MD, USA*

ANITA SCHAMBERGER • *Institute of Enzymology, Research Centre for Natural Sciences, Hungarian Academy of Sciences, Budapest, Hungary*

CARSTEN STICHT • *Medical Faculty Mannheim, Medical Research Center, University of Heidelberg, Mannheim, Germany*

LIANG SUN • *Key Laboratory of Intelligent Information Processing, Institute of Computing Technology, Chinese Academy of Sciences, Beijing, Republic of China*

OLGA TÖRÖK • *Department of Obstetrics and Gynecology, University of Debrecen Medical and Health Science Center, Debrecen, Hungary*

JAAK VILO • *Institute of Computer Science, University of Tartu, Tartu, Estonia*

URMO VÕSA • *Department of Biotechnology, Institute of Molecular and Cell Biology, University of Tartu, Tartu, Estonia*

YI XIAO • *Department of Physics, Huazhong University of Science and Technology, Wuhan, China*

YI ZHAO • *Key Laboratory of Intelligent Information Processing, Institute of Computing Technology, Chinese Academy of Sciences, Beijing, Republic of China*

VICTORIA ZISMANN • *Translational Genomics Research Institute, Phoenix, AZ, USA*

Part I

Structural RNA Mapping

Chapter 1

Full-Length Characterization of Transcribed Genomic Regions

Marc R. Reboll, M. Lucrecia Alvarez, and Mahtab Nourbakhsh

Abstract

In the last years, an enormous progress has been made in the identification of genomic sequences. Given that genomic sequences can have various functions (e.g., structural organization, gene regulation, transcriptional start, and protein coding), molecular characterization is essential for progressing from the initial identification of genomic sequences to the delineation of a specific biological mechanism. Mapping of transcribed sequences is the initial step in functional characterization of genomic sequences. Northern blot analysis allows for a direct and detailed characterization of transcribed sequences, like size and splicing variants, and provides a relative comparison of transcript abundance between different cellular conditions. This method includes separation of total cellular RNA by size via gel electrophoresis, RNA transfer to a membrane, and RNA hybridization with a complementary labeled genomic probe.

Key words Genomic sequences, RNA mapping, Northern blot analysis, Nonradioactive northern, Digoxigenin, DIG-labeling

1 Introduction

RNA synthesis is usually catalyzed by RNA polymerase using genomic DNA as a template. Transcription initiation starts with the binding of RNA polymerase to the regulatory DNA usually found "upstream" of the protein coding region. RNA polymerase progresses along the DNA template from 3′ to 5′ end synthesizing a complementary RNA molecule. When the RNA polymerase reaches a termination sequence, it dissociates from the complex and the generation of a primary RNA is accomplished. Depending on their functions, RNAs are often modified by enzymes after transcription. For example, a poly(A) tail and a 5′ cap structure are added to protein coding messenger RNA (mRNA) and introns are removed by the spliceosome. However, many RNAs do not encode a protein. Indeed, about 97 % of the transcribed sequences in mammalian cells are classified as non-protein-coding sequences. Some noncoding RNAs are transcribed by their own genes, the

M. Lucrecia Alvarez and Mahtab Nourbakhsh (eds.), *RNA Mapping: Methods and Protocols*, Methods in Molecular Biology, vol. 1182, DOI 10.1007/978-1-4939-1062-5_1, © Springer Science+Business Media New York 2014

so-called RNA genes. But some are derived from intron sequences of spliced RNAs. The most prominent noncoding RNAs are transfer (tRNA) or ribosomal RNAs (rRNA). They are involved in translation of mRNAs to proteins. Other non-protein-coding RNAs are involved in regulation of gene activity or RNA processing sequence directed digestion and ligation.

To date, bioinformatics approaches have been heavily utilized to obtain a hypothetical map of a genomic sequence, but functional characterization of all genes will require detailed and precise mapping of unknown cellular RNAs. While there are several laborious methods for identification and characterization of transcribed genomic regions, this chapter will focus on the best established and most comprehensive method: Northern blot analysis using cellular mRNA. Northern blot analysis allows for a direct and detailed characterization of transcribed sequences, and can be combined with different methods to detect cellular localization, splicing variants, and relative abundance. For instance, RNA stability can be monitored by blocking the overall RNA synthesis in cells treated with Actinomycin D. Thus, northern blot analysis can be used to monitor the stability of target RNA in the course of time. The first northern blot step, separation of cellular RNAs by size via gel electrophoresis, is very critical and should be adjusted to the expected RNA size. Otherwise, multiple gels need to be prepared to cover a broad range of RNAs from less than 20 nucleotides to over 10 kilo bases. The second step, RNA transfer to a membrane, needs to be adjusted to the gel characteristics. The third step, hybridization with a complementary genomic DNA, can be performed using radioactive or nonradioactive digoxigenin (DIG) labeled DNA probe.

2 Materials

2.1 Northern Blot Analysis

1. 10× MOPS electrophoresis buffer: 41.8 g MOPS in 700 ml of sterile DEPC-treated water. Adjust pH to 7.0 with 2 N NaOH. Add 20 ml of DEPC-treated 1 M sodium acetate and 20 ml of DEPC-treated 0.5 M EDTA (pH 8.0). Adjust volume to 1 L with DEPC-treated water.
2. Sample buffer: 1× MOPS, 6.85 % formaldehyde, 50 % deionized formamide, 4 % ficoll 400, 0.01 % bromophenol blue, and 0.01 % xylene cyanol.
3. SYBR green II RNA gel stain or ethidium bromide.
4. Horizontal electrophoresis unit (i.e., Sigma-Aldrich Corp).
5. Whatman 3MM paper.

2.1.1 Radioactive Method

1. Herring sperm DNA.
2. Yeast total RNA.

3. Dextran sulfate.
4. Sodium chloride (Sigma-Aldrich Corp).
5. Hybond-N+ membrane.
6. Rediprime II DNA labeling kit (GE Healthcare).
7. Illustra MicroSpin G-25 Columns (GE Healthcare).
8. Hybridization oven (i.e., Sigma-Aldrich Corp).
9. 20× SSC transfer buffer: 3 M sodium chloride, 300 mM trisodium citrate dehydrate, pH 7.0.
10. Buffer A: 2× SSC.
11. Buffer B: 1× SSC, 0.1 % SDS.
12. Buffer C: 0.2× SSC, 0.1 % SDS, preheat to 62 °C.
13. Buffer D: 0.1× SSC.

2.1.2 Nonradioactive Method

1. To test DIG-labeled DNA probe by agarose gel: agarose, bromophenol blue, xylene cyanol, SYBR® Safe DNA Gel Stain 10,000× concentrate (Invitrogen, Inc; Carlsbad, CA) or ethidium bromide.
2. 50× TAE (Tris-acetate-EDTA) buffer: 242 g/L Tris base, 5.7 % v/v, 0.05 M EDTA, pH 8.0.
3. Zeta probe GT membrane (Bio-Rad) or any positively charged nylon membrane.
4. UV crosslinker (Stratagene).
5. PCR DIG Probe Synthesis kit and Anti-DIG-AP antibody for detection (Roche Applied Science).
6. Oligonucleotide primers.
7. GeneAmp PCR system 9700 (Life Technologies) or any other thermocycler.
8. RNAse Zap (Life Technologies).
9. DIG East Hyb solution (Roche Applied Science).
10. Low Stringency Buffer: 2× SSC, 0.1 % (w/v) SDS.
11. High Stringency Buffer: 0.1× SSC, 0.1 % (w/v) SDS.
12. DIG block and wash buffer (Roche Applied Science).
13. CDP-star (Roche Applied Science).
14. KODAK BioMax XAR film.

3 Methods

Northern blotting is a straightforward procedure that offers opportunities to evaluate progress at various points during the protocol (e.g., assessing integrity of RNA samples, evaluating efficiency of membrane transfer, and so forth). In this protocol,

RNA that has been isolated from the tissue of choice is first separated by size via denaturing agarose gel electrophoresis. Following electrophoresis, RNA is transferred from the gel to a membrane by capillary action, after which it is cross-linked to the membrane for immobilization, and hybridized with a labeled probe for transcript detection. Northern blot analysis can be performed using either radiolabeled or non-isotopically labeled DNA as hybridization probes. The choice of label is generally dependent on a number of considerations, including sensitivity and resolution. Radioactive DNA probes can be quickly prepared, offer rapid detection, and, importantly, provide much higher sensitivity compared to nonradioactive probes. For example, nucleotides labeled with ^{32}P enable high specific activity probes up to 2×10^9 dpm/µg, which can detect as little as 10 fg of target RNA, and it is widely appreciated that sensitivity can be of paramount importance when investigating novel transcripts and potential splicing variants whose cellular abundance is not yet known. However, a drawback of the radioactive method is the need for licensing as well as the safety issues, both of which may limit the feasibility of radioisotope use in the laboratory. Further, half-lives of some radioisotopes (i.e., ^{32}P) are quite short, so reagents must be used quickly or purchased on a regular basis. In contrast, the nonradioactive system utilizing digoxigenin (DIG) is an alternative method for detection of cellular RNAs [1, 2] and offers unique advantages such as enhanced probe stability and shorter film exposure times for detection.

In this chapter we describe methods for northern blot analysis that utilize both radioactive and nonradioactive probe preparation. However, the initial steps of northern blot analysis, including gel electrophoresis transfer, and immobilization of RNA to membrane are independent of the type of probe used [3].

3.1 Denaturing Agarose Gel Electrophoresis

1. Prepare gel casting tray by sealing ends of gel chamber with tape or proper casting system. Place appropriate number of combs in gel tray.
2. Prepare 1 % (w/v) agarose in a glass flask, add water, and heat solution in a microwave until agarose is completely dissolved. For 120 ml gel volume, dissolve 1.2 g agarose in 86.4 ml water.
3. Cool agarose solution to 65–70 °C in a water bath. Swirl occasionally to prevent uneven cooling.
4. After cooling, add 21.6 ml formaldehyde and 12 ml of 10× MOPS to a final volume of 120 ml.
5. Add 12 µl of SYBR green II RNA gel stain to the gel mix. Alternatively, ethidium bromide can be added to the cooled agarose solution (final concentration 0.5 µg/ml) or directly to the RNA samples (*see* **step 13**).
6. Ensure that there is enough space between the bottom of the comb and the gel tray (0.5–1.0 mm) to allow for proper well formation and to avoid sample leakage (*see* **Note 1**).

7. Gently mix and pour the cooled agarose solution onto the gel tray in a fume hood to a thickness of 3–5 mm, being careful to avoid generation of bubbles on the gel surface. Insert comb either before or immediately after pouring; if inserted after pouring, make sure that no bubbles form along the teeth of the comb. Let the gel set for at least 60 min.
8. Leaving the comb in the gel, place the gel in the electrophoresis tank. Fill the tank with enough 1× MOPS gel running buffer to cover the gel with approximately 1 mm of liquid above the surface of the gel. If too much buffer is used, the electric current will flow through the buffer instead of the gel. Carefully remove the comb from the gel, being careful not to tear wells.
9. The amount of cellular RNA required to detect a specific transcript can vary over a wide range and strongly depends on the number of transcribed RNA molecules per cell. For radioactive northern blot analysis use 5–25 μg of total cellular RNA or 1–2 μg poly(A+) RNA in each lane. For the nonradioactive procedure, use 5 μg of total RNA. Dry down the appropriate amount of RNA in 1.5 ml tubes, if necessary, then reconstitute in buffer prior to gel loading.
10. Add DEPC-treated water to set all samples to an equal volume between 5 and 10 μl.
11. Add equal volume of sample buffer. Use a master mix of sample buffer to avoid concentration variations between the samples.
12. Denature samples 5 min at 90 °C, spin tubes for 10 s, then chill on ice.
13. If appropriate, add 0.5 μl ethidium bromide (0.5 μg/μl) to each sample (*see* **step 5**).
14. Slowly and carefully pipette samples into wells.
15. For a small electrophoresis chamber, perform a pre-electrophoresis run of 5–10 min at 55 V and a main run at 50–60 V. For a large electrophoresis chamber, do a pre-electrophoresis run for 5–10 min at 100 V and a main run at 100–120 V. Main run time depends on the size of the transcript that is going to be detected. The gel run can be adjusted by visualization of ribosomal RNA bands at any time during the run.

3.2 Transfer to Membrane and Immobilization

1. Cut a nylon membrane to the exact size of the gel. Use Hybond N+ or Zeta Probe membranes for radioactive and nonradioactive procedures, respectively. Both Hybond N+ and Zeta Probe are nylon membranes ideally suited for nucleic acids, but each has unique properties to enhance sensitivity to specific labeling techniques.
2. Cut four pieces of Whatman 3MM paper. Three of them should be the same size of the gel, and the fourth one should be twice as wide and long enough to fit under the gel and reach to the bottom of the dish on either side.

3. When electrophoresis is complete, wash the agarose gel twice in water and take a photograph under ultraviolet light. The integrity and size distribution of total RNA can assessed at this point (*see* **Note 2**).
4. Prepare 10× SSC and filter for long term storage. Carefully put the gel on a glass dish.
5. Wash gel in water twice for 10 min with gentle rocking. Wash once again in water for a minimum of 5 min. Take a picture to document RNA quality and parameters of the gel electrophoresis run.
6. Wash the gel in 10× SSC for 10 min.
7. Wet the nylon membrane in water. Then soak both the nylon membrane and the Whatman 3MM paper in 20× SSC for 1–2 min.
8. Assemble blotting unit as shown in Fig. 1 and allow for the transfer to proceed for 12–18 h.

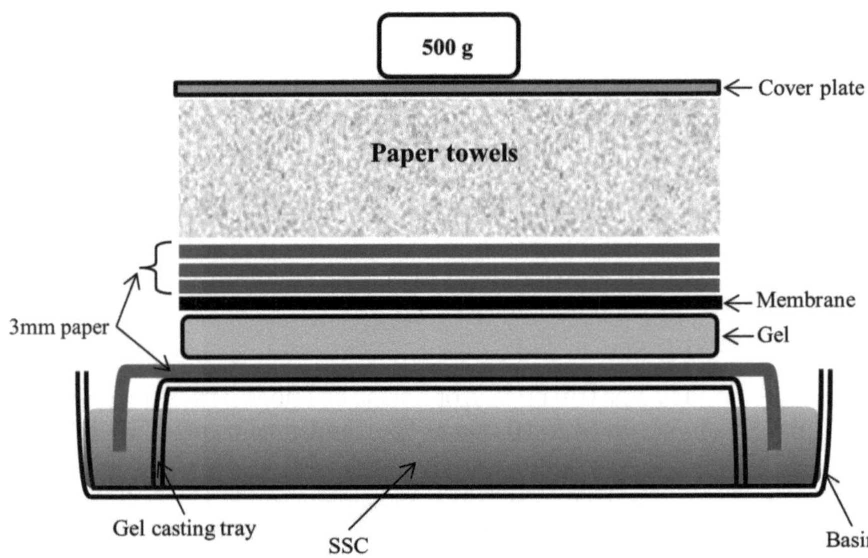

Fig. 1 Blot unit assembly. Fill the buffer tray with 1 L 20× SSC. Place a glass or Plexiglass plate across the tray or on top of a support. Place the two lengths of presoaked filter paper over the glass or Plexiglass plate so that the ends contact the bottom of the tray. Remove any air bubbles between the sheets of filter paper and the plate by rolling a pipet several times back and forth over the surface. Position the gel upside-down on the filter paper covering the plate. Ensure that the transfer buffer moves only through the gel and not around it. Place the presoaked nylon membrane on top of the gel so that it covers the entire surface. Do not move the nylon membrane once it has been placed on the gel. Remove any air bubbles between the membrane and the gel by gently rolling a pipet several times back and forth over the surface. Place the three presoaked sheets of Whatman 3MM paper on top of the nylon membrane. Again, remove any air bubbles by gently rolling a pipet several times back and forth over the surface. Place a 15–20 cm stack of dry paper towels on top of the filter paper. Place a second glass or Plexiglass plate on top of the paper towels. Place the 1 kg weight on top of the plate

9. After the transfer is complete, disassemble the blotting apparatus by removing the weight, paper towels, and two sheets of filter paper. Turn the gel and the nylon membrane over together, and lay them, gel side up, on a clean, dry sheet of filter paper.
10. Mark the positions of the gel lanes on the membrane using a ballpoint pen or a soft-lead pencil. Peel the gel from the membrane and discard it.
11. *Mark the gel lanes before removing the gel from the nylon membrane!* Without this marking, lanes will not be identifiable.
12. Fix the RNA to the blot by either baking 1 h at 80 °C in an oven or UV cross-linking. The latter method generally gives better results and enhanced sensitivity compared to baking. However, proper cross-linking requires prior optimization of the system (*see* **Note 3**).
13. If the blot will not be used immediately, it can be stored in plastic wrap at either 4 °C or room temperature for an indefinite period of time.

3.3 Radioactive DNA-Probe Labeling and Detection

The following steps vary depending on the probe labeling method, radioactive (Subheading 3.3) or nonradioactive (Subheading 3.4). However, both methods have the following general steps in common: probe generation, prehybridization and hybridization with a labeled DNA-probe, removal of non-hybridized probe (washing), detection, stripping, and reprobing.

3.3.1 Probe Generation

Both PCR amplification of genomic regions using high molecular weight DNA or isolation of genomic fragments from pre-cloned plasmids are suitable sources for generation of probes. In general, DNA fragments should be purified by gel electrophoresis prior to the labeling reaction. The random priming technique, which is based on extension of random hexanucleotide primers hybridized to a template by simultaneous incorporation of radioactive labeled nucleotides, is the most common method for generating a radioactive probe from an isolated DNA fragment. The ideal size of template DNA ranges from 450 bp to 2 kb. As a general rule, a minimum of 50 bp homology is required for hybridization with target RNA in this procedure. Prepare a reaction as follows:

1. Dilute DNA to be labeled to a concentration of 2.5–25 ng in 45 μl of 10 mM Tris–HCl pH 8.0, 1 mM EDTA.
2. Denature the DNA sample by heating to 95–100 °C for 5 min.
3. Chill DNA template on ice for 5 min.
4. Centrifuge briefly to bring the contents to the bottom of the tube.
5. Start labeling reaction using Rediprime II DNA labeling kit (GE Healthcare, Pittsburg, PA).

6. Add the template to a supplied ready-to-use reaction tube and carefully mix the components by gently flicking tube with finger.
7. Add 5 μl of [α-^{32}P] dCTP and carefully mix by pipetting up and down.
8. Incubate reaction at 37 °C for 10 min.
9. Stop the reaction by adding 5 μl of 0.2 M EDTA.
10. Purify the labeled probe using Illustra MicroSpin G-25 Columns (GE Healthcare, Pittsburg, PA).
11. Resuspend the resin in the supplied column by vortexing.
12. Detach the bottom closure by turning the cap.
13. Spin the column for 1 min at 735 ×*g*.
14. Place the column into a 1.5 ml tube.
15. Add labeling reaction to the center of the resin surface without disturbing the resin bed.
16. Spin for 2 min at 735 ×*g* to collect the sample at the bottom of the tube.
17. Store purified probe at −20 °C or use immediately.

3.3.2 Prehybridization and Hybridization with a Labeled DNA-Probe

1. Mix 0.58 g NaCl, 1 g dextran sulfate and 9.5 ml sterile water in 50 ml plastic tub for 10 min. Heat to 60 °C until solution turns clear, then add 0.5 ml 20 % SDS. Mix gently to avoid bubbles and incubate at 60 °C during next two steps.
2. Denaturate 1 % yeast RNA and 10 mg/ml herring sperm DNA by boiling in water bath for 5 min, then chilling on ice for 5 min.
3. Clean a hybridization tube, preheat to 60 °C, and insert blot into pre-warmed tube.
4. Add 0.5 ml herring sperm DNA and 0.5 ml RNA to the preheated hybridization buffer; mix and carefully pipet into the hybridization tube. Avoid bubbles!
5. Incubate the tube for prehybridization at 60 °C for a minimum of 30 min.
6. Radioactive probe can either be added directly to the tube or mixed with a new aliquot of hybridization buffer, which replaces the prehybridization buffer.
7. Incubate at 60 °C overnight or a minimum of 8 h.

3.3.3 Washing

1. Following hybridization, carefully remove the radioactive hybridization buffer by pipetting or draining from tube. Buffer can be directly disposed into radioactive waste or stored at −20 °C for a maximum of 1 week and used once more for hybridization (*see* **Note 4**).

2. Wash the blot in the tube with Buffer A for 2 min to remove residual, unbound radioactive probe.
3. Place blot in a glass dish and wash once again with Buffer A for 5 min with gentle agitation. Check counts using a hand counter and use this initial count as a baseline value of 100 % to estimate the efficiency of next wash steps.
4. Wash blot in Buffer B at room temperature for 15 min with gentle rocking. Check counts every 5 min. If the counts are reduced to 3 %, skip the next step.
5. Wash in Buffer C for 15 min at 62 °C. Check counts every 5 min. If the counts are lower than 10 %, stop the wash.
6. Wash in Buffer D for 5 min at room temperature.

3.3.4 Detection

1. Wrap blot in plastic wrap and expose to X-ray film.

3.3.5 Stripping and Reprobing

Typically, several hybridization experiments using different labeled genomic probes are required to characterize all transcripts from a single gene and provide visualization of an appropriate positive control. Therefore, a single blot is usually probed several times. However, prior to rehybridizing with new probes, the old label should be carefully removed from the blot. This procedure is called stripping.

1. First incubate the membrane in 0.1 % SDS at 95–100 °C for 10–20 min.
2. Next, wash the membrane in 2× SSC for 5 min at room temperature and start the next round of hybridization with a new probe. Expose blot to X-ray film to ensure that radioactive probe is completely removed.
3. Begin new hybridization.

3.4 Nonradioactive DNA-Probe Labeling and Detection

The digoxigenin (DIG) system—based on the steroid hapten, digoxigenin, which occurs in certain digitalis plants—is an effective method for nonradioactive labeling and detection of nucleic acids [4]. Digoxigenin is suitable for detection purposes based on three characteristics. First, high affinity antibodies can be easily generated if digoxigenin is coupled to a suitable carrier molecule. Second, there are no endogenous background problems with the anti-digoxigenin antibodies as in the case of other haptens, such as biotin, because digoxigenin occurs exclusively in digitalis plants. Third, digoxigenin can be coupled to nucleotides like dUTP and then incorporated into nucleic acids using Klenow, Taq, or RNA polymerases to generate DIG-labeled probes. These probes can be used in standard blotting and hybridization procedures and detected with anti-digoxigenin conjugates such as fluorescent or alkaline-phosphatase labeled antibodies [5].

Table 1
DIG-probe labeling by PCR using PCR DIG Probe Synthesis kit

Reagent	DIG-labeled probe	Un-labeled control[a]	Final concentration
Sterile double-distilled water	Variable volume	Variable volume	–
10× PCR buffer with $MgCl_2$ (vial 3)	5 μl	3 μl	1×
10× PCR DIG mix (vial 2)	5 μl	–	200 μM
10× dNTP stock solution (vial 4)	[b]	3 μl	200 μM
Forward and reverse primers [50 μM]	0.5 μl	0.3 μl	0.5 μM each primer
Enzyme mix (Expand High Fidelity, vial 1)	0.75 μl	0.45 μl	2.6 units total enzyme
Template DNA	Variable volume	Variable volume	10 ng genomic DNA or 10 pg plasmid DNA
Final volume	50 μl	30 μl[a]	30 or 50 μl

[a]The use of a lower total reaction volume for the un-labeled control than the DIG-labeled probe is only to save reagents of the PCR DIG Probe Synthesis kit (Roche Applied Science, Indianapolis, IN)
[b]For DIG-labeled probes < 1 kb long: use 1:3 DIG-dUTP–dTTP ratio and *do not* add 10× dNTP stock solution (vial 4). For DIG-labeled probes > 1 kb long use 1:6 DIG-dUTP–dTTP ratio and add 5 μl of 10× dNTP stock solution (vial 4)

3.4.1 Probe Generation

The nonradioactive DNA probe is labeled incorporating DIG-labeled dUTP by PCR using the PCR DIG Probe Synthesis kit and Anti-DIG-AP antibody for detection (*see* Subheading 2). This method is particularly recommended when template is available in limited quantity, is only partially purified, or is very short [6, 7]. The PCR DIG Probe Synthesis kit is especially designed for generation of highly sensitive hybridization probes suitable for detection of low (single) copy sequences.

1. In a 1.5 ml tube, prepare a DIG-labeled DNA probe that is going to be used as hybridization probe adding the reagents in the amounts shown in Table 1. Note that for DIG-labeled probes <1 kb long, a 1:3 DIG-dUTP–dTTP ratio (standard ratio) is recommended. However, for longer DIG-labeled probes (i.e., >1 kb long), a 1:6 DIG-dUTP–dTTP ratio should be used to avoid low yield of DIG-labeled PCR product.
2. In another 1.5 ml tube, prepare the unlabeled positive control adding the reagents and amounts according to Table 1. Mix the reagents and centrifuge briefly.
3. Transfer samples to 0.2 or 0.5 ml tubes suitable for PCR and mix thoroughly.
4. Place samples in thermocycler and begin cycling program. The optimal cycling conditions depend on the combination of template, primers, and thermocycler (*see* **Note 5**). The following conditions are a good starting point:

(a) Initial denaturation step (before the first cycle): 95 °C for 2 min

(b) PCR amplification (35 cycles): 95 °C for 30 s; 60 °C for 30 s; 72 °C for 40 s

(c) Final elongation step: 72 °C for 7 min

The labeling efficiency for PCR-labeled probes can be quickly estimated by gel electrophoresis as described below:

5. Prepare a 1–1.5 % w/v agarose gel in TAE buffer and add 10 μl SYBR® Safe DNA Gel Stain 10,000× concentrate (Invitrogen, Carlsbad, CA) per ml of agarose solution. Alternatively, gels can be stained with ethidium bromide, as described above.
6. Load 5 μl of each PCR amplification product on agarose gel. Include a DNA molecular weight marker. Electrophorese the samples at 60 V for 40 min or until the DIG-labeled and unlabeled probes are adequately separated. DIG-labeled probes are larger than unlabeled probes; thus, these probes migrate more slowly compared to unlabeled DNA of the same size. The intensity of the stained DIG-labeled probe should be equal to, or slightly less than, the intensity of the unlabeled probe DNA. If these conditions are met, then the DIG-probe was likely labeled efficiently and the standard amount of labeled probe (2 μl of PCR product per ml hybridization buffer) can be used in hybridization reactions. If the intensity of the labeled PCR product band is very strong on the gel, 0.5 μl probe per hybridization buffer should be used (*see* **Note 6**). If the signal is very faint, up to 4 μl probe per ml hybridization buffer should be used [5].

3.4.2 Prehybridization and Hybridization with a DIG-Labeled DNA-Probe

1. After fixing the RNA to a Zeta Probe membrane, or any similar positively charged nylon membrane, incubate the membrane in 2× SSC buffer to remove any agarose or salts remaining from the RNA capillary transfer. First incubate the membrane in 0.1 % SDS at 95–100 °C for 10–20 min.
2. Place membrane on 3MM paper with RNA face up.
3. For prehybridization and hybridization, use ready-to-use DIG Easy Hyb solution. In a hybridization oven at 50 °C, pre-warm a 50 ml tube containing 16 ml of DIG Easy Hyb solution.
4. Pre-warm an empty hybridization tube in the hybridization oven at 50 °C.
5. *Prehybridization step*: insert membrane into the pre-warmed hybridization tube and immediately add 10 ml of the pre-warmed DIG Easy Hyb solution prepared in **step 3** (*see* **Notes 7** and **8**). Incubate with rotation in a hybridization oven at 50 °C for at least 30 min.
6. In a 1.5 ml tube, mix 200 μl Easy Hyb solution with the DIG-labeled DNA probe (2 μl DNA probe labeled by PCR per ml of DIG Easy Hyb solution) (*see* **Note 6**).

7. Denature probe by boiling 10 min at 100 °C, then immediately quenching on ice.
8. Add the denatured probe to the remaining 6 ml of the prewarmed DIG Easy Hyb solution prepared in **step 3** to obtain 6 ml of hybridization solution. If a previous hybridization solution (containing probe from a previous northern blot experiment) is going to be used, incubate it at 68 °C for 10 min just before adding it to the hybridization tube with the membrane.
9. *Hybridization step*: immediately after the 30 min prehybridization step (**step 11**), replace the prehybridization solution in the hybridization tube with the 6 ml hybridization solution prepared in **step 8**. Incubate, rotating slowly, at 50 °C overnight (16–18 h).

3.4.3 Washes

1. After the hybridization is complete, save the hybridization solution at −20 °C.
2. Keep membrane in the hybridization tube and wash twice with 100 ml Low Stringency Buffer (high salt concentration and low temperature) at room temperature for 5 min each time.
3. Preheat High Stringency Buffer (low salt concentration and high temperature) at 50 °C.
4. Remove used Low Stringency Buffer and *immediately* add preheated High Stringency Buffer.
5. Wash with 100 ml High Stringency Buffer at 50 °C twice for 15 min each time.

3.4.4 Detection

Probe–target hybrids are detected with an enzyme-linked immunoassay, which may be more sensitive than radioactive detection procedures when the optimal experimental conditions are followed [4, 5]. Prepare the following solutions using DIG Wash and Block Buffer kit just before starting the detection:

- *500 ml 1× washing buffer*: 50 ml 10× Washing Buffer (shake vigorously before use) + 450 ml double-distilled RNAse-free water.
- *50 ml 1× maleic acid buffer*: 5 ml 10× Maleic Acid Buffer + 45 ml double-distilled RNAse-free water.
- *20 ml 1× blocking solution*: 2 ml 10× Blocking Solution + 18 ml 1× Maleic Acid buffer.
- *20 ml 1× detection buffer*: 2 ml 10× Detection Buffer + 18 ml double-distilled RNAse-free water.

1. Equilibrate the membrane in a clean tray (treated with anti-RNase spray) containing 50 ml 1× Washing Buffer for 1–5 min.
2. First wash the hybridization tube with soap and water, then treat it with RNAse Zap (Applied Biosystems, Carlsbad, CA) and rinse with RNAse-free water.

3. Place the membrane into hybridization tube and add 10 ml 1× Blocking Solution. Incubate for 30 min in hybridization oven at room temperature.
4. Prior to each use, centrifuge the anti-Digoxigenin-AP in the original vial, at 10,000×*g* for 5 min. Prepare the antibody solution adding 0.5 μl anti-Digoxigenin-AP taken carefully from the surface to 10 ml Blocking Solution to obtain the antibody solution (dilution 1:20,000) (*see* **Note 9**).
5. Remove Blocking Solution from hybridization tube and add 10 ml of antibody solution prepared in **step 5** of Subheading 3.3.3. Incubate for 30 min at room temperature in the hybridization oven with rotation. In a 0.2 ml tube, combine up to 5 μg total RNA, 1 μl primer (*see* **Note 4**), 1 μl 10 mM dNTP mix, and DEPC-treated water to a final volume of 10 μl. Make sure to include a negative control as well as a –RT control.
6. Transfer the membrane to a clean, RNase-free tray, and add 100 ml 1× Washing Buffer. Incubate for 15 min with shaking.
7. Repeat **step 6**.
8. Pour off washing buffer and add 50 ml 1× Detection Buffer (DIG block and wash buffer kit). Incubate at room temperature for 5 min.
9. Prepare substrate solution: add 10 μl CDP-star to 1 ml Detection Buffer and mix thoroughly.
10. Place the membrane with RNA-side facing up on a clean acetate sheet (*see* **Note 10**). Cover the membrane with 1–2 ml of substrate solution.
11. *Immediately* cover the membrane with a second acetate sheet to spread the substrate evenly over the membrane, *avoiding air bubbles*. Incubate for 5 min at room temperature.
12. Squeeze out excess liquid and seal edges of the development folder. Do not let the membrane dry completely.
13. Expose sealed envelope containing the membrane to one of the following at room temperature:
    (a) Lumi-Imager F1 Workstation (5–20 min). The Lumi-Imager allows for rapid, quantitative analysis of the chemiluminescent signal without X-ray film.
    (b) KODAK BioMax XAR film or Lumi-Film X-ray film (1–20 min) (*see* **Note 11**). Adjust the exposure time to get a darker or lighter band pattern, depending on the results.

Table 2 shows a summary of the most common problems that might arise during DIG labeling and detection and how to solve or prevent them.

Table 2
Troubleshooting parameters for DIG labeling and detection[a]

Problem	Possible cause	Recommendations
Low sensitivity	Inefficient probe labeling	Check labeling efficiency of your DIG-labeled probe by electrophoresis gel
	PCR not optimized	Always optimize PCR parameters (template, primers and $MgCl_2$ concentration and cycling conditions) for each template and primer set in the absence of DIG-dUTP before attempting incorporation of DIG
	Too much DIG-dUTP in reaction	Reduce the concentration of DIG-dUTP in the reaction. This is especially important for DIG-labeled probes > 1 kb long: use 1:6 DIG-dUTP–dTTP ratio
	Wrong type of membrane	The quality of the membrane used influences sensitivity and speed of detection. We recommend nylon membranes positively charged from Roche Molecular Biochemicals or Zeta probe GT membrane (Bio-Rad). Nitrocellulose membranes should not be used
	Inefficient hybridization	Increase concentration of DIG-labeled DNA probe in hybridization solution
	Low antibody concentration	Increase concentration of anti-DIG-AP conjugate
	Too short exposure time	Increase time of exposure to X-ray film. The type of film may also influence the sensitivity. We recommend KODAK BioMax XAR
Uniform high background	Wrong type of membrane	The protocols described in this chapter are optimized for the use of positively charged nylon membranes. However, some types of membrane are too highly charged and can cause background. Lot-to-lot variations in some membranes can also cause problems
	Too high concentration of labeled probe	The critical probe concentration limit (concerning background formation) can be determined by hybridization with increasing probe concentrations to unloaded membrane (Mock Hybridization). For example: use 12.5, 25, and 37.5 ng/ml probe to hybridize unloaded membrane (with no target RNA) to determine amount of probe giving highest signal with lowest background
	Too high antibody concentration	Decrease concentration of anti-DIG-AP conjugate. Increases volumes of the washing and blocking solutions and duration of the washing and blocking steps. Spotty background may be caused by precipitates in the anti-DIG-AP conjugate: do a short centrifugation step just before using it
	Membrane dried during hybridization or detection steps	Never let membrane dry at any stage of the prehybridization, hybridization or detection procedures. Always use enough liquid in each incubation to cover membrane completely (*see* **Note 3**). *Cloudy background*: membrane dried during the hybridization step. *Grainy background*: membrane dried during chemiluminescent detection procedure (CDP-star incubation)
Irregular and cloudy background	Uneven distribution of probe during hybridization	Do not add probe directly to the prehybridization solution Do not discard pre-hyb solution until hyb solution is ready to be used Shake or rotate the hybridization container during the hybridization incubation

[a]The information in this table was partially taken and modified from Roche Applied Science (2003) DIG Application Manual for Filter Hybridization (3rd edition)

3.4.5 Stripping and Reprobing

We recommend the blot be stripped soon after detection as follows:

1. Heat 0.1 % SDS solution to 100 °C.
2. Transfer the membrane into a clean, RNAse-free tray and cover immediately with plenty of the preheated 0.1 % SDS solution. Incubate for 10–60 min at 100 °C.
3. Wash the membrane with 100 ml 2× SSC for 5 min with shaking at room temperature.
4. Store stripped blot wet in Maleic acid Buffer at 4 °C.

4 Notes

1. Make sure that there are no air bubbles in the gel or trapped between the wells which could possibly connect single wells or lead to inconsistent sample runs. Air bubbles can be carefully removed with a plastic pipette tip before the gel sets.
2. The 18S and 28S ribosomal RNA bands should appear as sharp bands. If the ribosomal bands in a given lane are not sharp, but appear as a smear towards smaller-sized RNAs, it is likely that the RNA sample suffered major degradation during preparation. The 28S ribosomal RNA band should be present at approximately twice the intensity of the 18S rRNA band. As the 28S rRNA is more labile than the 18S rRNA, equal intensities of the two bands generally indicates that some degradation has occurred.
3. UV light can damage the eyes and skin. Always wear suitable eye and face protection.
4. Pre-used hybridization buffer should be denatured at 95 °C for 5 min before use.
5. To obtain a high yield of DIG-labeled PCR product, always optimize the PCR parameters (cycling conditions and concentrations of template, $MgCl_2$, and primers) for each template and primer set in the absence of DIG-dUTP *before* attempting incorporation of DIG.
6. An excessive probe concentration in the hybridization solution causes cloudy hybridization background. If this happens, reduce probe concentration to 0.5–1 μl per ml DIG Easy Hyb buffer.
7. Use enough buffer to completely cover the membrane during prehybridization and hybridization incubations. The amount needed will depend on the shape and capacity of the container used for the incubations. Uneven distribution of probe during hybridization produces an irregular and cloudy background and it is caused by using too little hybridization solution.

Use at least 3.5 ml of hybridization solution per 100 cm^2 of membrane. If roller bottles (hybridization tubes) are used for incubation, add at least 6 ml hybridization solution per bottle.

8. Do not allow the membrane to dry at any time from the beginning of prehybridization through the final detection. If the membrane dries or sticks to a second membrane, the assay will have a high background.
9. Several uses and centrifugation steps of the anti-digoxigenin-AP conjugate can cause a certain loss of material, which must be compensated by use of larger amounts.
10. Do not use plastic wrap to cover the membrane during the detection step: use hybridization bags, acetate sheet protectors, or two sheets of transparency film.
11. Luminescence continues for at least 24 h and signal intensity remains almost constant during the first hours. Multiple exposures at different times can be taken to achieve the desired signal strength.

References

1. Holtke HJ, Ankenbauer W, Muhlegger K, Rein R, Sagner G, Seibl R et al (1995) The digoxigenin (DIG) system for nonradioactive labeling and detection of nucleic acids—an overview. Cell Mol Biol 41:883–905
2. Rueger B, Thalhammer J, Obermaier I, Gruenewald-Janho S (1997) Experimental procedure for the detection of a rare human mRNA with the DIG System. Front Biosci 2:C1–C5
3. Sambrook J, Russell D (2001) Molecular cloning: a laboratory manual. Cold Spring Harbor Laboratory, Cold Spring Harbor, NY
4. Hloch P, Hoffmann K, Kruchen B, Rueger B (2001) The DIG system—a high sensitive substitute of radioactivity in northern blot analysis. Biochemica 2:24–25
5. Roche Applied Science (2003) DIG application manual for filter hybridization, 3rd edn, Roche Diagnostics GmbH, Roche Applied Science, Mannheim, Germany. http://lifescience.roche.com/wcsstore/RASCatalogAssetStore/Articles/05353149001_08.08.pdf
6. Rost A, Kohler T, Heilmann S, Lehmann J, Remke H, Rotzsch W (1995) A rapid and simple method to prepare digoxigenin-labeled DNA-probes by using PCR-generated DNA-fragments. Eur J Clin Chem Clin Biochem 33:A59
7. Finckh U, Lingenfelter PA, Myerson D (1991) Producing single-stranded DNA probes with the Taq DNA polymerase: a high yield protocol. Biotechniques 10:35–39

Chapter 2

Rapid Mapping of RNA 3′ and 5′ Ends

Victoria Zismann and Mahtab Nourbakhsh

Abstract

In recent years, an enormous progress has been made in applied genomics leading to identification and isolation of novel cDNAs. However, most attempts result in the acquisition of transcribed sequences that represent only a part of the mRNA's complete sequence. Rapid Amplification of cDNA Ends (RACE) is a technique used in molecular biology to obtain the full length sequence of an RNA transcript found within a cell. Since the first report of this technique, many significant improvements have been made on the basic approach. This chapter describes the most recent update of the relatively simple and versatile classic RACE protocol.

Key words Gene expression, mRNA mapping, Lambda cDNA library, cDNA end amplification, RACE amplification

1 Introduction

Rapid Amplification of cDNA Ends (RACE) is used to extend partial cDNAs by amplifying the 5′ or 3′ sequences of the corresponding mRNAs using gene-specific primers [1, 2]. The technique requires knowledge of only a small region of the sequence within the known partial cDNA (Fig. 1a). The first step begins with cDNA synthesis using cellular RNA. During this step, the Moloney Murine Leukemia Virus Reverse Transcriptase (MMLV RT) utilizes terminal transferase activity to add three to five residues to the 3′ end of the polymerized cDNA upon reaching the end of an RNA template. An additional primer, which contains a terminal stretch of annealing residues, is added to the reaction and serves as an extended template for reverse transcription. MMLV RT switches templates from the mRNA molecule to the primer generating a complete cDNA copy of the original RNA with the additional end sequence. This step is called Switching Mechanism at 5′ End of RNA Template (SMART). The relationship of the primers used in the SMART RACE reactions to the template and resulting RACE products is shown in detail in Fig. 1a.

M. Lucrecia Alvarez and Mahtab Nourbakhsh (eds.), *RNA Mapping: Methods and Protocols*, Methods in Molecular Biology, vol. 1182, DOI 10.1007/978-1-4939-1062-5_2, © Springer Science+Business Media New York 2014

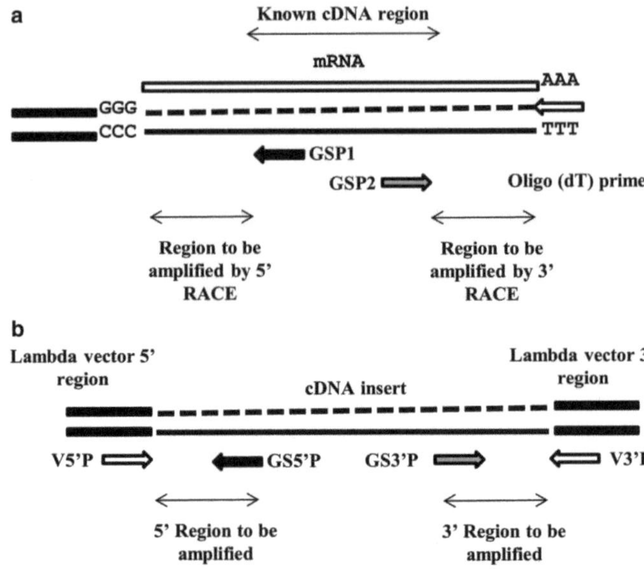

Fig. 1 The relationship of gene-specific primers to the cDNA template by PCR-based amplification of cDNA ends. (**a**) This diagram shows a generalized first-strand cDNA template and direction and position of gene-specific primers, GSP1 and GSP2, required for *R*apid *A*mplification of *c*DNA *E*nds (RACE). (**b**) This diagram shows direction and position of gene-specific primers, GS5′P and GS3′P, and lambda-specific primers, V5′P and V3′P, for identification of cDNA ends using premade cDNA libraries

A straightforward alternative approach to RACE relies on amplification of cDNA ends from commercially available cDNA libraries. The cDNA libraries represent the total RNA from a cell line or tissue integrated into lambda vectors as single copies. This strategy takes advantage of both stability and integrity of pre-made cDNA libraries. Commercially available cDNA libraries are designed and evaluated to preserve highest independent copy numbers of full-length cDNAs. In addition, the 3′ and 5′ adjunct arms of the vector allow for more stringent design for 3′ or 5′ amplification primers (Fig. 1a).

2 Materials

2.1 Rapid Amplification of cDNA Ends (RACE)

1. SMART RACE cDNA Amplification (Clontech, Inc; Mountain View, CA).
2. Advantage PCR Kit and Polymerase Mix (Clontech).
3. PCR clean-up Gel extraction NucleoTraPCR or NucleoTrap (Clontech).
4. February.
5. Thermocycler (e.g., Applied Biosystems, Inc).

2.2 Identification of cDNA Ends Using Premade Lambda cDNA Libraries

1. Storage medium: 100 mM NaCl, 50 mM Tris–Cl, pH 7.5 (25 °C), 10 mM $MgSO_4$, 2 % (w/v) gelatin, 7 % (w/v) DMSO.
2. 10× PCR buffer: 500 mM Tris–Cl, pH 9.2 (25 °C), 160 mM $(NH_4)_2SO_4$, 22.5 mM $MgCl_2$.

3 Methods

3.1 Rapid Amplification of cDNA Ends (RACE)

The following reaction is capable of converting 0.1–1 μg of poly(A+) RNA into first-strand cDNA. The use of poly(A+) RNA is recommended if there is an evidence for polyadenylation, such as poly(A) signal in genomic sequence or poly(A) tail at cDNA 3′ end. The use of total RNA may increase background noise and should only be considered if the target RNA is not polyadenylated.

1. Combine appropriate amount of poly(A+) RNA with 1 μl of a 12 μM 5′-antisense or 3′-sense cDNA primer, and 1 μl of SMART II A oligo included in the SMART RACE cDNA Amplification kit (Clontech). The appropriate amount of poly(A+) RNA strongly depends on the specific abundance of a transcript and should be determined using different amounts of poly(A+) RNA in parallel reactions.
2. Add RNase-free water to a final volume of 5.0 μl for each reaction.
3. Mix contents and spin tubes briefly in a microcentrifuge.
4. Incubate the tubes at 70 °C for 2 min.
5. Cool the tubes on ice for 2 min.
6. Spin briefly to collect contents to bottom of tubes.
7. Add 2 μl 5× First-Strand Buffer (SMART RACE cDNA Amplification Kit) and 1 μl MMLV Reverse Transcriptase to a total volume of 10 μl.
8. Mix reaction mixture gently by pipetting.
9. Spin the tubes briefly.
10. Incubate the tubes at 42 °C for 1.5 h.
11. Heat reaction at 72 °C for 10 min and store at −20 °C.

The reaction contains extended 3′ or 5′ cDNA sequences. Only a fraction of this material should be used for amplification of the cDNA end of interest (*see* **Note 1**).

12. Mix the following reagents for each PCR reaction:
    (a) 34.5 μl PCR-Grade Water
    (b) 5 μl 10× Advantage 2 PCR Buffer
    (c) 1 μl dNTP Mix (10 mM)
    (d) 1 μl 50× Advantage 2 Polymerase Mix

13. Mix well by vortexing (without introducing bubbles), and briefly spin in a microcentrifuge.
    For 5′-RACE add:
    (a) 2.5 μl 5′-RACE-Ready cDNA
    (b) 5 μl Universal Primer A Mix
    (c) 1 μl gene-specific 5′ primer GSP2 (10 μM)

    For 3′-RACE add:
    (a) 2.5 μl 3′-RACE-Ready cDNA
    (b) 5 μl Universal Primer A Mix
    (c) 1 μl gene-specific 3′ primer GSP2 (10 μM)
14. Perform 20–25 cycles using the following program (*see* **Note 2**)
    (a) 94 °C for 30 s
    (b) 78 °C for 30 s
    (c) 72 °C for 3 min (*see* **Note 3**)
15. Prepare a 1–1.5 % w/v agarose gel in TAE buffer and ethidium bromide (EtBr) to a final concentration of approximately 0.2–0.5 μg/ml (*see* **Note 4**).
16. Load 5 μl of each PCR amplification product on the gel; include a DNA molecular weight marker for size estimation, and begin electrophoresis at 60 V for 40 min or until the amplified fragment and unincorporated primers are clearly separated.
17. Locate the position of your fragment under UV light. If you find no or too many fragments, adjust the PCR conditions as outlined in Table 1.
18. Excise the DNA fragment of interest using a clean scalpel and transfer it to a clean 1.5-ml microcentrifuge tube (*see* **Note 5**).
19. For every 100 mg of agarose, add 300 μl of Buffer NE (NucleoTraPCR) and vortex the NucleoTrap Suspension thoroughly until the beads are completely resuspended.
20. Add 10 μl of NucleoTrap Suspension or more (4 μl of NucleoTrap Suspension for each 1 μg of DNA) and incubate the sample at 50 °C for 5–15 min. Vortex briefly several times during the incubation period.
21. Centrifuge the sample at 10,000 × *g* for 30 s at room temperature; discard supernatant.
22. Add 500 μl of Buffer NT2 to the pellet. Vortex briefly and centrifuge at 10,000 × *g* for 30 s at room temperature. Remove supernatant completely, and repeat **steps 20** and **21**.
23. Add 500 μl of Buffer NT3 to the sample. Vortex briefly and centrifuge the sample at 10,000 × *g* for 30 s at room temperature. Remove the supernatant completely and repeat this step.

Table 1
Troubleshooting guide for SMART RACE cDNA Amplification kit[a]

Problem	Recommendations
5′ or 3′-RACE product is not the expected size or is absent	The cause may be a GCrich template. Use Clontech Mixes for efficient amplification of GC-rich templates. PCR parameters may need to be optimized for these templates
No amplified 3′ or 5′ products after the minimum number of cycles at 68 °C	Return tube(s) to the thermal cycler and run five additional cycles. If the product still does not appear, add an additional three to five cycles at 68 °C. If you are still unsuccessful, run a new PCR experiment, changing the annealing temperature in the third set of cycles from 68 °C to 65 °C. This last program is especially useful if Tm close to 70 °C
Neither 3′ nor 5′ amplified products	Check the quality of first-strand cDNA (if generated from poly A+ RNA) using a 32P-labeling procedure. Repeat the first-strand synthesis, substituting 1 μl of 0.1 μCi/μl [α-32P] dATP or dCTP for 1 μl of water. Run the reaction products on an alkaline agarose gel, and examine the banding pattern by autoradiography. If the first-strand reaction was successful, you should see a banding pattern similar to that produced by your RNA. Mammalian poly A+ RNA typically produces a smear from 0.5 to 12 kb. Mammalian total RNA usually exhibits two bright bands at 1.9 kb and 4.5 kb
Multiple 5′- and/or 3′-RACE products	By multiple fragments you can generally start with the largest fragment from each RACE reaction, because it is most likely to be a true, complete RACE product. However, in the long run you should try to eliminate nonspecific fragments by troubleshooting the reactions

[a]The information in this table was partially taken and modified from SMART RACE cDNA Amplification kit manual from Clontech; Mountain View, CA

24. Air-dry DNA pellet for 10–15 min.
25. Add 20–50 μl of Buffer NE and resuspend the pellet by vortexing.
26. Elute DNA by incubating the sample at room temperature for 10–15 min.
27. Centrifuge the sample at 10,000 × *g* for 30 s at room temperature, and then transfer the supernatant containing the purified DNA to a fresh tube. The isolated fragment(s) can now be directly cloned into a T/A-type PCR cloning vector (e.g., Life Technologies).

3.2 Identification of cDNA Ends Using Premade cDNA Libraries

This approach relies on amplification of a cDNA sequences from commercially available human cDNA libraries which represent the total RNA from a cell line or tissue. The main advantage of premade libraries is that they contain high-quality, full-length cDNAs inserted into a self-replicating lambda vector. Once the sequence is available in the form of a cDNA library, individual processed segments of the original cDNA can be isolated and examined with relative ease.

For this method gene-specific and lambda vector-specific primers are utilized. The relationship of the required primers is shown in detail in Fig. 1b. In general, primers should be:

- 23–28 nt
- 50–70 % GC
- Tm ≥ 65 °C; the best results are obtained when Tm > 70 °C

For amplification of 5′ or 3′ ends, lambda vector-specific primers should hybridize 50 nt upstream or downstream to the cDNA cloning site, respectively.

1. Combine 107 plaque forming units of a λ cDNA library in 1–5 μl of storage medium.
2. Add 4.5 μl of 10× PCR buffer.
3. Add 5 μl of 10 mM dNTPs and 50 pmol of each lambda (V5′P or V3′P) and gene-specific primer (GS5′P or GS3′P).
4. Add sterile water to a final volume of 45 μl.
5. Incubate reaction at 95 °C for 10 min in a thermocycler to denature phage particles.
6. Incubate reaction 5 min at 75 °C. Within this step, add 5 μl of a pre-made master polymerase mix: 0.5 μl 10× PCR buffer, 0.5 μl polymerase, and 4 μl sterile water.
7. Continue with standard PCR reaction cycles. For example, 30 cycles of denaturation at 94 °C for 45 s, annealing at 63 °C for 30 s, and extension at 72 °C for 3 min.
8. Continue with the **steps 15–17** of Subheading 3.1 for detection of cDNA fragment of interest.
9. If you find no or too many cDNA fragments, adjust the PCR conditions as outlined in Table 2.
10. Continue with the **steps 18–27** of Subheading 3.1 for cloning and characterization of cDNA fragment of interest.

4 Notes

1. Prepare enough PCR Master Mix for all PCR reactions, plus one extra reaction, to ensure sufficient volume. The same Master Mix can be used for both 5′- and 3′-RACE reactions.
2. Because the necessary number of cycles depends on the abundance of the transcript, you may need to determine the optimal cycling parameters for your gene empirically.
3. If fragments >3 kb are expected, add 1 min for each additional 1 kb.

Table 2
Troubleshooting guide for cDNA end amplification using cDNA libraries[a]

Problem	Recommendations
5′ or 3′ RACE product is not the expected size or is absent	The cause may be a GC-rich template. Use buffer conditions for efficient amplification of GC-rich templates. PCR parameters may need to be optimized for these templates Depending on cDNA abundance or the quality of the cDNA library you may use up to 10^8 plaque forming units
No amplified 3′ or 5′ products after the minimum number of cycles at 68 °C	Change the annealing temperature in the third set of cycles from 72 °C to 68 °C. This last program is especially useful if Tm close to 70 °C
Neither 3′ nor 5′ amplified products	Check the quality of cDNA library using primers for abundant cDNAs
Multiple 5′- and/or 3′-RACE products	By multiple fragments you can generally start with the largest fragment from each PCR reaction. If you cannot isolate a single band, you should try to eliminate nonspecific fragments by troubleshooting the reactions by changing the annealing temperature, reducing cycle number, or using high stringent buffer conditions

[a]Although most convenient, the use of cDNA libraries in PCR reactions might cause artifacts based on high copy number of high abundant cDNAs in a library. Thus, you may require optimizing the PCR conditions step by step

4. Prepare and use a 1,000-fold master EtBr solution (0.2–0.5 mg/ml) and 1 μl/ml agarose gel volume.
5. EtBr solution must be handled with extreme caution and decontaminated prior to disposal. Caution is required by handling with UV which can cause serious damage to your eyes and skin.

References

1. Yeku O, Frohman MA (2011) Rapid amplification of cDNA ends (RACE). Methods Mol Biol 703: 107–122
2. Frohman MA, Dush MK, Martin GR (1988) Rapid production of full-length cDNAs from rare transcripts: amplification using a single gene-specific oligonucleotide primer. Proc Natl Acad Sci U S A 85: 8998–9002

Chapter 3

Single Nucleotide Mapping of RNA 5′ and 3′ Ends

Mahtab Nourbakhsh

Abstract

Nuclease protection assay is a sensitive method for detection, quantitation, and mapping of a specific RNA in an extremely heterogeneous mixture of RNAs, such as total cellular RNA. The assay is based on a small volume solution hybridization of a single-stranded synthetic antisense and labeled RNA probe to a RNA sample. Thus, it is much more efficient than the common immobilized hybridization on a membrane, such as in northern-blot analysis. After solution hybridization, different nucleases are used to remove any remaining single-stranded nucleotides within the probe and sample RNA by digestion. Then, the remaining probe-target hybrids are purified and separated on a denaturing polyacrylamide gel. Using a radioactive labeled probe, the protected probe can be visualized by direct autoradiography and the copy number can be calculated based on the specific radioactivity of the RNA probe and the length of protected fragment. Because of its high sensitivity and resolution, nuclease protection assay is the most effective procedure for mapping internal and external boundaries in mRNA compared to other RNA detection methods such as RT-PCR.

Key words Intron, 5′ UTR, 3′ UTR, Double stranded RNA, Hybridization, Nuclease, S1 RNA mapping

1 Introduction

Molecular characterization of genomic sequences involves the qualitative and quantitative analysis of possibly transcribed sequences. Transcribed regions can be hypothetically determined by aligning Expressed Sequence Tags (ESTs) with genome sequences [1]. However, available ESTs databases represent only a fraction of transcribed regions. This makes further experimental analysis indispensable.

The current book includes three most frequently used technical procedures for mapping, detecting, and quantifying a particular RNA in a total RNA sample: Northern blot analysis, nuclease protection assays, and reverse transcription-polymerase chain reaction (RT-PCR). Although each of these techniques can be used for qualitative and quantitative RNA detection, each procedure has decisive advantages and/or limitations. Northern-blot analysis is

M. Lucrecia Alvarez and Mahtab Nourbakhsh (eds.), *RNA Mapping: Methods and Protocols*, Methods in Molecular Biology, vol. 1182, DOI 10.1007/978-1-4939-1062-5_3, © Springer Science+Business Media New York 2014

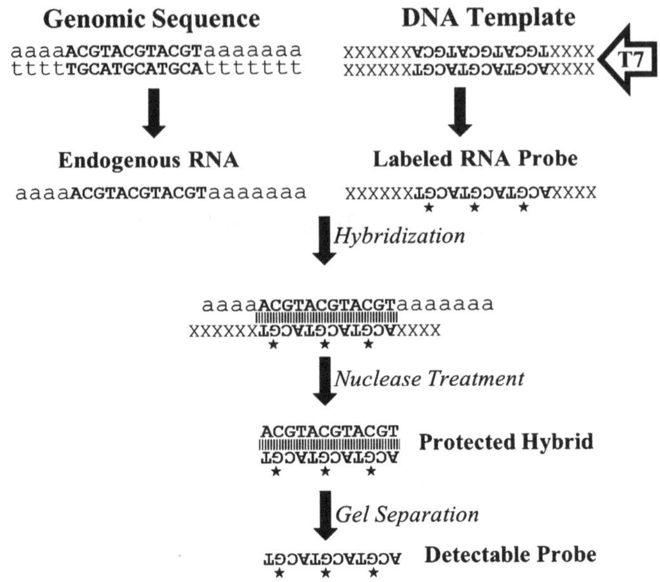

Fig. 1 Nuclease protection assay principle. The method is based on hybridization of a labeled synthetic antisense RNA probes to a target RNA which forms a double-stranded RNA fragment. S1 or similar nucleases remove single-stranded sequences. The remaining fragment signal corresponds to the length and the copy number of a target RNA. Digestion of the full-length probe is an indicative step. Therefore, unspecific vector sequences should be co-transcribed which overspans a possible homology region

the only method that can provide information about the size of the mature transcripts. RT-PCR is the most convenient, fast, and sensitive method for comparative detection and estimation of relative abundance of transcripts in different samples. The principle of nuclease protection assays is based on hybridization of a labeled synthetic antisense RNA probes to a target RNA which forms a double-stranded RNA fragment protected against specific single strand nucleases (*see* Fig. 1). Following nuclease treatment and removing of single-stranded sequences, the length and the copy number of protected probe fragments correspond to the boundaries and number of target RNA in a sample. As truncation of the full-length probe is the indicative step, it is crucial to use a probe which overspans the possible homology region. In most cases, probes are designed to convey unrelated vector sequences (*see* Fig. 1).

The potential use of different antisense probes in nuclease protection experiments allows for simultaneous quantification of different RNA species in a single sample [2]. Thus, several different target RNAs or several different fragments of a single target RNA can be simultaneously detected in samples. First, this approach can discriminate between closely related targets if probes are designed to span the regions where the related genes differ at the most. Second, probes can be designed to protect fragments of different sizes of target RNAs. Nuclease protection assay is thereby suitable for precise single nucleotide mapping of external and internal junctions in RNA including transcription initiation or termination sites as well as intron and exon boundaries. This method is an important tool for analysis of alternative splicing [3].

Although DNA can serve as a probe as well, RNA probes are mostly preferred because RNA–RNA duplexes are more stable than RNA–DNA duplexes in solution hybridization. The antisense RNA probe is usually synthesized using available in vitro Transcription kits and radioactive nucleotides. This step requires a DNA template containing antisense sequence downstream to a synthetic promoter, such as T7. Although end-labeling of RNA probe is feasible as well, incorporation of numerous radioactive nucleotides during transcription reaction reveals a probe with significantly higher specific activity. Using the MAXIscript Kit described in this chapter, radiolabeled RNA probes can be synthesized in a 10 min reaction. The Kit can be used to incorporate any labeled nucleotide into RNA using Sp6, T3 or T7 polymerases.

Following in solution hybridization of RNA target and antisense probe a nuclease treatment removes all single-stranded sequences, precisely at the nucleotide 3′ and 5′ to double stranded sequence. The last stage of the assay involves the separation of protected probe and RNA fragments in a denaturing gel. A short gel run (15 cm) is sufficient for most quantification purposes. For precise mapping experiments and exact determination of the size of the protected fragments, they may be resolved on a denaturing sequencing gel (60 cm) in combination with a RNA marker or "sequencing ladder" reaction [4].

This chapter provides the complete RNase protection assay protocol including support protocols for synthesis of labeled probes, and quantitation and mapping of target mRNA.

2 Materials

2.1 Synthesis of RNA Antisense Probe

1. MAXIscript® kit (Life Technologies).
2. DNA template (*see* **Note 1**).
3. Labeled nucleotide (*see* **Note 2**).
4. Trichloroacetic acid: molecular biology grade.
5. Ethanol: ACS reagent grade.
6. 0.5 M EDTA.
7. Centri-Spin™ 40 Columns (Life Technologies).

2.2 Probe-Target Hybridization and RNAse Digestion

1. RPA III™ ribonuclease protection assay kit (Life Technologies).
2. Heat block (42–45 °C and 85–95 °C).
3. RNAse-free polypropylene microfuge tubes and pipette tips.
4. Microcentrifuge (10,000 × *g*).
5. 100 % ethanol (ACS grade).
6. Trichloroacetic acid (molecular biology grade).

2.3 Denaturing Acrylamide Gel

1. Urea (high quality).
2. 40 % Acrylamide (acryl–bis-acryl = 19:1).
3. 10× TBE (0.9 M Tris base, 0.9 M Boric Acid, 20 mM 0.5 M EDTA).
4. 10 % ammonium persulfate.
5. EMED.
6. Vertical S2 Gel Electrophoresis Apparatus Life Technologies.
7. Power supply.
8. Gel Dryer Model 583 Bio-Rad.

3 Methods

3.1 Synthesis of RNA Antisense Probe

1. Thaw the frozen reagents, mix, and microfuge briefly to prevent loss and/or contamination of material by opening the lid.
2. Keep all reagents on ice except the 10× Transcription Buffer.
3. Vortex the 10× Transcription Buffer several times at room temperature until it is completely in solution (*see* **Note 3**).
4. Assemble transcription reaction according to manufacturer's recommendations at room temperature by adding the DNA, water, nucleotides, and 10× Transcription Buffer (*see* **Note 4**).
5. Mix thoroughly by pipetting the mixture up and down gently.
6. Incubate the reaction for 10 min to 1 h at 37 °C (*see* **Note 5**).
7. Add 1 μl TURBO DNase, mix well, and incubate at 37 °C for 15 min (*see* **Note 6**).
8. Add 1 μl of 0.5 M EDTA to stop the reaction to inactivate DNase and block the heat-induced RNA degradation.
9. Purify the transcripts using Centri-Spin™ 40 Columns (*see* **Note 7**).

3.2 Hybridization and RNase Digestion

The amount of RNA probe required will depend on the abundance of the mRNA being detected and on the specific activity of the probe. 5–20 μg of total RNA is sufficient for most purposes. It is important to set up the hybridization with threefold to tenfold molar excess of the probe over the target mRNA.

Further detailed guidelines for optimizing amounts of probe and sample RNA are given in manufacturer's protocol.

1. For each experimental tube, mix about 150–600 pg (2–8 × 10^4 cpm) of RNA antisense probe per 10 μg total sample RNA (0.6 μg poly(A)).
2. For each different probe used, include two control tubes containing the same amount of labeled probe used for the experimental tubes in **step 1**, plus Yeast RNA equivalent to the highest amount of sample RNA (*see* **Note 8**).
3. Add 1/10th volume of 5 M NH_4OAc.
4. Add 2.5 volumes of ethanol, mix thoroughly, and allow RNA to precipitate at −20 °C for at least 15 min or overnight (*see* **Note 9**).
5. Centrifuge at maximum speed in a microcentrifuge (≥10,000 × *g*) for 15 min at 4 °C (*see* **Note 10**).
6. Remove the supernatants carefully and air-dry the pellets for 5 min.
7. Add 10 μl of Hybridization Buffer III to each pellet, vortex each tube briefly, then microfuge for a few seconds.
8. Heat samples to 90–95 °C for 3–4 min to denature the RNA.
9. Vortex tubes after the incubation and microfuge briefly.
10. Incubate at 42 °C overnight for hybridization.
11. Prepare a master mix dilution of RNase in RNase Digestion III Buffer (150 μl buffer and 1.5 μl RNAse A/T1 mixture per reaction).
12. Briefly centrifuge the sample tubes to remove condensation in the tube.
13. Add 150 μl of the RNase mix to each reaction.
14. Vortex and microfuge tubes briefly (*see* **Note 11**).
15. Incubate the tubes for 30 min at 37 °C.
16. Add 225 μl RNase Inactivation Solution III and incubate for 15 min at −20 °C.
17. Centrifuge the tubes for 15 min at maximum speed at 4 °C.
18. Carefully remove all supernatant from each tube and air-dry the pellets for 15 min.

3.3 Separation and Detection of Protected Fragments

The gel size and acrylamide concentration will be dictated by the experiment; specifically, the number and sizes of probes, and their relation to each other. A 5 % acrylamide gel will effectively resolve fragments of about 50–1,000 nucleotides.

1. Prepare 45 ml 5 % acrylamide gel using 21.6 g urea, 4.5 ml 10× TBE, 5.7 ml 40 % acrylamide (acryl–bis-acryl = 19:1) and water to 45 ml.

2. Stir at room temperature until the urea is completely dissolved.
3. Follow the manufacturer's instructions for the details of attaching the glass plates to pour the gel.
4. Add 60 μl 10 % ammonium persulfate to gel solution.
5. Start the polymerization by adding 48 μl TEMED and mix briefly.
6. Start to pour gel immediately and wait 60 min to complete polymerization.
7. Follow the manufacturer's instructions to set up the gel and S2 apparatus.
8. Use 1× TBE as the gel running buffer.
9. It is very important to rinse the wells of urea-containing gels immediately before loading the samples.
10. Heat the samples at 85 °C for 5 min and chill on ice before loading to gel.
11. Add appropriate RNA size marker or sequence marker to a single well for RNA mapping experiments (*see* **Note 12**).
12. Run the gel about 60 W.
13. Dry gel after the run using a vacuum gel dryer.
14. Use autoradiography or digital radioactive imaging system to assess the recovered product.
15. If you detect no or too many fragments like a smear, adjust assay conditions according to Table 1.

4 Notes

1. To synthesize a labeled antisense RNA probe from DNA templates in vitro, SP6, T3, and T7 phage RNA polymerases are widely used. The template must have a double-stranded 19–23 base promoter upstream of the sequence to be transcribed. Many commercially available cloning vectors contain two or more separate phage promoters flanking a multiple cloning site.
2. The MAXIscript Kit can be used to incorporate virtually any labeled nucleotide into RNA. Traditionally 32P labeled UTP or CTP has been used in the MAXIscript Kit, but other isotopically labeled nucleotides (33P, 35S, 3H) can be used with this kit as well.
3. The spermidine in the 10× Transcription Buffer can precipitate.
4. 10× Transcription Buffer can coprecipitate the template DNA if the reaction is assembled on ice.
5. Incubate reactions with 3–10 μM limiting nucleotide for 10 min and reactions with >10 μM limiting nucleotide for 1 h.

Table 1
Troubleshooting guide

Problem	Recommendations
The target is not present in the sample	Overdigestion with RNase is rarely seen and would lead to smearing of the signal below the expected position of the protected fragment Confirm that absence of signal is a legitimate result. Use a separate sample known to contain detectable levels of the target RNA. If this is not possible, in vitro synthesized sense-strand RNA can be added to a RNA samples to serve as a positive control. Increase the sensitivity of the assay using a longer probe with a higher specific activity, or increase the amount of target RNA up to about 50 μg of total RNA per hybridization reaction in the assay. In some cases better results can be achieved by higher hybridization temperatures
Smear or ladder in the no-target/no-RNase control lane	Gel wells might be overloaded which reduces the resolution of the bands. Use wider gel wells Degradation of probe is possible. This is most often from radiolysis, but it can also be due to RNase contamination of the probe solution, your tubes, or pipette tips. Resynthesize the probe, avoid contamination and do not store the prober at −20 °C longer than a week
Full-length probe is seen in all lanes	RNase(s) were completely inactive or were omitted. Use a new batch of enzvme Too much probe might be added to the reaction. No more than $2–8 \times 10^4$ cpm of high specific activity probe should be used for up to 10 μg of RNA Residual DNA is protecting the probe from digestion. Use less DNA template in transcription reaction. Alternatively, transcription reaction can be treated with RNase-free DNase I before hybridization
Aberrant, pointed, or smeared bands appear	The supernatant from the final precipitation step might be not completely removed. Salt may lead to aberrant migration ("tunneling") of the protected fragment. We recommend removing the residual supernatant thoroughly using a very fine-tipped pipette

6. DNase digestion is important to remove the template as it can hybridize to the probe and cause false positive signals.
7. By heterogeneous length of probe, it is necessary to isolate primarily full length probe. We recommend purification of probe using polyacrylamide gel as described in manufacturer's protocol.
8. Yeast RNA is not an appropriate control if the probe is expected to hybridize with sequences found in yeast RNA.
9. Extended storage of radiolabeled probes will result in radiolysis.
10. To make it easier to locate the pellets, it is helpful to position all the tubes with the hinges of the lids facing away from the center of rotation. The pellets will all form directly below the hinges.

11. It is important to have no-RNase control tube(s) which serves as a control for probe integrity. It will also show the gel migration of the full-length probe. If there is any unexpected degradation of the probe, it will be seen in this control. Ideally, this lane should show a single band.

12. The basic requirement for RNA mapping is that the probe spans the region to be mapped. This usually means that the probe is derived from a genomic clone, as opposed to a cDNA clone. For example, in order to map the transcription initiation site for a given mRNA, a probe is prepared by subcloning and transcribing a genomic fragment that extends from upstream of the gene of interest to some point in the first exon. Probe synthesis, purification, hybridization, and RNase digestion are carried out using the standard RNase protection assay. The transcription start site is mapped by comparing the size of the protected fragment to the size of the undigested probe. For exact determination of the size of the protected fragment, the sample is analyzed on a gel in conjunction with a sequencing reaction of the RNA probe or RNA maker.

References

1. Ma C, Wang J, Li L, Duan MJ, Zhou YH (2011) Identification of true EST alignments for recognising transcribed regions. Int J Data Min Bioinform 5:465–484
2. Hobbs MV, Weigle WO, Noonan DJ, Torbett BE, McEvilly RJ, Koch RJ, Cardenas GJ, Ernst DN (1993) Patterns of cytokine gene expression by CD4+ T cells from young and old mice. J Immunol 150:3602–3614
3. Kekule AS, Lauer U, Meyer M, Caselmann WH, Hofschneider PM, Koshy R (1990) The pre S2/S region of integrated hepatitis B virus DNA encodes a transcriptional transactivator. Nature 343:457–461
4. Gelfman S, Ast G (2013) When epigenetics meets alternative splicing: the roles of DNA methylation and GC architecture. Epigenomics 5:351–353

Chapter 4

Analysis of RNA Secondary Structure

Mahtab Nourbakhsh

Abstract

RNA has different levels of structural organization. The primary structure is the linear order of the nucleotide monomers, the RNA sequence. During transcription process, the partially synthesized RNA is folded by base-pairing and thermodynamic intramolecular or intermolecular interactions. This results in a dynamic spreading of a secondary structure along the length of the transcribed section of the RNA. The analysis of both primary or secondary structures requires the RNA end-labeling either at its 5′ end using a kinase reaction with [gamma-32P]ATP, or at its 3′ end using an RNA ligation reaction with [32P]pCp. End-labeled RNAs are then gradually breakdown using hydrolysing chemicals or a variety of enzymes targeting specific RNA sequences and secondary structure. The most commonly used enzymes are RNase A, T1, and V1. The partial digestion of the RNA reveals a mix of truncated RNA fragments of different lengths, called RNA ladder. The products are then separates through a high resolution gel system and subjected to autoradiographic analysis. Each visible fragment is labeled at one end, but comprises an enzyme specific sequence at the other end. Final comparison of the detected RNA ladders reveals a hypothetical model of the secondary RNA structure under assay conditions.

Key words Secondary structure, Double-stranded RNA, Single-stranded RNA, RNAse, Mfold

1 Introduction

RNA molecules possess a variety of single-stranded and double-stranded regions that lead to complex three-dimensional structures. These structures are mostly crucial for the molecule's interactions with other regulatory molecules like nucleic acids and proteins. Thus, RNA structure plays a central role in many cellular processes, including transcription initiation, elongation and termination, regulation of gene expression, and protein translation. Thus, elucidating the mechanistic aspects of RNA interactions often requires a detailed understanding of the underlying RNA structure.

Analysis of RNA structure is traditionally based on successive enzymatic cleavage of folded RNA using specific RNases. First, RNA of interest needs to be synthesized in vitro using available in vitro transcription systems. This step requires a DNA template containing the RNA sequence of interest localized downstream to

M. Lucrecia Alvarez and Mahtab Nourbakhsh (eds.), *RNA Mapping: Methods and Protocols*, Methods in Molecular Biology, vol. 1182, DOI 10.1007/978-1-4939-1062-5_4, © Springer Science+Business Media New York 2014

a synthetic promoter, T3, T7, or SP6 (*see* **Note 1**). Using the MAXIscript Kit described in this chapter, high amount of RNA can be synthesized in a 10-min reaction using corresponding polymerase. Next, the RNA product needs to be labeled at 5′ or 3′ end using a kinase reaction with [gamma-32P]ATP or an RNA ligation reaction with [32P]pCp, respectively.

T4 Polynucleotide Kinase (PNK) catalyzes the transfer of the gamma-phosphate of ATP to the 5′-hydroxyl termini of RNA. This phosphate transfer is commonly referred to as a kinase or phosphorylation reaction. RNAs with a 5′-hydroxyl (OH) group can be added directly to a kinase reaction. However, RNAs with a 5′-phosphate should be dephosphorylated first using Calf Intestinal Phosphatase prior to labeling with PNK and [gamma-32P]ATP.

T4 RNA ligase catalyzes the ligation of the 5′ phosphate terminus of a nucleic acid donor to the 3′ OH terminus of a nucleic acid acceptor. The reaction is ATP dependent, and the 3′ end of target RNA is labeled by adding [32P]pCp to the reaction.

The most commonly used RNases for performing RNA structural analysis are RNases A, V1, and T1. These enzymes bind to specific sequences and cleave folded RNA at specific sequence patterns. RNase A is a pancreatic ribonuclease that cleaves the target RNA 3′ of single-stranded C and U residues. It cleaves the phosphodiester bond between the 5′-ribose of a nucleotide and the phosphate group attached to the 3′-ribose of an adjacent pyrimidine nucleotide. The resulting 2′, 3′-cyclic phosphate is hydrolysed to corresponding 3′-nucleoside phosphate. RNase V1 is a metal-dependent ribonuclease specific for dsRNA regions of 4 nucleotides or more. Cleavage occurs between the 3′-hydroxyl of any ribonucleotide and the 5′-phosphate group of the adjacent ribonucleotide. Commercially available RNase T1 is isolated by a series of purification steps from recombinant *E. coli* strains overexpressing the RNase T1 gene of *Aspergillus oryzae*. The purified enzyme specifically cleaves the target RNA 3′ to single-stranded guanosine residues, producing 3′-phosphorylated ends.

To facilitate RNA structural studies, the exposure time of RNA to RNase is strictly limited allowing for a single cleavage per RNA strand. The end products of RNase reactions comprise of labeled RNA ladders which can be analyzed using high resolution gel electrophoresis. To help identify the cleavage site locations, another ladder is generated by alkaline hydrolysis which cleaves RNA strand by single nucleotides.

In addition to the described experimental analysis here, we recommend the use of available software for predicting RNA structure [1–3]. Secondary structure is the set of the canonical base pairs, and secondary structure can be predicted by comparative sequence analysis in silico. The most commonly used method is free energy minimization. The accuracy of structure prediction is then improved either by using experimental mapping data or by predicting a structure conserved in a set of homologous sequences. Additionally, tertiary structure, the three-dimensional arrangement of atoms, can be modeled with guidance from comparative analysis and experimental techniques. New approaches are also available for predicting tertiary structure.

2 Materials

2.1 RNA Synthesis

1. MAXIscript® kit (Life Technologies).
2. DNA template (*see* **Note 1**).
3. Ribonucleotides.
4. Trichloroacetic acid: molecular biology grade.
5. Ethanol: ACS reagent grade.
6. 0.5 M EDTA.
7. Centri-Spin™ 40 Columns (Life Technologies).

2.2 RNA 3′ End Labeling

8. T4 RNA Ligase (Ambion Cat #2140).
9. 10× T4 RNA Ligase Buffer (supplied with Ambion's T4 RNA Ligase: 0.5 M Tris–HCl, pH 7.8, 0.1 M $MgCl_2$, 0.1 M DTT, 10 mM ATP).
10. [32P]pCp.
11. RNase-free Sephadex G-25 or G-50 spin columns such as Ambion's NucAway™ Spin Columns (Cat #10070).

2.3 RNA 5′ Labeling

1. Nuclease-free Water.
2. 10× Dephosphorylation buffer (0.5 M Tris–HCl, pH 8.5, 1 mM EDTA, pH 8).
3. Calf Intestinal phosphatase (CIP; 0.1 U/μl).
4. Phosphatase removal reagent (available in the KinaseMax™ Kit).
5. [gamma-32P]ATP (7,000 Ci/mmol, 150 mCi/ml).
6. 10× Kinase buffer (500 mM Tris, pH 7.5, 100 mM $MgCl_2$, 50 mM DTT).
7. T4 Polynucleotide Kinase (10 U/ml).
8. RNase-free Sephadex G-25 or G-50 spin columns such as Ambion's NucAway™ Spin Columns (Cat #10070).

2.4 RNA Hydrolysis

1. 0.1–3 μg end-labeled RNA.
2. Yeast RNA (10 mg/ml, Ambion Cat # 7118).
3. 1× Alkaline Hydrolysis Buffer (supplied with Ambion's RNA Grade Ribonucleases: 50 mM Sodium Carbonate [$NaHCO_3$/Na_2CO_3] pH 9.2, 1 mM EDTA).

2.5 RNA Enzymatic Digestion

4. 0.2–4 μg end-labeled RNA.
5. Yeast RNA (10 mg/ml, Ambion Cat # 7118).
6. 10× RNA structure buffer (supplied with Ambion's RNA Grade Nucleases: 100 mM Tris pH 7.0, 1 M KCl, 100 mM $MgCl_2$).
7. RNase T1 (1 U/μl, Ambion Cat # 2283).
8. Inactivation/precipitation buffer (supplied with Ambion's RNA Grade Nucleases).
9. 100 % ethanol.

2.6 High Resolution Gel Analysis

1. Acrylamide Gel Loading Buffer (Supplied with Ambion's RNA Grade Ribonucleases: 95 % Formamide, 18 mM EDTA, 0.025 % SDS, 0.025 % Xylene Cyanol, 0.025 % Bromophenol Blue; or Gel Loading Buffer II Cat #8546G).
2. Urea (high quality).
3. 40 % Acrylamide (acryl–bis-acryl = 19:1).
4. 10× TBE (0.9 M Tris base, 0.9 M Boric Acid, 20 mM 0.5 M EDTA).
5. 10 % ammonium persulfate.
6. EMED.
7. Vertical S2 Gel Electrophoresis Apparatus Life Technologies.
8. Power supply.
9. Gel Dryer Model 583 Bio-Rad.

3 Methods

3.1 RNA Synthesis

1. Thaw the frozen reagents, mix, and microfuge briefly to prevent loss and/or contamination of material by opening the lid.
2. Keep all the reagents on ice except the 10× Transcription Buffer.
3. Vortex the 10× Transcription Buffer several times at room temperature until it is completely in solution (*see* **Note 2**).
4. Assemble transcription reaction according to manufacturer's recommendations at room temperature by adding the DNA, water, nucleotides, and 10× Transcription Buffer (*see* **Note 3**).
5. Mix thoroughly by pipetting the mixture up and down gently.
6. Incubate the reaction for 10 min to 1 h at 37 °C.
7. Add 1 μl TURBO DNase, mix well, and incubate at 37 °C for 15 min (*see* **Note 4**).
8. Add 1 μl of 0.5 M EDTA to stop the reaction to inactivate DNase and block the heat-induced RNA degradation.
9. Purify the transcripts using Centri-Spin™ 40 Columns (*see* **Note 5**).

3.2 RNA 3′ End Labeling

1. Combine 2 μl 10× T4 RNA Ligase Buffer, 50–100 pmol RNA, equimolar amount (50–100 pmol) [32P]pCp and RNase-free water to a final volume of 18 μl in a single RNase-free microfuge tube.
2. Mix thoroughly by pipetting the mixture up and down gently.
3. Add 2 μl T4 RNA Ligase (10 U) and mix thoroughly by pipetting the mixture up and down gently.
4. Incubate at 4 °C overnight (10–12 h).
5. Remove unincorporated nucleotides by applying the mixture to an RNase-free Sephadex G-25 or G-50 spin column (e.g., NucAway Spin Columns) following the manufacturer's recommendations.
6. If the RNA is not labeled efficiently, follow the instructions in Table 1.

3.3 RNA 5′ End Dephosphorylation and 5′ End Labeling

1. Combine Nuclease-free Water to make a final volume of 10 μl, 0.1–10 pmol RNA, 1 μl 10× dephosphorylation buffer (0.5 M Tris–HCl, pH 8.5, 1 mM EDTA, pH 8) and 1 μl Calf Intestinal Phosphatase (CIP; 0.1 U/μl) in a single RNase-free microfuge tube.
2. Mix thoroughly by pipetting the mixture up and down gently.
3. Incubate for 1 h at 37 °C.
4. Remove the Calf Intestine Alkaline Phosphatase by use of the Phosphatase Removal Reagent (available in the KinaseMax™ Kit).
5. Combine nuclease-free water to make a final volume of 20 μl, 25 pmol [gamma-32P]ATP (7,000 Ci/mmol, 150 mCi/ml), 2 μl 10× Kinase Buffer (500 mM Tris, pH 7.5, 100 mM $MgCl_2$, 50 mM DTT), and 1 μl T4 Polynucleotide Kinase (10 U/ml).
6. Mix thoroughly by pipetting the mixture up and down gently.
7. Incubate at 37 °C for 1 h.
8. Remove unincorporated nucleotides by applying the mixture to an RNase-free Sephadex G-25 or G-50 spin column (e.g., NucAway Spin Columns) following the manufacturer's recommendations.
9. If the RNA is not labeled efficiently, follow the instructions in Table 1.

3.4 RNA Secondary Structure Analysis

3.4.1 Alkaline Hydrolysis

This procedure provides a gel electrophoresis "ladder" of hydrolyzed RNA fragments. In the procedure, three different hydrolysis times are used. After the experiment, select the ladder that provides the best distribution of nucleic acids over the range of lengths needed for your experiments.

1. Mix 0.1–3 μg of end-labeled RNA and 3 μg of yeast tRNA in 5 μl or less.

Table 1
Troubleshooting guide

Problem	Recommendations
Samples do not label well	The quality of the RNA preparation is a crucial factor in the labeling reaction. Make sure your RNA does not contain small RNA or DNA fragments, residual salts, or monovalent cation concentrations ≥100 mM. You may clean up the RNA by spin-column purification Your RNA might form a complex or strong tertiary structure decreasing the efficiency of the CIP or kinase reaction. Preheating the RNA to 90 °C for 2 min and immediate transfer on ice might dissolve any secondary structure in the RNA
The abundance of detected fragments is very low	The RNAses are optimized and purified for RNA structure analysis, RNA end mapping experiments. The enzymes cannot be contaminated with other nucleases causing unexpected cleavage sites. Thus, we recommend that the concentration of target RNA and enzymes need to be optimized in individual experiments. Reduce the concentration of the nucleases to increase the abundance of cleavage products
Identical fragments observed by 5′ labeled RNAs independent of nuclease concentration or enzyme	In most cases, this is caused by unexpected termination of transcription during probe synthesis based on truncated templates. Examine the integrity and length of the DNA template In some cases a strong tertiary structure can cause a nick in the RNA which might affçct the run pattern of fragments in gel analysis. Examine the denaturation step and temperature. You may denaturate the samples at higher temperature or for an extended time
No RNA fragments can be observed in all reactions	This indicates the RNAse contamination in the synthesized RNA. In this case you will need to start all over again and prepare new DNA template

2. Add sufficient 1× Alkaline hydrolysis buffer to bring the final volume to 15 μl.
3. Aliquot 5 μl of the RNA–buffer mixture into three tubes labeled 1–3.
4. Heat the tubes to 95 °C to denaturation.
5. After 2 min, remove Tube #1 to an ice bucket.
6. After 5 min, remove Tube #2 to an ice bucket.
7. After 15 min, remove Tube #3 to an ice bucket.
8. Add 10 μl of Acrylamide Gel Loading Buffer to each of the three tubes. For an untreated control, mix 1 μl of 5′ end-labeled RNA with 8 μl of Acrylamide Gel Loading Buffer.

Table 2
Selectivity of RNases

RNase	Cleavage site
A	3′ of single-stranded cytosine or uracil nucleotides
V1	Any base-paired nucleotide
T1	3′ of single-stranded guanosine nucleotide

3.4.2 RNA Enzymatic Digestion

This procedure uses RNase A, T1, or V1 to provide information about the structure of an RNA molecule. The sequence selectivity of these enzymes is listed in Table 2. In the procedure, four tubes are prepared. Tube 1 contains a negative control of end-labeled RNA without nuclease treatment. Non-full length bands in this sample represent cleavage products present within the labeled RNA itself. These bands will also be present in the nuclease treated samples and should be disregarded in your analysis. Tubes 2, 3, and 4 contain decreasing amounts of ribonuclease. Bands evident in the samples with lower amounts of ribonuclease typically represent nucleotides that are most accessible to the nuclease. Further enzyme dilution may be necessary to achieve the optimal digestion ladder.

1. Add 3.2 ml of 100 % ethanol to the Inactivation/Precipitation Buffer.
2. Thaw the 10× RNA Structure Buffer and end-labeled RNA by incubating at room temperature.
3. Mix 4 μl of 10× RNA Structure Buffer, 0.2–4 μg end-labeled RNA, 4 μg yeast RNA, and Nuclease-Free Water to bring the final volume to 36 μl.
4. Aliquot 9 μl of the RNA–buffer–water mixture into four tubes numbered 1–4. To Tube #2 add 1 μl of ribonuclease. Mix thoroughly by pipetting.
5. Transfer 1 μl from Tube #2 to Tube #3. Mix thoroughly by pipetting.
6. Transfer 1 μl from Tube #3 to Tube #4. Mix thoroughly by pipetting.
7. Incubate all tubes at room temperature for 15 min.
8. Add 20 μl of Inactivation/Precipitation Buffer and vortex.
9. Incubate at −20 °C for 15 min, centrifuge at 1,600 × *g* for 15 min, aspirate, and wash the pellet with 70 % ethanol.
10. Dry the pellet at the room temperature and dissolve in 7 μl of Acrylamide Gel Loading Buffer.

11. For denaturation incubate the RNA at 95 °C for 5 min.
12. Analyze 3 μl of the product, concomitantly with hydrolized RNA ladder using a 6–20 % acrylamide/7 M Urea high resolution sequencing gel.
13. Dry gel after the run using a vacuum gel dryer.
14. Use autoradiography or digital radioactive imaging system to assess the digestion product.
15. If you detect no or too many identical fragments in all reactions, follow the instructions in Table 1.

4 Notes

1. SP6, T3, and T7 phage RNA polymerases are widely used to transcribe RNA from DNA templates in vitro. The template must have a double-stranded 19–23 base promoter upstream of the sequence to be transcribed. Many commercially available cloning vectors contain two or more separate synthetic promoters flanking a multiple cloning site.
2. The spermidine in the 10× Transcription Buffer can precipitate.
3. 10× Transcription Buffer can coprecipitate the template DNA if the reaction is assembled on ice.
4. DNase digestion is important to remove the template as it can hybridize to the probe and cause false positive signals.
5. By heterogeneous length of probe, it is necessary to isolate primarily full length probe. We recommend purification of probe using polyacrylamide gel as described in manufacturer's protocol.

References

1. Mathews DH, Turner DH, Zuker M (2007) RNA secondary structure prediction. In: Beaucage S, Bergstrom DE, Glick GD, Jones RA (eds) Current protocols in nucleic acid chemistry. Wiley, New York, pp 1–11
2. Zuker M (2003) Mfold web server for nucleic acid folding and hybridization prediction. Nucleic Acids Res 31:3406–3415
3. Hofacker IL, Lorenz R (2014) Predicting RNA structure: advances and limitations. Methods Mol Biol 1086:1–19

Chapter 5

Tertiary Structure Mapping of the Pri-miRNA miR-17~92

Steven G. Chaulk and Richard P. Fahlman

Abstract

The understanding of RNA in regulating gene expression has exploded over the past 15 years. MicroRNAs (miRNAs) have vastly expanded the role of RNA in gene regulation beyond spliceosomal, ribosomal, and messenger RNAs. Approximately one half of miRNAs are polycistronic, where two or more miRNAs are encoded on a single pri-miRNA transcript, termed a miRNA cluster. The six miRNAs of the miR-17~92 cluster are contained within a ~800 nucleotide region within intron 3 of the *c113orf25* ~7 kb pri-miRNA transcript. We recently reported on the tertiary structured domain of miR-17~92 and its role in modulating miRNA biogenesis. The key finding was that the cluster structure explained the differential processing of the miRNA hairpins by Drosha. This work demonstrated the need to consider pri-miRNA tertiary structure in miRNA biogenesis. Since biochemical structure probing is typically performed on relatively short RNAs (≤200 nucleotides), we had to adapt these methodologies for application on large RNAs (~800 nucleotide miR-17~92 pri-miRNA). We present here our adaptation of a protection footprinting method using ribonucleases to probe the structure of the ~800 nucleotide miR-17~92 pri-miRNA. We outline the technical difficulties involved in probing large RNAs and data visualization using denaturing polyacrylamide gel electrophoresis and how we adapted the existing approaches to probe large RNAs. The methodology outlined here is generally applicable to large RNAs including long noncoding RNAs (lncRNA).

Key words Pri-miRNA, miR-17~92, RNA footprinting, Ribonuclease, miRNA biogenesis, RNA structure

1 Introduction

RNA footprinting has been a standard method for investigating protein–nucleic acid interaction and RNA tertiary structure (Fig. 1). Some examples of methods used to probe tertiary structure of both large and small ribozymes include the use of hydroxyl radicals generated from Fe-EDTA Fenton chemistry, peroxynitrous acid decay or from UV photolysis of thiones [1–4]. Cleavages of end-labeled RNA are resolved by denaturing PAGE (polyacrylamide gel electrophoresis) and visualized by autoradiography. The typical 31 × 38.5 cm gel used for this experiment allows for nucleotide resolution of up to ~170 nucleotides of sequence. For RNAs less than ~200 nucleotides

M. Lucrecia Alvarez and Mahtab Nourbakhsh (eds.), *RNA Mapping: Methods and Protocols*, Methods in Molecular Biology, vol. 1182, DOI 10.1007/978-1-4939-1062-5_5, © Springer Science+Business Media New York 2014

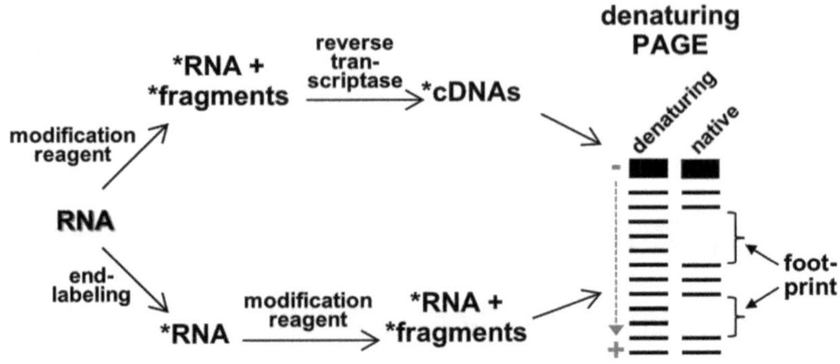

Fig. 1 Visualization of RNA footprinting experiments can use either a direct end-labeling or a reverse transcription based approach. RNA fragments from a footprinting experiment can be reverse-transcribed into end-labeled cDNAs using an end-labeled DNA probe in the reverse transcription reaction. The end-labeled cDNAs are then resolved by denaturing PAGE for subsequent data analysis. Alternatively, if end-labeled RNA is used in the footprinting experiment, the resulting end-labeled RNA fragments are resolved by denaturing PAGE. In a footprinting experiment, RNA is probed under denaturing and native conditions. Footprints are areas of reduced intensity in the native lane versus the denaturing lane and indicate regions of the RNA that are internalized by tertiary structure formation and are thus inaccessible to the cleavage reagent. The direction of electrophoresis is indicated by an *arrow*

in length, one gel is sufficient to analyze the probing data of a 5′ or 3′ end-labeled RNA giving information along the entire length of the RNA. Larger RNAs (200–400 nucleotides in length) require probing of both 5′ and 3′ end-labeled RNA to cover the entire sequence [5].

Reverse transcription-based methods (Fig. 1) are an alternative to direct end-labeling of RNA for structure probing, where RNA structure probing is performed by chemical cleavage of the phosphodiester backbone, or modification of the Watson-Crick base-pairing face of the nucleotides [1, 6–8]. The RNA cleavage or modification pattern is inferred from a reverse transcriptase assay; where the reverse transcriptase stops correspond to sites of RNA cleavage or modification. Reverse transcription-based methods are not limited by the size of the RNA and have been successfully applied to large RNAs like ribosomal RNA [9]. We attempted to apply this approach to the study of the miR-17~92 cluster but, unfortunately, we were unable to establish conditions where the reverse transcriptase read-through was efficient enough to enable visualization of a footprint pattern. Extensive stable secondary structure elements in the miR-17~92 cluster, depicted schematically in Fig. 2, created severe premature stops for the transcriptase even when using elevated extension temperatures (55 °C) and a thermal stable reverse transcriptase (Invitrogen Superscript III) (Fig. 3).

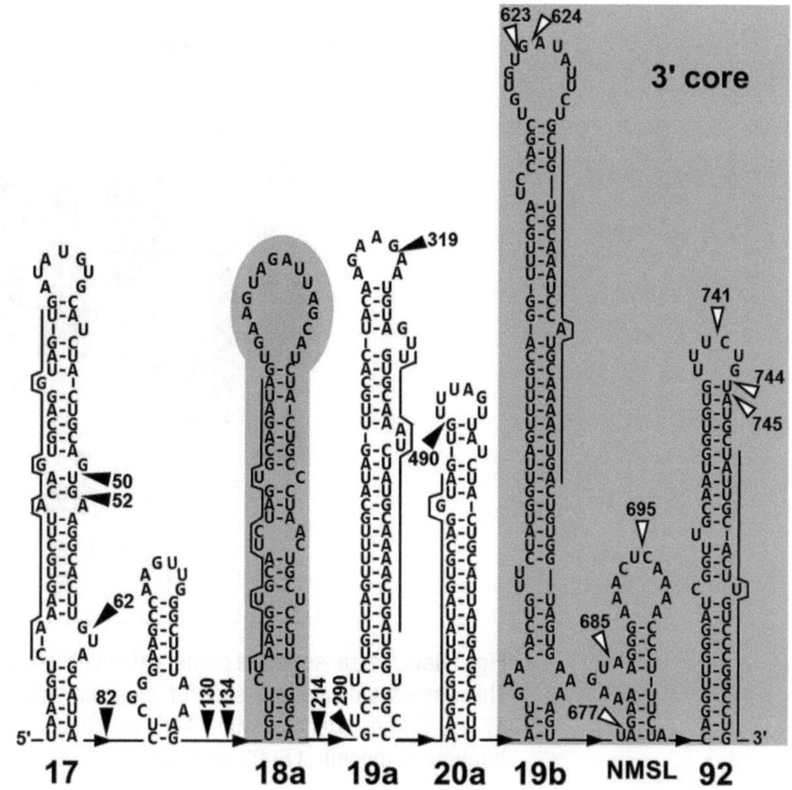

Fig. 2 Summary of miR-17~92 pri-miRNA footprinting [10]. The predicted secondary structure of miR-17~92 with RNAse T1 and A cleavages is indicated. The position of the mature miRNA sequences are indicated with *black lines*. The protected miRNA hairpins are *shaded grey*. The *white triangles* indicate positions that are cleaved when the ~275 nucleotide 3′ core domain itself is probed as a separate RNA, and are protected when the full length miR-17~92 is probed

Thus, the cleavages induced during a footprinting experiment are difficult to identify because of the high background of premature stop sites arising from the stable RNA secondary structure.

In the standard style of tertiary structure probing, hydroxyl radical cleavages are assigned to a specific nucleotide in the RNA by comparison to a hydroxide ladder and chemical or enzymatic sequencing (cleavage under denaturing conditions) reactions at nucleotide resolution. Hydroxide cleavage, under denaturing conditions (elevated temperature or in the presence of denaturants), cleaves RNA at every nucleotide position in a sequence independent manner. The resulting cleavage pattern, or ladder, serves as a polynucleotide ruler and indicates the distance (number of nucleotides) between any two given cleavages in the sequencing lane. This allows for the annotation of cleavages in the sequencing lane with nucleotides in the RNA sequence. Once the sequencing lane has

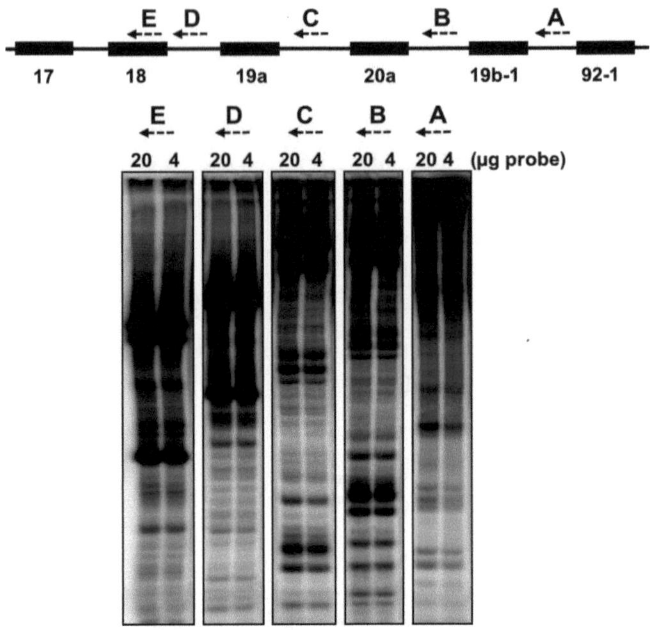

Fig. 3 Reverse transcriptase primer extension using miR-17~92 as RNA template. The *arrows* indicate the sites of complementarity between probe and miR-17~92. Primer extension with all probes exhibited extensive premature stoppage throughout the miR-17~92 sequence

been annotated by comparison with the hydroxide ladder, cleavages in the structure probing lanes can be annotated unambiguously. Often, only a ~2-fold difference in band intensity is seen between protected and unprotected nucleotides [2, 4, 5]. Thus, nucleotide resolution is essential for determining whether a particular nucleotide is protected or not. Generally, the smaller the RNA, the better the nucleotide resolution visible by denaturing PAGE. Conversely, the larger the RNA, the poorer the nucleotide resolution attainable, and often, the lane background is higher due to the greater complexity of RNA fragments resolved on the gel from large RNAs. In the case of the ~800 nucleotide miR-17~92 pri-miRNA, in our hands, only the bottom third of the denaturing PAGE (spanning ~100 nucleotides of sequence) is at nucleotide resolution. The combination of high background and relatively weak signal between cleavage and protection with hydroxyl radicals renders the hydroxyl radical method of probing of limited utility for investigating large RNAs (>400 nucleotides).

The sequence selectivity of the single strand specific ribonucleases RNase T1 and RNase A results in fewer cleavages and a simpler cleavage pattern to annotate, and obviates the need for nucleotide resolution. The combination of results coming from RNAse T1 cleaving 3′ of single-stranded guanosines and RNase A cleaving 3′ of single-stranded cytosines and uridines allows for the probing of all residues in the RNA with the exception of adenosine. Here, we outline the experimental procedures involved in ribonuclease structural probing of the ~800 nucleotide miRNA cluster miR-17~92 that were used in our previous report of regulation of miRNA biogenesis by pri-miRNA tertiary structure (Fig. 1) [10]. A summary of the ribonuclease probing data is summarized in Fig. 2. This approach can be utilized for any large RNA (lncRNA, pri-miRNA, etc.) to obtain global information on RNA folding.

2 Materials

2.1 RNA Transcription

1. T7 RNA polymerase [50 U/μl] (*see* **Note 1**).
2. 5× transcription buffer: 200 mM Tris–HCl pH 8.0, 30 mM $MgCl_2$, 10 mM spermine, 50 mM DTT, 50 mM NaCl.
3. cDNA transcription template [10].
4. 100 mM nucleotide triphosphates: 100 mM total NTP, 25 mM each NTP.
5. SUPERase-IN RNase inhibitor [20 U/μl] (Life Technologies).
6. 4 % (19:1 acrylamide–bis-acrylamide) 8 M urea for denaturing PAGE (10 % ammonium persulfate and TEMED for acrylamide polymerization) purification of RNA transcript.
7. 1× TBE: 10× stock: 1.8 M Tris, 1.8 M borate, 40 mM EDTA, pH 8.0 for denaturating PAGE.
8. 5× denaturing PAGE loading dye: 8 M urea, 1× TBE, 0.04 μg/ml bromophenol blue, 0.04 μg/ml xylene cyanol as tracking dyes.
9. Hand held UV lamp (Entela UVG-54) for UV shadowing visualization of purified transcript on gel.
10. Silica based thin layer chromatography plate (10 cm × 10 cm).
11. A razor blade for excising the gel slice containing the RNA transcript.
12. Sodium acetate 3 M pH 5.5.
13. Phenol–chloroform–isoamyl alcohol (25:24:1).
14. Chloroform–isoamyl alcohol (24:1).
15. 100 % ethanol.

2.2 RNA Labeling

1. γ-[^{32}P]-ATP (Perkin Elmer, 6,000 Ci/mmol).
2. α-[^{32}P]-Cordycepin 5′ triphosphate 5,000 Ci/mmol (Perkin Elmer).
3. T4 polynucleotide kinase 10 U/μl (New England Biolabs).
4. 10× T4 polynucleotide kinase buffer.
5. PolyA polymerase 5 U/μl (New England Biolabs).

6. 5× polyA polymerase buffer.
7. Calf intestinal alkaline phosphatase 10 U/μl (New England Biolabs).
8. 10× NEB Buffer 3.
9. 4 % (19:1 acrylamide–bis-acrylamide) 8 M urea for denaturing PAGE (10 % ammonium persulfate and TEMED for acrylamide polymerization) purification of end labeled RNA.
10. X-ray film for localization of radiolabeled RNA to be excised and purified.
11. Razor blade for excising the radiolabeled RNA gel slice.
12. Sodium acetate 3 M, pH 5.5.
13. Phenol–chloroform–isoamyl alcohol (25:24:1).
14. Chloroform–isoamyl alcohol (24:1).
15. 100 % ethanol.

2.3 RNA Folding

1. RNA folding buffer: 10 mM sodium cacodylate pH 6.8, 10 mM NaCl, 10 mM $MgCl_2$, and 0.1 mM EDTA.

2.4 Ribonuclease Digestion

1. RNAse T1 [100 U/μl] (Life Technologies), used to prepare 1 U/μl and 5 U/μl dilutions in ddH_2O.
2. RNase A [100 mg/ml] (Life Technologies) used to prepare a 1 mg/ml dilution in ddH_2O.
3. Phenol–chloroform–isoamyl alcohol (25:24:1).
4. Chloroform–isoamyl alcohol (24:1).
5. 100 % ethanol.
6. tRNA [10 mg/ml] (Life Technologies) competitor RNA used to prepare a 1 mg/ml dilution in ddH_2O.
7. RNAse T1 1× sequencing buffer: 7 M urea, 20 mM citrate (pH 5), 1 mM EDTA, 120 mM HCl.

2.5 Denaturing PAGE Analysis of Ribonuclease Digestion Reactions

1. 6 % (19:1 acrylamide–bis-acrylamide) 8 M urea for denaturing PAGE (10 % ammonium persulfate and TEMED for acrylamide polymerization) of digestion reactions.
2. Model S2 Sequencing Gel electrophoresis apparatus.
3. Autoradiography storage screen and image analyzer (Molecular Dynamics GE Healthcare).

2.6 Data Analysis

1. Typhoon Imager (GE Health Sciences), and autoradiography screen and cassette.
2. ImageQuant and Microsoft Excel software for image and data analysis, respectively.

3 Methods

All electrophoresis is performed in 1× TBE.

3.1 RNA Preparation

1. To generate the DNA template for run-off transcription of miR-17~92, digest with Xho I the pcDNA 3.1 (+) plasmid harboring the ~800 nucleotide sequence of the miR-17~92 pri-miRNA downstream of the T7 promoter [10].
2. Perform miR-17~92 transcription at 37 °C for 4 h in 0.4 ml containing 1× transcription buffer, 8 mM of each nucleoside triphosphate (32 mM total NTP concentration), 20 U of SUPERase-In RNase inhibitor, 800 U T7 RNA polymerase, and 20 μg of Xho I digested plasmid [11, 12].
3. Add 1 ml of 5× denaturing PAGE loading dye to the RNA prepared in **step 1** and heat it for 2 min at 95 °C, place it on ice for 1 min, and then load it onto a 4 % (19:1) 8 M urea PAGE gel for purification (gel dimensions 20×15 cm, 3 mm thick) (*see* **Note 2**).
4. Remove the gel from the glass plates, wrap it in Saran wrap, and place it on a TLC plate to visualize the RNA by UV shadowing with a hand held UV lamp. Excise the RNA transcript with a razor blade and place it in a 15 ml plastic centrifuge tube.
5. Crush the gel slice with a glass or plastic rod thin enough to reach the bottom of the tube. Extract the RNA out of the gel by adding 50 μl of phenol–chloroform–isoamyl alcohol (25:24:1) and 1.6 ml of 0.3 M NaOAc pH 5.5, and incubate it with shaking or rotation at 37 °C for 4 h.
6. Centrifuge the RNA extraction-gel mixture at 4,000×*g* for 30 min and remove the RNA containing supernatant using a pipette.
7. Divide the RNA extraction solution into 400 μl aliquots and precipitate the RNA by adding 1 ml of 100 % ethanol, briefly vortex, and place it on dry ice for ~15 min.
8. Centrifuge the frozen RNA solution at 12,000×*g* for 15 min to pellet the precipitated RNA.
9. Remove the ethanol supernatant by pouring off or pipetting.
10. Wash the RNA pellets with 200 μl of 75 % ethanol and centrifuge for 2 min at 12,000×*g*.
11. Remove the supernatant and let the RNA pellets air-dry.
12. Resuspend the RNA in ~50 μl ddH_2O.

3.2 RNA 5′ End-Labeling

1. Prior to 5′ end-labeling, 5′ dephosphorylate the purified pri-miRNA transcript (120 pmol) from the previous section by adding 5 μl of calf intestinal alkaline phosphatase, 2 μl of SUPERase-In RNase Inhibitor, 20 μl of 10× NEB Buffer 3, and ddH_2O to bring the final reaction volume to 200 μl and incubate for 20 min at 37 °C. When the reaction is complete, add 22 μl of 3 M NaOAc.
2. Add 200 μl phenol–chloroform–isoamyl alcohol (25:24:1), vortex for ~10 s, and centrifuge for 2 min at 12,000 × *g* to separate the organic and aqueous layers.
3. Collect the aqueous layer using a pipette.
4. Add 200 μl chloroform–isoamyl alcohol (24:1), vortex for ~10 s, and then centrifuge for 2 min at 12,000 × *g* to separate the organic and aqueous layers.
5. Collect the aqueous layer and add 500 μl of 100 % ethanol, vortex for ~10 s, and place on dry ice for ~15 min.
6. Centrifuge the frozen RNA solution at 12,000 × *g* for 15 min to pellet the precipitated RNA. Remove the ethanol supernatant by pouring off or pipetting.
7. Wash the RNA pellet with 200 μl of 75 % ethanol, centrifuge for 2 min at 12,000 × *g*.
8. Remove the supernatant and let the RNA pellet air-dry.
9. Resuspend the RNA in 35 μl nuclease-free water and perform the 5′ labeling reaction by adding 5 μl of γ-[^{32}P]-ATP, 4 μl T4 polynucleotide kinase, 2 μl SUPERase-In RNase Inhibitor, and 4 μl 10× T4 PNK reaction buffer. Incubate the reaction at 37 °C for 20 min.
10. Add 160 μl of 5× loading dye, heat it to 95 °C for 2 min, and cool it on ice prior to loading on a 4 % (19:1) 8 M urea PAGE for purification.
11. Locate and excise the purified radiolabeled RNA using autoradiography, and place the RNA in a 1.5 ml plastic centrifuge tube.
12. Crush the gel slice with a glass or plastic rod thin enough to reach the bottom of the tube.
13. Extract the RNA out of the gel by adding 25 μl of phenol–chloroform–isoamyl alcohol (25:24:1) and 0.8 ml of 0.3 M NaOAc pH 5.5. Incubate with shaking or rotation at 37 °C for 4 h.
14. Centrifuge the RNA extraction-gel mixture at 12,000 × *g* for 20 min and remove the RNA containing supernatant using a pipette.
15. Divide the RNA extraction solution into 400 μl aliquots and precipitate the RNA by adding 1 ml of 100 % ethanol, briefly vortex, and place on dry ice ~15 min.

16. Centrifuge the frozen RNA solution at 12,000 × *g* for 15 min to pellet the precipitated RNA. Remove the ethanol supernatant by pouring off or pipetting it away.
17. Wash the RNA pellets with 200 μl of 75 % ethanol, centrifuge for 2 min at 12,000 × *g*, and remove the supernatant and let the RNA pellets air-dry.
18. Resuspend the purified 5′ end labeled RNA pellets in ~20 μl nuclease-free water with up to 50 million cpm (counts per minute) total radioactivity.

3.3 RNA 3′ End-Labeling

1. Carry out the 3′ end labeling of pri-miRNA (100 pmol) by adding 50 μl α-[^{32}P]-Cordycepin 5′ triphosphate, 3 μl polyA polymerase, and 24 μl 5× polymerase buffer in a 120 μl reaction and incubate at 37 °C for 15 min.
2. When the reaction is over, add 200 μl of 5× loading dye, denature at 95 °C for 2 min, cool on ice, and purify by denaturing PAGE as outlined for the 5′ end-labeling procedure (Subheading 3.2, **steps 10–18**). Resuspend the end-labeled RNA in ~20 μl ddH_2O. Typically ~25 million cpm of total radioactivity is isolated (*see* **Note 3**).

3.4 RNA Folding

Folding or renaturation of the 5′ or 3′ end-labeled RNA in 1× RNA folding buffer is done by thermal denaturation followed by a slow cool down to room temperature that allows for RNA folding. This heating and slow cool procedure can be programmed into most current PCR machines.

1. Prepare an end-labled RNA solution by adding 2 μl of 10× RNA folding buffer and 10×10^6 cpm of radiolabeled RNA from Subheading 3.2 or Subheading 3.3 and ddH_2O to a final volume of 20 μl.
2. Heat the RNA solution to 90 °C for 30 s for denaturation (*see* **Note 4**).
   Cool it down slowly from 75 °C to 25 °C over 20 min.

3.5 Ribonuclease Footprinting Reactions

We use unlabeled competitor tRNA in molar excess over radiolabeled pri-miRNA to minimize variability in extent of cleavage between experiments. The competitor tRNA also slows the RNase reactions to the extent that time dependent differences in cleavage can be seen on a convenient minute time scale. Figure 4 shows an RNase T1 footprinting experiment with too little competitor tRNA, which results in over digestion of the pri-miRNA and the loss of the miR-19b footprint (Fig. 5) (*see* **Note 5**).

1. To the 20 μl annealed RNA solution from Subheading 3.4, add 1 μl of 1 mg/ml tRNA, and 79 μl of ddH_2O.
2. Immediately prior to addition of 2 μl of 1 U/μl RNAse T1, remove 20 μl for the "time zero" aliquot and quench it by adding 200 μl of phenol–chloroform–isoamyl alcohol (25:24:1)

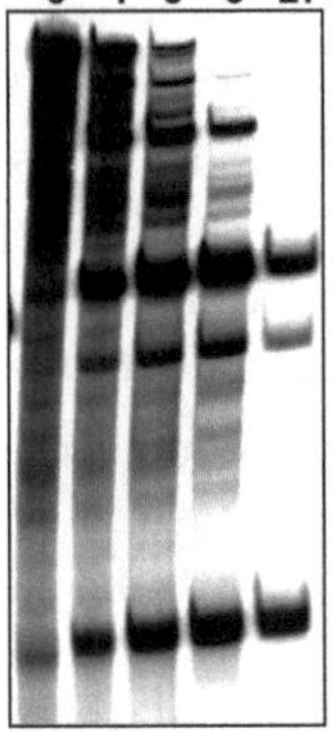

Fig. 4 Ribonuclease probing of miR-17~92 structure. Denaturing PAGE analysis of RNAse T1 footprinting reactions of 3′ [^{32}P] end-labeled miR-17~92. An RNase T1 or A cleavage site is defined as a band or region having a greater than twofold increase in intensity over the same band or region in the no RNase treatment lane (*arrowheads* indicate the sites of cleavage). The protection of miR-19b relative to miR-20a (as seen in Fig. 5) is already reduced in the first time point generating a weaker footprint signal. In subsequent time points, the protection of miR-19b relative to miR-20a is lost completely

and vortex immediately (*see* **Note 5**). After RNase T1 is added, remove 20 μl for each time point aliquot and quench them as previously indicated. Once the last time point aliquot is removed and quenched, add 40 μl of 3 M NaOAc and 320 μl of nuclease-free water to each time point aliquot, vortex again, and then proceed with the standard phenol–chloroform extraction and ethanol precipitation as outlined in Subheading 3.2. Resuspend the precipitated RNA in 20 μl of 5× denaturing PAGE loading dye.

3. Resuspend the footprinting reaction cleavage products in 20 μl of 5× loading dye, heat to 95 °C for 2 min, then cool on ice prior to loading on a 6 % (19:1) 8 M urea denaturing PAGE (*see* **Note 6**).
4. RNase T1 sequencing reactions are RNase T1 reactions performed under denaturing conditions. Incubate 4×10^6 cpm of end-labeled RNA from Subheading 3.2 or Subheading 3.3, with 1.2 μl of yeast tRNA 10 mg/ml and add 1× sequencing buffer to bring the final volume to 80 μl. Heat the solution at 65 °C for 5 min and then cool on ice.
5. Remove the 20 μl zero time point aliquot and quench as described in Subheading 3.5.
6. Then add 12 μl of 5 U/μl RNaseT1, perform the digestion reaction at 65 °C, and remove three 20 μl time point aliquots over a 10–20 min time course (*see* **Note 7**). Quench and purify the time point aliquots in the same way as the footprinting reactions in Subheading 3.5.

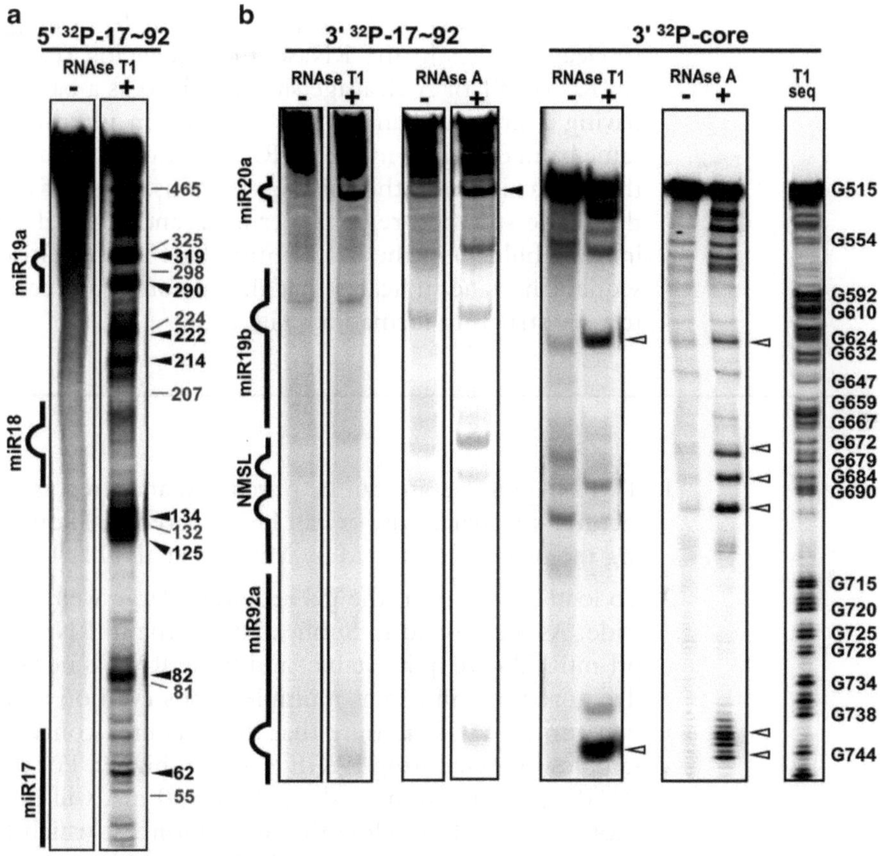

Fig. 5 Denaturing PAGE analysis of RNAse T1 and RNase A footprinting reactions of 5′ (**a**) and 3′ [^{32}P]-end-labeled miR-17~92 and 3′ [^{32}P]-end-labeled 3′ core domain (**b**). The sites of cleavage in the 3′ core domain are indicated by *white triangles*. *Grey lines* in (**a**) indicate positions of size standards generated from an RNase T1 sequencing reaction of 3′ ^{32}P end-labeled 3′ core domain used to assign approximate positions of RNAse T1 cleavage (*black triangles*) in the miR-17~92 cluster. Positions of miRNA stems and loops are indicated schematically adjacent to the gel panels

3.6 Denaturing PAGE of Sequencing and Footprinting Reactions

1. Load the RNAse T1 sequencing reactions, footprinting reactions, and respective zero time points onto a 31 × 38.5 cm 6 % (19:1) 8 M urea denaturing PAGE (1 mm thick) gel. Run the gel at low amperage (20–25 mA) to minimize heat gradient "smiling." Run the gel until the bromophenol blue dye is within an inch of the bottom of the gel.
2. Dismantle the gel and transfer it to filter paper for drying in a gel dryer for 2 h at 80 °C.
3. Wrap the dried gel in Saran wrap and place it in an autoradiography cassette and expose it to an autoradiography screen overnight.
4. Scan the autoradiography screen on an imager; visualize and analyze the gel image using ImageQuant software. Analyze the data with Microsoft Excel (*see* **Note 8**).

5. Assign sites of cleavage in the footprinting lanes by comparing to cleavage sites in the RNase T1 sequencing reaction lanes. An RNase T1 or A cleavage site is defined as a band or region having a greater than twofold increase in intensity over the same band or region in the no RNase treatment lane. Annotate the gel to indicate the miRNA hairpins. Loss of cleavage in the single-stranded regions (terminal and internal loops, and internal bulges) in the footprinting lane in comparison to the sequencing lane indicates that the region is protected due to tertiary structure formation (Fig. 5).

4 Notes

1. T7 RNA polymerase can be purchased and used as per manufacturer's instructions, or can be prepared "in-house" following the protocol outlined by Milligan et al. [11, 12].
2. To load 1.4 ml onto the gel requires a large well, about 5 cm wide. As mentioned in Subheading 1, the miRNA hairpins in pri-miRNAs are quite stable, and if the RNA is not sufficiently denatured, it will run as multiple bands even on an 8 M urea denaturing PAGE. Ensure that the final urea concentration in samples for denaturing PAGE is 6 M or higher. For preparative PAGE gels, pre-running the gels at ~80 mA will heat the gel (hot to touch, but below the temperature at which the plates will crack) and help denature the RNA within the gel.
3. Typically, the 3′-end labeling is not as robust as 5′-end labeling, giving about 50 % as much total radioactivity due to a lower specific activity of α-[^{32}P]-Cordycepin 5′ triphosphate compared to γ-[^{32}P]-ATP as well as lower efficiency of polyA polymerase compared to polynucleotide kinase.
4. Heating end-labeled RNA at 90 °C longer than 30 s in 10 mM $MgCl_2$ will result in nonspecific hydrolytic cleavage of the RNA backbone, generating a high background of cleavages visible by autoradiography, which will complicate analysis of the gel image and reduce the amount of full length RNA that is required in the footprinting experiments.
5. When first performing footprinting experiments, a time course of RNase cleavage is required to establish the optimal reaction time for digestion of the end-labeled RNA, such that cleavage products as well as regions of reduced cleavage are clearly visible (footprints) (Figs. 1 and 5). Figure 4 shows an example of over-digested RNA with little full length RNA remaining.

   In subsequent rounds, the amount of cleavage can be controlled by decreasing the reaction time or by increasing the cold-competitor yeast tRNA (twofold to fivefold) or adding less RNase T1 (twofold to fivefold). Figure 5 shows optimized RNase T1 and RNase A footprinting experiments [10].

6. For footprinting with RNase A, simply replace RNase T1 [1 U/μl] with RNase A [1 mg/ml] in the above protocol.
7. Like for the footprinting reactions, the reaction time for the sequencing reactions should be optimized using the same parameters. An optimized RNAse T1 sequencing reaction is shown in Fig. 5.
8. We use a Typhoon gel imager and an autoradiography cassette and screen from Molecular Dynamics (GE Healthcare) and ImageQuant 5.2 software which links directly to Excel when quantifying gel features.

Acknowledgement

This work was supported by a research grant from the Canadian Breast Cancer Foundation (CBCF to R.P. Fahlman).

References

1. Ehresmann C, Baudin F, Mougel M et al (1987) Probing the structure of RNAs in solution. Nucleic Acids Res 15:9109–9128
2. Chaulk SG, MacMillan AM (2000) Kinetic footprinting of an RNA-folding pathway using peroxynitrous acid. Angew Chem Int Ed Engl 39:521–523
3. Kent O, Chaulk SG, MacMillan AM (2000) Kinetic analysis of the M1 RNA folding pathway. J Mol Biol 304:699–705
4. Chaulk SG, Pezacki JP, MacMillan AM (2000) Studies of RNA cleavage by photolysis of N-hydroxypyridine-2(1H)-thione. A new photochemical footprinting method. Biochemistry 39:10448–10453
5. Chaulk SG, MacMillan AM (2000) Characterization of the Tetrahymena ribozyme folding pathway using the kinetic footprinting reagent peroxynitrous acid. Biochemistry 39:2–8
6. Merryman C, Moazed D, Daubresse G et al (1999) Nucleotides in 23S rRNA protected by the association of 30S and 50S ribosomal subunits. J Mol Biol 285:107–113
7. Merryman C, Moazed D, McWhirter J et al (1999) Nucleotides in 16S rRNA protected by the association of 30S and 50S ribosomal subunits. J Mol Biol 285:97–105
8. Brunel C, Romby P (2000) Probing RNA structure and RNA-ligand complexes with chemical probes. Methods Enzymol 318: 3–21
9. Wilson KS, Noller HF (1998) Mapping the position of translational elongation factor EF-G in the ribosome by directed hydroxyl radical probing. Cell 92:131–139
10. Chaulk SG, Thede GL, Kent OA et al (2011) Role of pri-miRNA tertiary structure in miR-17~92 miRNA biogenesis. RNA Biol 8: 1105–1114
11. Milligan JF, Groebe DR, Witherell GW et al (1987) Oligoribonucleotide synthesis using T7 RNA polymerase and synthetic DNA templates. Nucleic Acids Res 15:8783–8798
12. Milligan JF, Uhlenbeck OC (1989) Synthesis of small RNAs using T7 RNA polymerase. Methods Enzymol 180:51–62

Chapter 6

In Situ Hybridization Detection of miRNA Using LNA™ Oligonucleotides

Stefania Cotta Doné and Olga Beltcheva

Abstract

MicroRNAs are a family of small noncoding ribonucleic acids involved in regulation of gene activity. They have been implicated in both normal cellular pathways related to proliferation, differentiation, and apoptosis and pathological processes leading to disease. It is believed that better understanding of their structure and function will shed more light on a number of cellular functions while at the same time providing the basis for development of novel therapeutic applications. That is why identification and quantification of miRNAs are of great scientific interest. Several techniques have been developed which allow accurate, fast, and easy detection of these RNA species. This chapter focuses on in situ hybridization (ISH), a method which combines identification of miRNAs with histochemistry (ICH). We describe in detail a protocol for ISH in formalin-fixed paraffin-embedded tissue with the help of synthetic nonradioactive LNA oligonucleotide probes.

Key words In situ hybridization, miRNA, LNA oligonucleotides, Nonradioactive miRNA ISH

1 Introduction

MicroRNAs (miRNAs) belong to the heterogeneous class of noncoding RNAs with important structural, catalytic, or regulatory functions. Lately, miRNAs have come into the focus of attention largely due to their potentially key role in numerous cell processes both under normal physiological conditions and as part of pathological changes in the organism.

Studying the structure, expression, and function of these RNA molecules serves both the fundamental biological sciences and the medical practice [1–4]. miRNAs are investigated for their potential application as marker of disease progression or therapeutic tool [5].

Different techniques have been designed or redesigned for the purpose of miRNA research. In situ hybridization (ISH) is probably the most time and labor consuming amongst them as well as the one requiring most skill. However, it has certain advantages over the more simple methods for identification and detection of

M. Lucrecia Alvarez and Mahtab Nourbakhsh (eds.), *RNA Mapping: Methods and Protocols*, Methods in Molecular Biology, vol. 1182, DOI 10.1007/978-1-4939-1062-5_6, © Springer Science+Business Media New York 2014

Table 1
Comparison of qPCR, Northern blot, and in situ hybridization

Technique	Cost[a]	Handling	Time[b]	Quantitative[c]	Qualitative[c]	Multiplex	Localization
qPCR	Medium	Easy	Short	Yes	Yes	Yes	No
Northern blot	Low	Difficult	Long	No	Yes	No	No
In situ hybridization	Medium	Difficult	Medium	No[d]	Yes	Yes	Yes

[a]It includes consumables and equipment
[b]Time required from the start until the results are ready for analysis
[c]This is considering the technology available today
[d]ISH can be considered semiquantitative

these small noncoding molecules and in many cases successfully complements them allowing the researchers to gain better insight into the biological processes in question. This is especially relevant when whole embryos, organs, or tissue sections comprising a mixture of different cell types are studied. Using real-time PCR, Northern blot, or array applications it is possible to identify the miRNA molecules involved in certain biological process and determine their absolute or relative expression levels. These techniques are pretty straightforward, reliable, and fast, which accounts for their widespread use when studying the role of miRNA molecules in different model organisms. However, approaches that rely on the use of cell and tissue homogenates, as well as purified miRNAs, cannot provide information about the type of cells expressing the target miRNAs, if changes in the expression levels are due to expansion or decrease in the population of specific cell types, whether the expression levels correlated with changes in cell morphology, etc. ISH is a powerful technique, which allows one to study the macroscopic distribution and cellular localization of DNA and RNA sequences in a heterogeneous cell population. This kind of knowledge is vital when the aim of the study is to understand the role miRNAs play in the biology of more complex tissues like tumors and brain [6, 7]. Table 1 summarizes the advantages and disadvantages of each technique.

ISH allows the detection of RNA or DNA species in cells or tissue sections using labeled complementary probes. It is a distinct form of histochemistry, which allows the researcher to detect certain nucleic acids and determine their cellular and tissue-specific localization at the same time. Different approaches for ISH have been developed and are used depending on the type of target molecule and target cells. It can be used for both paraffin-embedded and cryosections; the probes can be double- or single-stranded, DNA or RNA probes and the labeling radioactive or nonradioactive. Efforts have been made for development of standardized quantitative ISH protocols [8].

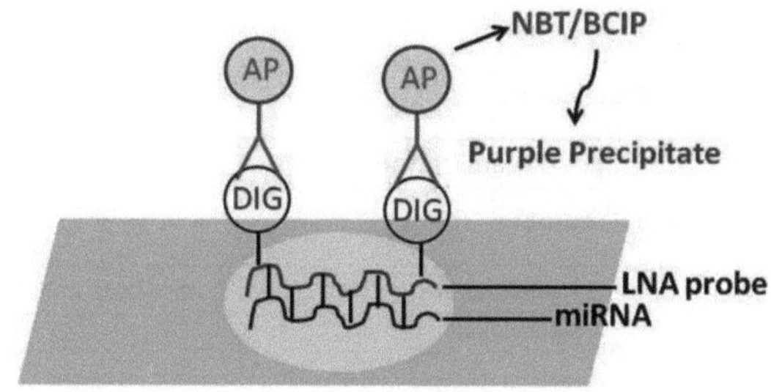

Fig. 1 General overview of ISH using LNA oligonucleotides. DIG-labeled LNA oligonucleotides are hybridized to miRNA present in the tissue. Signal is visualized as color deposit by histochemistry

ISH for detection of miRNA has certain peculiarities largely due to the distinct nature of the target molecules. Unlike mRNA, miRNAs are very short, approx. 22 nt, which decreases the specificity of their sequence and hinders the use of standard DNA or RNA probes for detection due to generation of high background noise. This problem has been overcome by the development of locked nucleic acid (LNA) probes. LNA is a nucleic acid analogue containing nucleotides in which an O2′, C4′-methylene bridge is introduced within the ribose ring, thus chemically locking it in C3′-endo (North type) conformation. This conformation enhances base stacking and backbone pre-organization, thus increasing the melting temperature and specificity of oligonucleotides [9–11].

The application of LNA probes for the purpose of miRNA ISH was first adapted by Kloosterman and coauthors [12]. Figure 1 shows the general ISH procedure using labeled LNA probes. Today, the implementation of LNA-modified DNA probes has become a standard in miRNA ISH assays for animal and plant, embryo and adult sections [13].

Here we describe a basic protocol for ISH in formalin-fixed paraffin-embedded tissue using nonradioactive LNA oligonucleotide probes. This protocol can be used as a base for optimization for new probes or as an experimental procedure for already established probes. Figure 2 summarizes the steps explained in the following sections.

2 Materials

1. Gloves (*see* **Note 1**).
2. High-quality, nuclease-free water (*see* **Note 2**).
3. Calibrated single- and multichannel pipettes.
4. RNase/DNase-free pipette tips and tubes.
5. Silane-coated glass slides (Lab Scientific Inc.) (*see* **Note 3**).
6. Glass histology jars with racks.

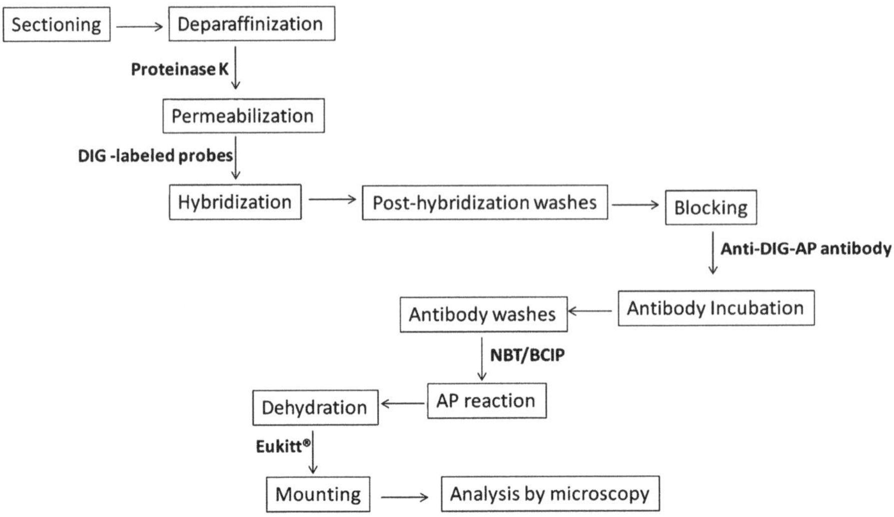

Fig. 2 Schematic flow-through illustrating the steps for miRNA ISH

7. Molecular biology-grade xylene (*see* **Note 4**).
8. Molecular biology-grade ethanol (*see* **Note 5**).
9. Nuclease-free PBS.
10. Ultrapure SSC buffer (20×) (Life Technologies).
11. Tissue sections in glass slides.
12. Sheep-anti-DIG-AP antibody (# 11093274910, Roche) (*see* **Note 6**).
13. NBT/BCIP ready-to-use tables (#11697471001, Roche).
14. Levamisole stock 100 mM (*see* **Note 7**).
15. Proteinase K stock solution (20 mg/mL) (*see* **Note 8**).
16. Proteinase K buffer:
    (a) 50 mM Tris–HCl pH 7.5.
    (b) 5 mM EDTA.
17. miRCURRY LNA™ microRNA Detection Probes for ISH (Exiqon) (*see* **Note 9**).
18. miRCURRY LNA™ microRNA ISH Buffer Set (Exiqon).
19. RNase ZAP.
20. Hybridizer (*see* **Note 10**).
21. Humidifying chamber (*see* **Note 11**).
22. Hybridization oven (*see* **Note 10**).
23. Immunohistochemistry blocking solution:
    (a) 1× PBS.
    (b) 0.1 % Tween-20.
    (c) 2 % Sheep serum.
    (d) 1 % BSA.

24. Hydrophobic barrier (PAP pen, DAKO).
25. Antibody diluent:
    (a) 1× PBS.
    (b) 0.05 % Tween-20.
    (c) 1 % Sheep serum.
    (d) 1 % BSA.
26. Alkaline phosphatase stop solution:
    (a) 50 mM Tris–HCl pH 9.0 (*see* **Note 22**).
    (b) 150 mM NaCl.
    (c) 10 mM KCl.
27. Nuclear Fast Red (Vector Laboratories).
28. Eukitt® mounting medium (Sigma).

3 Methods

The main concern when performing ISH is to work in an environment free of RNases. Therefore, it is important to clean all tools and surfaces before starting the procedure. Moreover, because of the nature of the sample, it is essential that tissue samples from both human and animal sources are handled with care to avoid damage and contamination. For this, fixation should be done immediately after harvest, followed by embedding and proper storage.

3.1 Sample Preparation

1. The tissue should be fixed in neutral formalin and embedded in paraffin. Alternatively, tissue fixed in 4 % PFA or fresh-frozen tissue can be used (*see* **Note 12**).
2. Spray all surfaces with RNase Zap or similar before the experiments. This includes benches, tweezers, forceps, microtome, blades, water baths, and brushes.
3. Prepare two water baths containing RNase-free water, and set one of them at room temperature and the other one to 40–50 °C.
4. Place paraffin blocks at 4 °C for a few minutes before sectioning. This step will facilitate the sectioning, especially in warm environments.
5. Prepare the block to be cut, adjusting it in the microtome. Make sure that the microtome is equipped with new RNase-free blade!
6. Cut 6 mM serial sections, place them at the room-temperature water bath, and gently uncurl the sections (*see* **Note 13**).
7. Transfer the sections to the warm bath, and let them stretch for a while.
8. Place the section in a silane-coated glass slide, and allow section to dry at room temperature for 2–3 h (*see* **Note 3**).

3.2 In Situ Hybridization

3.2.1 Deparaffinization of the Samples

1. Remove the sample slides from storage, allow them to reach room temperature, and place them in an oven heated to 60 °C for 30 min to remove part of the paraffin.
2. Let the slides cool at room temperature for 5 min.
3. Proceed with the deparaffinization dipping the slides in histological jars as follows:

Xylene	2× 15 min
Ethanol 99 %	2× 15 min
Ethanol 95 %	2× 10 min
Ethanol 80 %	1× 10 min
Ethanol 70 %	1× 10 min
PBS	2× 5 min

3.2.2 Proteinase K Treatment

1. Immediately before use, dilute the proteinase K to a final concentration of 15 mg/mL in proteinase K buffer (*see* **Note 8**).
2. Remove the slides from the PBR, slightly dry the glass around the section, and add 300 μL of proteinase K 15 mg/mL per slide. Make sure that the sections are fully covered.
3. Incubate the slides for 10 min at 37 °C (*see* **Note 8**).
4. Discard the proteinase K by tilting the slide sidewise into an absorbent tissue paper.
5. Wash slides twice with PBS using the histology jars.
6. Dehydrate the slides in new ethanol solutions as follows:

70 % Ethanol	2× 5 min
95 % Ethanol	2× 5 min
100 % Ethanol	2× 5 min

7. Air-dry the slides at room temperature for 15 min.

3.2.3 Hybridization

1. Preheat the hybridization oven to 55 °C (*see* **Note 14**).
2. Prepare the RT cocktail (amounts are for one reaction, final volume 2 mL):
   (a) Dilute the 2× miRNA ISH buffer 1:1 with nuclease-free water.
   (b) Label 2 mL non-stick RNase-free tubes, one tube per probe to be used.
   (c) Pipette the amount of probes for the chosen concentration in the tubes (*see* **Note 15**).

   (d) Denature the probes at 90 °C for 4 min using a heating block.

   (e) Spin down the tubes shortly, and *immediately* add 2 mL 1× miRNA ISH buffer to each tube.

3. Add 15 μL of RT cocktail to each section, making sure that the tissue is covered with hybridization solution (*see* **Note 16**).
4. Cover the sections with an RNase-free glass cover slip, but *do not press the glass* or the samples will be squeezed. This step is to avoid evaporation during incubation.
5. Place the slides into the hybridization chamber (*see* **Note 10**).
6. Incubate slides at 55 °C for 1 h.
7. While the samples are incubating, dilute the 20× SSC buffer in nuclease-free water to the following final concentration:

   (a) 5× SSC buffer enough for one wash.

   (b) 1× SSC buffer enough for two washes.

   (c) 0.2× SSC buffer enough for four washes.

8. Warm up the SSC buffers to 55 °C, but for one jar of 0.2× SSC. All the washes, save the last one, will be performed at the hybridization temperature (*see* **Note 17**).
9. Remove hybridization chamber from the oven, and open it carefully.
10. Carefully remove the cover slips from the sections. For such, press one end of the cover slip with your fingernails and lift the other end with tweezers. Do not press too hard or the cover slip will break. And do not slide the cover slip, because you are going to damage the tissue. Another way is to dip the slides into a histological jar containing warm 5× SSC buffer and gently shake the cover slip away from the slides.
11. Place the slides in a histological jar containing 5× SSC buffer warmed to 55 °C and wash for 5 min. The jars should be placed at the 55 °C water bath to keep the buffer warm (*see* **Note 17**).
12. Proceed with the washing steps as follows:

5× SSC	5 min	55 °C
1× SSC	5 min	55 °C
1× SSC	5 min	55 °C
0.2× SSC	5 min	55 °C
0.2× SSC	5 min	55 °C
0.2× SSC	5 min	RT

13. Transfer the slides to a jar containing PBS. *Do not* let the tissue sections dry at any time.

3.2.4 Detection of miRNA by Immunohistochemistry

1. Prepare the IHC blocking solution.
2. Isolate the tissue sections by circling it around with hydrophobic pen. The tissue needs to be moist at all times, so work swiftly.
3. Place the sections flat in a clean humidifying chamber (*see* **Note 10**).
4. Add 300 μL of blocking solution to each section, making sure that the volume stays inside the hydrophobic circle and that all the sections are fully covered.
5. Incubate slides for 15 min at room temperature (*see* **Note 18**).
6. In the meantime, prepare the antibody dilutant.
7. Dilute the anti-DIG antibody with antibody dilutant to the correct concentration. Here we will use 1:800 dilution (*see* **Note 19**).
8. Remove blocking solution, and dry the slides outside the hydrophobic circle, making sure that the tissue is always wet and you do not remove the barrier.
9. Replace the slides flat into the humidifying chamber.
10. Add 150 μL antibody to each section, making sure that the sections are all covered and the volume is inside the hydrophobic barrier.
11. Incubate the slides for 1 h at room temperature.
12. Remove antibody from the slides by gently tilting the slides sidewise in an absorbent paper towel.
13. Immediately transfer the slides to a histological jar containing 1× PBS.
14. Wash the slides three times, 5 min each in 1× PBS (*see* **Note 20**).
15. Prepare the AP substrate solution, by dissolving one tablet of substrate in ultrapure water, following the manufacturer's instructions, and add levamisole to a final concentration of 200 μM. This solution should be prepared *immediately* before use!
16. Replace the slides flat in the humidifying chamber, drying carefully outside the hydrophobic barrier. Remember not to let the tissue dry!
17. Dispense 150 μL of AP substrate solution to each section, always making sure that the sections are completely covered and the volume is all inside the hydrophobic barrier.
18. Incubate protected from light for 2 h or until purple spots appear in the tissue (*see* **Note 21**).
19. Prepare AP stop solution.
20. Stop the reaction by removing the AP substrate from the slides and immersing them in a histological jar containing AP stop solution.

21. Wash the slides 3× 5 min with AP stop solution.
22. Wash the slides 2× 2 min with water (*see* **Note 23**).
23. Remove the slides from the jars, dry around the hydrophobic barrier, and place the slides flat in the humidifier chamber. Again, make sure that the tissue is kept wet during the process.
24. Dispense 150 μL of Nuclear Fast Red to each slide and let incubate for 1 min at room temperature (*see* **Note 24**).
25. Remove the dye, tilting the slides sidewise in an absorbent tissue, and place the slides in a histological jar with water.
26. Wash the slides under tap water for 10 min to properly remove all traces of dye (*see* **Note 25**).
27. Dehydrate tissue as follows:

70 % Ethanol	Immerse slides 10× with gentle movements
70 % Ethanol	1 min
96 % Ethanol	Immerse slides 10× with gentle movements
96 % Ethanol	1 min
99 % Ethanol	Immerse slides 10× with gentle movements
99 % Ethanol	1 min

28. Place slides in a paper towel, and proceed with the mounting (*see* **Note 26**).
29. Mount slides using EuKitt.
30. Allow precipitate to settle overnight.
31. Analyze results by light microscopy. Positive ISH signal will appear as purple dots along the tissue, whereas nuclear staining will show light pink.

4 Notes

1. Use of gloves is essential for ISH success. RNA is extremely sensitive to degradation and skin contact with samples and material should be avoided at all costs.
2. It is essential to use nuclease-free ultrapure (RNase-free) water for the ISH steps. We recommend using nuclease-free water that is not DEPC treated offered by some companies such as Life Technologies, Qiagen, and Abnova. After the RNA hybridization is complete, MilliQ purified water may be used.
3. Glass slides for histology are usually coated with positively charged substances to increase tissue adherence and prevent tissue tearing and loss during the process. Here we recommend the use of silane-coated slides, as also recommended by the

ISH probe manufacturer. However, other coating may be used. Make sure that the treatment is compatible with the chemistry used during the ISH protocol.

4. Xylene is an extremely toxic organic compound; therefore use it always under chemical hood, protected from fumes. There are substitutions for xylene, which are less toxic and considered as efficient. I used K-Clear from Kaltek (http://www.kaltek.it). Millipore also offers a xylene substitute called *Neo-Clear*®. All those chemicals can be used successfully in the deparaffinization of the tissue.
5. The ethanol used during the ISH procedure *must* be of molecular biology grade and RNase free. Make the working dilutions using RNase-free water. This will prevent RNA degradation during hydration and dehydration steps. As well, *never* re-utilize the solutions in the following steps. All the solutions must be *fresh* to avoid contamination during the experiment.
6. Digoxigenin (DIG) is a steroid found exclusively in plants. It makes a perfect antigen for detection by immunostaining experiments, as DIG-specific antibodies generate little to no background staining. Many companies offer antibodies against DIG. Here we use the antibody from Roche as well as the developing agent from the same manufacturer. Roche offers many products for DIG ISH protocols. For more information, consult its homepage at http://www.roche-applied-science.com.
7. Levamisole is an AP inhibitor, used during the experiment to inhibit endogenous AP. The working concentration varies from 200 μM to 5 mM. Prepare a 100 mM stock solution, and dilute to the final concentration as needed. In tissues with high endogenous AP concentration, perform a 2-min wash with PBS containing 200 μM levamisole just before adding the AP substrate.
8. Proteinase K digestion is necessary for the probe to reach its target. However, proteinase K treatment inevitably leads to tissue damage and loss. Thus, the degree of the treatment depends on the nature of the tissue and fixation used (i.e., harder fixatives will call for longer or more concentrated proteinase K treatment, while fresh tissues will call for more gentle conditions). Proteinase K, as all enzymes, has an optimal condition for digestion. It depends on time, temperature, buffer composition, and enzyme concentration; therefore, it is highly recommended to do an optimization for proteinase K treatment for your tissue in hands. For this, make several dilutions of proteinase K (i.e., 0.5–20 μg/mL) and incubate for 10 min to adjust the concentration. Then, set up a time-course

experiment (5–30 min) to adjust the proper duration for the process. Use a housekeeping gene as probe or any other highly expressed gene.

9. miRNA ISH probes are commercially available from http://www.exiqon.com. This is the inventor and manufacturer of LNA miRNAs, a technology which increased greatly the research in miRNAs. Exiqon offers a number of LNA oligonucleotides for different applications, including ISH. As well, the manufacturer offers different labeling for the LNA oligonucleotides, allowing not only single-probe experiments but also multiplex capabilities. For every experiment you must include, besides your probe of interest, a positive control and a scrambled miRNA negative control. Those will be your comparative standards. For more information on the probes, consult the manufacturer website or contact local representative. In addition to the probes, Exiqon also offers a buffer system, optimized for the probes. I recommend the use of the whole system.
10. For labs that own a hybridizer, this is the best method for ISH. However, if your laboratory does not own a hybridizer, ISH can be performed using a humidifying histological chamber or any closed slide holder that allows the slides to lay flat and allows a humid tissue to be placed on the bottom. The incubation step will be performed in a hybridization oven. Remember that your tissue cannot dry during the process, so do not forget to place a soaked tissue on the bottom of the chamber, just below the slides. For the high temperatures utilized for ISH, I would recommend a thick tissue layer, heavily soaked in water or PBS.
11. Humidifying chambers can be bought from different manufacturers. However, if your laboratory does not own one, you can use a slide box with a humid tissue paper, as described in **Note 9**. Again, make sure that the chamber is kept humid at all times and that the sections do not dry during incubation.
12. Harvest and processing of tissues vary greatly according to laboratories, tissue origin, and treatment protocol. In addition, protocols on handling human and animal tissues vary from institution to institution, from country to country. Therefore, consult your department on the proper rules for handling of tissues before setting a harvest/fixing protocol.
    (a) Tissue can be fixated by immersion (most common) or by perfusion (done with small laboratory animals such as mice and rats). Perfusion protocols may be cleared with animal care personnel before implementation.
    (b) ISH tissues can be fixed with neutral formalin or 4 % paraformaldehyde (PFA) and then embedded in paraffin.

(c) PFA-fixed tissues can also be embedded in cryogenic gel (O.C.T. medium). If you are using such protocols, before embedding the tissue, incubate overnight in a 30 % sucrose solution at 4 °C. This step will protect the tissue from the low temperatures and preserve its morphology.

(d) Paraffin-embedded tissues have a more preserved morphology than O.C.T.-embedded ones, but O.C.T. tends to preserve better the proteins and nucleic acid due to its gentler protocol.

(e) Fresh-frozen tissues may also be used for ISH, despite the low preservation of morphology. When using fresh-frozen sections, harvest the tissue as fast as possible and snap freeze it in liquid nitrogen. Then proceed with O.C.T. embedding. Tissue may also be immersed in 30 % sucrose solution overnight at 4 °C. But make sure that all the solutions are RNase free.

13. Thickness of the sections varies with the origin of tissue, experimental protocol, and fixation/embedding. Usually paraffin sections are between 4 and 6 mM. Paraffin makes a sturdy bed, and sectioning is easily performed. Frozen species, on the other hand, require more care during sectioning. Cryopreserved tissues should be sectioned in thicker sections in a cryostat. Start with 10 μM sections. However, do not use too thick sections, or the ISH will not be optimal, as the probe will not be able to penetrate the tissue and reach its target. Make sure that your cryostat is set for the optimal sectioning for your tissue or your sections will not be perfect. Also make sure that the instrument and all the accessories used are RNase free and all the accessories are kept inside the cryostat chamber, so that they are as cold as the tissue at all times.

14. Hybridization temperature depends on the type and length of the probe being used. The mercury LNA™ miRNA ISH probe manufacturer recommends a temperature of 55 °C for its probes. However, when using self-made or other manufacturer's product, you should determine empirically the optimal hybridization temperature for your probe of interest.

15. Likewise, the optimal probe concentration depends on the representation of target miRNA on the tissue of interest. Some miRNAs are more abundant in one tissue, but not in another. You must do dose curve to determine optimal probe concentration for your experiment. But remember that some probes as U6 lnRNA, which is usually employed as positive control, will give a higher signal than some double-DIG miRNA probes. Therefore, set different concentration protocols for the probes (i.e., a curve from 0.01 nM to 2 mM for the

positive control, and a curve from 5 to 40 mM for double-DIG miRNA probe).

16. The volume of probe cocktail used in each experiment is very small due to costs. So, you must spread the volume equally along the section, making sure that the whole tissue is covered with probe. A good wait to do it is to use a pipette tip. Place the side of the tip on the top of the volume, and use surface tension force to spread the volume. But be very careful not to touch the tissue with the tip in any moment.
17. Stringency of the washes will determine whether you will have high or low background. This means that washes are most important for the success of your experiments. Wash temperatures are set to remove as much unspecific signal as possible without removing the positive signal. I recommend to start washes at hybridizing temperature and adjust conditions as you get the results. Some probes will ask for less stringent washes (i.e., lower SSC concentration or lower temperatures), whereas other probes will call for more stringency. There is no rule, so the conditions should be determined empirically.
18. There is no rule for how long you should block your sections. However, keep in mind that insufficient blocking may result in high background. Thus, make sure to block your tissue for at least 15 min.
19. Antibody dilutions depend on the amount of antigen present on the tissue to be probed. Thus, the entire signal will come from the probe used. This way, high hybridization will lead to higher signal, while low hybridization will result in low signal. In the ISH, we use DIG, which is not present endogenously in the tissue. Nonetheless, nonspecific binding occurs. Therefore, optimal antibody dilutions for your particular assay should be determined by the final user. The probe manufacturer recommends a dilution of 1:800. I would suggest optimizing the antibody dilution using at least one dilution up and one dilution down. Too much antibody may result in high background signal, whereas too little antibody may result in false-negative results.
20. Washes after antibody incubation are essential to remove unspecific bound antibody. Make sure that your tissue is well washed before proceeding to detection step.
21. The AP staining reaction may take longer or shorter time depending on the amount of AP present in the sample. Some reactions develop in 5–10 min, while some reactions may take up to 5 h to show any signal. Therefore, I recommend tracking the staining reaction development as it occurs. Be careful to protect the slides from direct light as much as possible. BCIP/NBT are light sensitive and tend to form precipitates

when exposed to light. It is also very important to remember that too long incubations not only do not increase significantly signal but may increase precipitate formation and high background. So, be sure to clock correctly your reaction time.

22. Optimal pH for AP staining buffer is between pH 9 and pH 9.5. Make sure that the pH of your buffer is in this range or your reaction efficiency will be decreased.
23. ISH protocols also allow different probes to be detected in the same tissue, at the same time, as well as a combination between ISH and IHC protocols (i.e., multiplex FISH). We do not discuss these protocols, but those are easily found in the literature. If you are planning on multiplexing your assay, I recommend designing the experiment carefully, choosing the colors and techniques that are compatible with each other (i.e., protocols for fluorescence-based probes are different than the ones for IHC probes such as AP and peroxidase). Exiqon, Affymetrix, and other companies offer platforms that allow multiplexing as well as a wide array of detection probes. For more information, consult their websites for products and protocols.
24. To facilitate identification of positive signal, it is common to use nuclear dye after the AP reaction is finished. Here we suggested Nuclear Fast Red from Vector Laboratories. Other dyes may be used, but make sure that the nuclei are stained in a distinct color from the ISH probes so that you are able to differentiate between the signals.
25. When washing your slides under tap water, make sure that the flow is gentle enough to wash but does not damage the sections. Moreover, avoid direct flow on the tissue; always direct your flow to the back of the slides.
26. Eukitt® is resin-based mounting medium widely used in immunohistochemistry. Eukitt® dries clear and very fast in room temperature, so you should work swiftly to achieve a good result. Because of its organic nature, tissue should be free of water before the mounting step. For this, some perform a quick wash with xylene before mounting; others use toluene instead, which is less toxic than xylene. In the protocol described here, we use a 100 % ethanol wash before mounting. Make sure that you proceed *immediately* from the ethanol wash to the mounting. If time and/or speed are an issue, I recommend a xylene or a toluene wash before mounting. Also very important, *do not* let the tissue dry at any moment or the mounting will not be adequate.

References

1. Ambros V (2011) MicroRNAs and developmental timing. Curr Opin Genet Dev 21(4): 511–517
2. Hoffmann TW, Duverlie G, Bengrine A (2012) MicroRNAs and hepatitis C virus: toward the end of miR-122 supremacy. Virol J 9:109
3. Suh N, Blelloch R (2011) Small RNAs in early mammalian development: from gametes to gastrulation. Development 138(9):1653–1661
4. Volinia S, Galasso M, Costinean S, Tagliavini L, Gamberoni G, Drusco A et al (2010) Reprogramming of miRNA networks in cancer and leukemia. Genome Res 20(5):589–599
5. Nana-Sinkam SP, Croce CM (2013) Clinical applications for microRNAs in cancer. Clin Pharmacol Ther 93(1):98–104
6. Li J, Li X, Li Y, Yang H, Wang L, Qin Y et al (2013) Cell-specific detection of miR-375 downregulation for predicting the prognosis of esophageal squamous cell carcinoma by miRNA in situ hybridization. PLoS One 8(1):e53582
7. Nelson PT, Abner EL, Scheff SW, Schmitt FA, Kryscio RJ, Jicha GA et al (2009) Alzheimer's-type neuropathology in the precuneus is not increased relative to other areas of neocortex across a range of cognitive impairment. Neurosci Lett 450(3):336–339
8. Broide RS, Trembleau A, Ellison JA, Cooper J, Lo D, Young WG et al (2004) Standardized quantitative in situ hybridization using radioactive oligonucleotide probes for detecting relative levels of mRNA transcripts verified by real-time PCR. Brain Res 1000(1–2): 211–222
9. Kaur H, Arora A, Wengel J, Maiti S (2006) Thermodynamic, counterion, and hydration effects for the incorporation of locked nucleic acid nucleotides into DNA duplexes. Biochemistry 45(23):7347–7355
10. Owczarzy R, You Y, Groth CL, Tataurov AV (2011) Stability and mismatch discrimination of locked nucleic acid-DNA duplexes. Biochemistry 50(43):9352–9367
11. Vester B, Wengel J (2004) LNA (locked nucleic acid): high-affinity targeting of complementary RNA and DNA. Biochemistry 43(42):13233–13241
12. Kloosterman WP, Wienholds E, de Bruijn E, Kauppinen S, Plasterk RH (2006) In situ detection of miRNAs in animal embryos using LNA-modified oligonucleotide probes. Nat Methods 3(1):27–29
13. Wheeler G, Valoczi A, Havelda Z, Dalmay T (2007) In situ detection of animal and plant microRNAs. DNA Cell Biol 26(4):251–255

Chapter 7

Quantification of miRNAs by a Simple and Specific qPCR Method

Susanna Cirera and Peter K. Busk

Abstract

MicroRNAs (miRNAs) are powerful regulators of gene expression at posttranscriptional level and play important roles in many biological processes and in disease. The rapid pace of the emerging field of miRNAs has opened new avenues for development of techniques to quantitatively determine miRNA expression levels in different systems.

In this chapter we describe a PCR method for quantification of miRNAs based on a single reverse transcription reaction for all miRNAs combined with real-time PCR with two miRNA-specific DNA primers. This method quantifies synthetic templates over eight orders of magnitude and successfully discriminates miRNAs that differ by one single nucleotide. Due to the usage of DNA primers this method allows higher amplification efficiencies than a similar method based on locked nucleic acid-spiked primers. The high efficiency translates into higher sensitivity and precision in miRNA quantification. Furthermore, the method is easy to perform with common laboratory reagents, which allows miRNA quantification at low cost.

Key words MicroRNA, qPCR, DNA primers, Specificity, Quantification, Noncoding RNA

1 Introduction

MicroRNAs (miRNAs) are important regulators of many cellular processes in a broad range of organisms. Due to the important biological role of miRNAs there is a high interest in measuring their expression levels and determine their biological functions in many physiological scenarios. Furthermore, miRNAs have been involved in many disease processes and quantification of their expression can readily contribute to improving diagnostics and prognostics of many human diseases. The presence of stable miRNAs in blood holds great promise as a diagnostic tool for human disease.

The three most commonly used platforms for profiling miRNAs are high-throughput sequencing (miRNA), microarray profiling, and reverse transcription quantitative PCR (RT qPCR).

M. Lucrecia Alvarez and Mahtab Nourbakhsh (eds.), *RNA Mapping: Methods and Protocols*, Methods in Molecular Biology, vol. 1182, DOI 10.1007/978-1-4939-1062-5_7, © Springer Science+Business Media New York 2014

The latter method is used alone and for validating data obtained with the other two techniques.

Accurate measurement of miRNAs by qPCR has some inherent challenges. One of them is the design of PCR primers because miRNAs are only around 22 bases long, which is about the size of a conventional PCR primer. Several methods have been developed to overcome this problem. Chen and co-workers [1] developed a stem–loop RT-qPCR in which reverse transcription is done with a special loop primer followed by a PCR with one specific primer and a universal primer. The PCR product is detected with a TaqMan probe. Unfortunately, this method does not allow control of the specificity of the assay by melting curve analysis.

A recently published method based on circularization of the miRNA also depends on a specific primer for reverse transcription [2]. This method uses RNA ligase, which is sensitive to sequence bias.

Another way to perform qPCR on miRNAs is to add a poly(A) tail to the miRNA sequence and use an anchored poly(T) primer for reverse transcription [3]. This method has the advantage that one reverse transcription reaction per sample is sufficient to measure expression of all miRNAs. Subsequently, PCR is performed with a miRNA-specific primer and a universal primer. However, this method relies on only one specific primer for PCR and there is no detection of a sequence-specific probe. In addition, the assays are difficult to design and often the specificity of the PCR is low [4]. An important drawback of this method is the difficulty to discriminate between members of a miRNA family that differ by only one or a few nucleotides.

Universal RT miRNA PCR combines the benefits of a universal RT reaction with the specificity of two miRNA-specific PCR primers [5]. The PCR product is detected with SYBR Green dye, and the specificity of the assay can be assessed by melting curve analysis. The method relies on poly(A) tailing of the miRNAs followed by reverse transcription with an anchored poly(T) primer. The high specificity of this assay arises from the use of two specific primers for PCR. Originally, it was thought that it was necessary to spike the primers with locked nucleic acid (LNA) [5]. However, later it was found that LNA-spiked primers often lead to low amplification efficiency and DNA primers perform better than LNA primers in this assay and are easier to design [6].

In this chapter we describe how to measure miRNA expression by miRNA-specific qPCR. The protocol consists of reverse transcription with a poly(T) adaptor primer followed by qPCR with two miRNA-specific DNA primers (Fig. 1) and includes instructions on PCR primer design. The method has all the benefits regarding freedom of primer design and specificity of the LNA-based method but requires normal DNA primers, which lowers the cost and makes the method accessible to everybody.

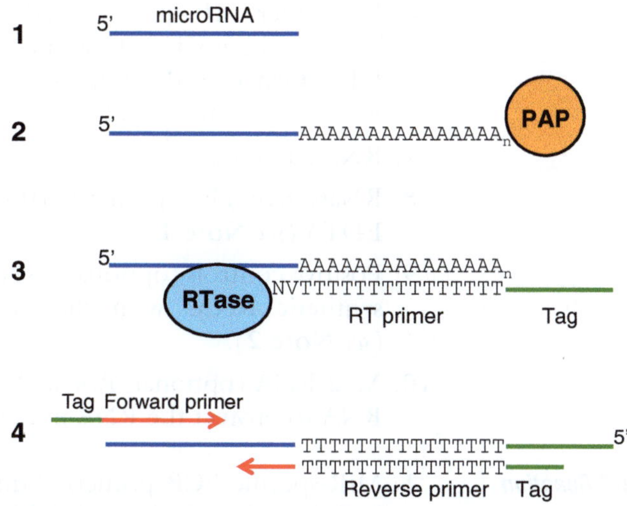

Fig. 1 Scheme of miRNA-specific qPCR. 1. Start with total RNA or purified miRNA. 2. Use poly(A) polymerase to incorporate a tail of adenosines. 3. Make cDNA with reverse transcriptase, RT primer, and dNTPs. Steps 2 and 3 are performed simultaneously in one reaction. *4.* Perform real-time PCR with two miRNA-specific primers. *PAP* Poly(A) polymerase, *RTase* MuLV reverse transcriptase. Modified from Balcells et al. [6]

2 Materials

All materials and samples should be handled with gloves to avoid contamination with RNases and with nucleic acids that can lead to false-positive results.

All materials should be kept on ice unless indicated otherwise.

2.1 Reverse Transcription

1. Poly(A) polymerase: *E. coli* poly(A) polymerase 5,000 U/ml (New England Biolabs, USA).
2. Reaction buffer: 10× *E. coli* poly(A) polymerase buffer (included with the *E. coli* poly(A) polymerase).
3. Reverse transcriptase: M-MuLV Reverse transcriptase 200,000 U/ml (New England Biolabs, USA).
4. 1 mM RNase-free adenosine 5′-triphosphate (ATP): Make it by mixing 1 part 10 mM ATP (included with the *E. coli* poly(A) polymerase) with 9 parts of RNase-free water.
5. 1 mM RNase-free dNTP mix: Make it by 100 times dilution of a mix of 100 mM deoxy-ATP (dATP), 100 mM deoxy-Cytosine 5′-triphosphate (dCTP), 100 mM deoxy-Guanosine 5′-triphosphate (dGTP), and 100 mM deoxy-Thymidine 5′-triphosphate (dTTP) (Stratagene, USA) with RNase-free water.

6. RT-primer: 10 μM 5′-CAGGTCCAGTTTTTTTTTTTTTTVN, where V is A, C, and G and N is A, C, G, and T (Sigma, UK). Order as desalted primer, and dissolve and dilute in RNase-free water or TE.
7. RNase-free water.
8. RNase-free TE (optional): 10 mM Tris–HCl, pH 8.0, 1 mM EDTA (*see* **Note 1**).
9. Positive controls (optional): RNA positive controls are made as synthetic RNAs with the same sequence as the miRNA (*see* **Note 2**).
10. MS2 RNA (optional, Roche Applied Science) or other carrier RNA to protect the RNA in dilute samples (*see* **Note 2**).

2.2 QPCR Reaction

1. MiR-specific PCR primers: Primers are designed as described in Subheading 3. They should be desalted after synthesis and dissolved and diluted to 10 μM in water or TE.
2. PCR mix: Quantifast SYBR Green PCR kit (Qiagen) and Brilliant III Ultra-Fast QPCR Master Mix (Agilent) perform well, but other master mixes may also work (*see* **Note 3**).
3. Positive controls for PCR (optional): Synthetic, positive control DNA templates are designed with the sequence 5′-CAGGTCCAG(T_{15}) followed by the sequence complementary to the miRNA. This sequence is identical to the DNA sequence generated by the reverse transcription step (Fig. 1) (*see* **Note 4**).
4. RNase-free water.
5. RNase-free TE (optional): 10 mM Tris–HCl, pH 8.0, 1 mM EDTA.
6. Salmon sperm DNA (optional; Sigma, USA): Use it as a carrier DNA for dilution of synthetic templates (*see* **Note 4**).

3 Methods

3.1 Poly(A) Tailing and Reverse Transcription

1. Mix on ice (*see* **Note 5**):
   (a) 1 μl 10× Reaction buffer *E. coli* poly(A) polymerase.
   (b) 1 μl 1 mM ATP.
   (c) 1 μl 10 μM RT primer 5′-CAGGTCCAGTTTTTTTTTTTTTTVN.
   (d) 1 μl dNTP mix (1 mM dATP, 1 mM dCTP, 1 mM dGTP, 1 mM dTTP).
   (e) 0.5 μl (200 U/μl) M-MuLV reverse transcriptase.
   (f) 0.2 μl *E. coli* poly (A) polymerase 5,000 U/ml.

2. Add the RNA sample (*see* **Note 6**), 10 pg to 100 ng (*see* **Note** 7), and complete with RNase-free water up to 10 μl.
3. Incubate at 42 °C for 1 h.
4. Inactivate the reaction by heating at 95 °C for 5 min.
5. The resulting cDNA may be diluted 4–8 times in RNase-free water before being used in qPCR (*see* **Note 8**).

3.2 QPCR Primer Design

Design miR-specific primers according to the previously described rules [6]. They can be made with the software miRprimer [7] or manually by following these steps:

1. Delete all A's from the 3′ end of the miRNA sequence until the most 3′ base is C, G, or T (*see* **Note 9**).
2. Make the longest possible forward primer (12–20 bases long) that leaves at least four bases at the 3′ end of the miRNA for design of the reverse primer and where the 3′ end of the primer fulfills as many as possible of the following criteria (*see* **Note 10**):
   1. The primer includes one A or T residue in the two bases closest to the 3′ end.
   2. The primer includes one to two A or T residues in the three bases closest to the 3′ end.
   3. The primer includes two to three A or T residues in the five bases closest to the 3′ end.
3. Calculate the Tm by the nearest neighbor method [8] with a NaCl concentration of 115 mM (*see* **Note 11**).
4. If the Tm of the forward primer is below 59 °C, add the following bases G, A, C, G, and C to the 5′ end of the primer one at a time until the Tm is 59 °C (*see* **Note 12**).
5. If the Tm of the forward primer is above 59 °C, remove bases from the 5′ end one at a time until the Tm is 59 °C.
6. Choose the longest possible reverse primer (4–8 bases long) that is not complementary to the 3′ end of the forward primer.
7. Choose the reverse primer with the best 3′ end according to criteria 1, 2, and 3 as in **step 2**.
8. Add 15 T's at the 5′ end of the reverse primer.
9. Calculate the Tm of the reverse primer as in **step 4**.
10. If the Tm of the reverse primer is below 59 °C, add the bases G, A, C, C, T, G, G, A, and C to the 5′ end, one at a time, until the Tm is 59 °C (*see* **Note 13**).

3.3 Quantitative PCR

1. Mix on ice for each sample (*see* **Note 14**):
   (a) 0.25–1 μl 10 μM Forward primer (*see* **Note 15**).
   (b) 0.25–1 μl 10 μM Reverse primer (*see* **Note 16**).
   (c) 5 μl 2× PCR master mix.
   (d) RNase-free water to a final volume of 9 μl.
2. Aliquot 9 μl in PCR plates or tubes on ice or in a cooling block.
3. Add 1 μl of diluted cDNA from previous step. Seal the plate or tubes and spin down (*see* **Note 17**).
4. Immediately run the real-time PCR (*see* **Note 18**): 95 °C for 5 min followed by 40 cycles of 95 °C for 10–30 s and 60 °C for 30–60 s (*see* **Note 18**).
5. Run melting curve analysis from 55 °C to 99 °C (*see* **Note 19**).

Analyze results according to the software in the user's qPCR machine or in MS Excel. For a more thorough statistical analysis we recommend GenEx Pro software (Multid Analyses AB).

4 Notes

1. The RT primer can be diluted in TE instead of diluting in RNase-free water. The primer is stable for longer time in TE than in water.
2. Positive control miRNAs can be used for controlling the specificity and the efficiency of the reverse transcription. They can be used as spike in the samples or as pure templates diluted to the desired concentration in 2 ng/μl MS2 carrier RNA. RNA oligonucleotides are kept on ice during use and should be stored in aliquots at −80 °C to avoid degradation.
3. Solutions containing SYBR Green should be kept in the dark as the dye is degraded by light.
4. Synthetic DNA can be used for measuring the specificity, detection limit, and amplification efficiency of the qPCR reaction but should be handled with care. Positive controls are extremely prone to contaminate the lab. Therefore, positive controls should be purchased from a different oligonucleotide provider than the one that provides the primers and diluted to 10^6–10^8 copies per μl in a background of salmon sperm DNA (carrier DNA) in a different lab and using different pipettes than for the RT and PCR samples. We strongly discourage working with synthetic DNA templates if only one set of pipettes is available. However, it can be done if the pipettes are properly decontaminated according to the manufacturer's guidelines after handling the positive controls. Please refer to http://www.pipettecalibration.net/pipette-cleaning.html for some general guidelines.

5. It is preferable to make a master mix for all the samples to minimize the effect of pipetting inaccuracy between samples. Make a 10 % excess of the master mix. For example make 47 μl mix for nine samples. The mix can be scaled up or down for larger or smaller reverse transcription sample volumes, respectively, by changing the amounts of all components proportionally.
6. Several commercial methods are available for total RNA sample preparation. We have tried two of them with success: miRNeasy (Qiagen) and TriReagent™ (MRC Research Inc.).

   miRNeasy is a column-based method that according to the manufacturer purifies RNA molecules down to 18 base pairs long. The problem of this method is that it is necessary to start with small amounts of tissue (50–100 mg), which results in low amounts of purified RNA. This is especially problematic in tissues with low content of RNA, i.e., fat tissues.

   TriReagent™ is a single-step total RNA isolation reagent, which is not based on columns, and, therefore, it is less expensive and allows larger amounts of tissue than miRNeasy resulting in higher yields of pure RNA. The drawback of this method is that it is difficult to completely remove the phenol used for purification. Often small rests of phenol remain in samples isolated with this method resulting in 260/230 ratios below 1.8. However, this is usually not a big issue for samples analyzed by qPCR, although it might be a problem for techniques that require starting from a highly pure RNA sample.
7. The amount of RNA can vary but should not exceed 100 ng/10 μl reaction as too much RNA can inhibit the reaction. The amount of RNA necessary for detection depends on the concentration of the target. For samples containing less than 1 ng/μl RNA, add 1–2 ng/μl MS2 carrier RNA to avoid losing the sample by absorption to the walls of plasticware.
8. Dilution of the cDNA depends on the amount of input sample and the expression level of the specific miRNA. To be able to detect miRNAs in cDNA made from minute amounts of total RNA, samples should not be diluted, whereas cDNA made from 100 ng of total RNA can be diluted ten times or more.
9. One or more A residues at the 3′ end of the miRNA will be part of the poly(A) tail added by the poly(A) polymerase. Regardless of the length of the tail, 15 T residues will be incorporated during the reverse transcription as the RT primer is anchored to the first non-A base at the 3′ end of the miRNA (Fig. 1).
10. It is far from always possible to comply with the three criteria 1, 2, and 3. Criterion 1 is more important than criterion 2, which is more important than criterion 3.

11. Several web services for calculating the Tm of an oligonucleotide are available (i.e., Integrated DNA Technologies website http://www.idtdna.com/analyzer/Applications/OligoAnalyzer/). Make sure that the site uses the nearest neighbor model for calculation and that the NaCl concentration is set to 115 mM.
12. The longest possible forward primer is $CGCAGN_{20}$, where "N_{20}" are 20 miRNA-specific bases and "CGCAG" is a tail sequence that is not complementary to the miRNA (Fig. 1). More bases can be added to the 5′ end of the primer if necessary.
13. The longest possible reverse primer is $CAGGTCCAGT_{15}N_8$, where "N_8" are eight miRNA-specific bases, "T_{15}" are 15T's, and "CAGGTCCAG" is a sequence complementary to the tail of the RT primer (Fig. 1). More bases can be added to the 5′ end of the primer if necessary.
14. It is preferable to make a master mix for all the samples to minimize the effect of pipetting inaccuracy between samples. A 10 % excess of mix is recommended; for example make 90 μl mix for nine samples. Furthermore, the mix can be scaled up or down for larger or smaller sample volumes by changing the amounts of all components proportionally.
15. Optimal primer concentration depends on the sequences of the primers and miRNA. Lowering the primer concentration can solve problems with unwanted products but may decrease the amplification efficiency, whereas higher primer concentration may give higher efficiency but increases the risk of unwanted products. For most primer sets and applications, any concentration within the range of 250–1,000 nM will work.
16. Spin the tubes or the plates to assure that the samples are located at the bottom of the reaction vessel.
17. Although the PCR samples are placed on ice, start the real-time PCR step as soon as possible after mixing all the reagents because some primer–template combinations create artifacts if they are stored on ice for a prolonged time.
18. The incubation time at the different steps in the PCR depends on the thermocycler and should follow the manufacturer's recommendations, whereas the incubation temperatures are fixed. The Tm of the PCR primers is optimized to an annealing temperature of 60 °C. If it is changed, it may also be necessary to design primers with different Tms or change the primer concentration.
19. The melting curve analysis should give a single peak (the Tm of the PCR product), which is normally located between 70 and 76 °C. However, very AT- or GC-rich templates may result in products with higher or lower Tms.

References

1. Chen C, Ridzon DA, Broomer AJ et al (2005) Real-time quantification of microRNAs by stem-loop RT-PCR. Nucleic Acids Res 33:e179
2. Kumar P, Johnston BH, Kazakov SA (2011) miR-ID: a novel, circularization-based platform for detection of microRNAs. RNA 17:365–380
3. Shi R, Chiang VL (2005) Facile means for quantifying microRNA expression by real-time PCR. Biotechniques 39:519–525
4. Chugh P, Dittmer DP (2012) Potential pitfalls in microRNA profiling, Wiley interdisciplinary reviews. RNA 3:601–616
5. Busk PK (2010) Method for quantification of small RNA species. WO2010085966(A2)
6. Balcells I, Cirera S, Busk PK (2011) Specific and sensitive quantitative RT-PCR of miRNAs with DNA primers. BMC Biotechnol 11:70
7. Busk PK (2014) A tool for design of primers for microRNA-specific quantitative RT-qPCR. BMC Bioinformatics 15:29
8. SantaLucia J (1998) A unified view of polymer, dumbbell, and oligonucleotide DNA nearest-neighbor thermodynamics. Proc Natl Acad Sci U S A 95:1460–1465

Chapter 8

RNA Isolation for Small RNA Next-Generation Sequencing from Acellular Biofluids

Kasandra L. Burgos and Kendall Van Keuren-Jensen

Abstract

There are a number of considerations when choosing protocols both upstream and downstream of Next-Generation Sequencing experiments. On the front end, purification methods, additives, and residuum can often inhibit the sensitive chemistries by which sequencing-by-synthesis is performed. On the back end, data handling, analysis software packages, and pipelines can also impact sequencing outcomes. The current chapter will describe stepwise how acellular biofluid samples are prepared for small RNA sequencing.

With regard to purification methods, we found that small RNA yield can be improved considerably by following the total RNA isolation protocol included with Ambion's mirVana PARIS Kit but modifying the organic extraction step. Specifically, after transferring the upper aqueous phase to a fresh tube, water is added to the residual material (interphase and lower organic layer) and again phase-separated. In contrast, all the protocols provided with the commercially available kits at the time of this chapter publication require only one organic extraction. This simple yet, as it turns out, quite useful modification allows access to previously inaccessible material. Potential benefits from these changes are a more comprehensive sample profiling of small RNA, as well as wider access to small volume samples, such as is typically available for acellular biofluids, which now can be prepared for small RNA sequencing on the Illumina platform.

Key words RNA purification, miRNA, Next-generation sequencing, NGS, CSF, Extracellular miRNA

1 Introduction

Scientists looking to perform next-generation sequencing (NGS) must consider the manner and method of sample preparation. The way that DNA or RNA is isolated from tissue, the preparation chosen to construct sequencing libraries, and the type of sequencing that is being performed all become crucial factors in the experimental design [1].

For RNA sequencing in particular, classes of molecules are, at least in part, defined and sequenced by their size. MicroRNAs (miRNAs; 16–27 nucleotides or nt), small interfering RNAs (siRNAs; 16–27 nt), and PIWI interacting RNAs (piRNA; ~30 nt) are all part of a class of small noncoding RNA involved in sequence-specific

M. Lucrecia Alvarez and Mahtab Nourbakhsh (eds.), *RNA Mapping: Methods and Protocols*, Methods in Molecular Biology, vol. 1182, DOI 10.1007/978-1-4939-1062-5_8, © Springer Science+Business Media New York 2014

gene silencing [2]. While currently known as the smallest functional class, the depth of small RNA's biological significance to regulate gene expression is still being uncovered some 15 years after discovery [3]. NGS has recently been used to obtain small RNA differential expression profiles for a number of developmental stages [4], tissue types [5], and pathological conditions such as cancer [6] and mental illness [7] with the potential for novel biomarker discovery.

Until recently, methods for isolating RNA from tissues of origin had been thought to recover all RNA species. Roughly from large to small, RNA as a family of molecules includes coding RNA (mRNA), long noncoding RNA (lncRNA), transfer RNA (tRNA), small nucleolar RNA (snoRNA), PIWI Interacting RNA (piRNA), and microRNA (miRNA) [2]. The purification of all species of RNA is implied in the description of many commercially available kits and methods touting "total" RNA isolation. In fact, it had been used for methods that do not recover small RNA at all, such as column-based kits that washed the small RNA off the column during the cleaning steps. In addition, other kits used ratios of salt and alcohol that are too low to precipitate small RNA out of solution. There are now many commercially available kits for small RNA purification from which to choose. While the solutions have some differences, the majority of them use guanidinium thiocyanate to denature RNases and other proteins, followed by phenol-chloroform extraction to isolate RNA [8, 9]. Systematic testing has shown that the performance of RNA extraction kits varies quite a bit depending on the type of sample [10]. Reasonably, different kits may deal with a particular sample type better than another. For example, a fibrous tissue such as muscle has to be handled differently than lipid-rich nervous tissue. When available, the best option may be to choose a Kit specifically designed to deal with the challenges of a particular type of tissue. Regardless, it is worth the effort to seek to improve upon what is available and to make that resource available to other scientists.

We tested different commercially available RNA extraction kits and found that some of them were more efficient at isolating small RNA from biofluids than others. Common protocol changes that produced a higher yield of RNA were also tested in all kits. The best conditions to obtain high small RNA yield from cell-free biofluids are outlined in this chapter, as we believe it has value to researchers looking to perform small RNA NGS. The current protocol was specifically developed and tested for small RNA isolation from human plasma, serum, and cerebrospinal fluid for the purposes of Illumina-based NGS (Illumina, San Fransico, CA, USA). It has since been further applied to human saliva and urine samples. This method potentially expands the sample types and amounts used for human small RNA profiling.

From among the top four kits for isolation of total and small RNA, MaxRecovery BiooPure RNA Isolation Reagent

(Bioo Scientific, Austin, TX, USA) was not selected because the invisible final pellet caused some loss of RNA in some samples, and the miRNeasy Kit (Qiagen, Valencia, CA, USA) was not selected either because it has an 18 nt lower size limit cutoff for RNA recovery, precluding 67 of 2,578 or ~2.6 % of all mature miRNAs [11]. The standard mirVana Kit (Life Technologies), which does not offer researchers the option for protein isolation from the original lysate, performed well but was not chosen because the first buffer is added at 10 times the sample volume. Therefore, more than 50 individual centrifugation steps would be required for each 1 mL of sample, making this method logistically unreasonable for biofluid RNA isolation. The mirVana PARIS (Protein and RNA Isolation) Kit (Life Technologies) performed the best for RNA yield, ease, and application when systematically compared with the other commercially available kits and methods [10].

The mirVana PARIS Kit includes use of a proprietary lysis buffer with β-mercaptoethanol which serves to denature biofluid proteins, an acidic phenol:chloroform extraction to isolate RNA from the protein, lipid, and DNA content, followed by an alcohol/column-based cleaning step before RNA elution. In this chapter, we describe an off-label method for optimized miRNA extraction from acellular biofluids [10]. The main changes are in addition to the standard protocol provided by the manufacturer, and include re-extracting RNA from, instead of disposing of, the organic residual phenol:chloroform by adding a volume of water, remixing, and separating another aqueous volume. These changes are summarized in Subheading 3 of this chapter from **steps 9** to **11**, Subheading 3.3. Although the level of improvement in small RNA yield using the modifications proposed in this chapter will vary depending upon the particular Kit this method is applied to, it has been shown to have cross platform applicability [10]. Kits using a phenol:chloroform RNA isolation may benefit by adding the extra steps that we used for the mirVana PARIS Kit. The RNA yield from all kits that were tested benefitted from a second aqueous extraction from the phenol:chloroform residual material [10].

A notable finding was that the best kits for recovery of large RNA molecules (quantified fluorometrically using Quant-iT Ribogreen RNA, Life Technologies) were not the best for recovery of small RNA (quantified by TaqMan qRT-PCR, Life Technologies). In fact, of the top four kits in each category of either the best small RNA recovery or the best large RNA recovery, only two kits were shared across them; therefore, some kits recovered one size RNA better than another [10]. Hence, this chapter will focus on the description of methods that will enable researchers to maximize small RNA recovery. Since current methods of NGS on small RNA are performed separately from large RNA, the fact that the best kits for extraction of small or large RNA molecules are different does not pose an issue at the time of this publication.

The method described here was tested and shown to improve small RNA recovery from plasma, serum, and cerebrospinal fluid. However, this method is not limited to these sample types and can reasonably be applied to other types of acellular biofluids. In addition, the Illumina Small RNA Sample Preparation Kit and Illumina HiSeq 2000 were used for NGS downstream of the purification (Life Technologies). Since each type of sample may introduce unique challenges, we strongly recommend testing different methods for each particular type of biofluid sample and downstream application.

2 Materials

1. Ambion mirVana PARIS Kit (*see* **Note 1**): miRNA Wash Solution 1, Wash Solution 2/3 (*see* **Note 2**), Collection Tubes and Filter Cartridges (*see* **Note 3**), Cell Disruption Buffer (*see* **Note 4**), 2× Denaturing Solution, Acid-Phenol:Chloroform (*see* **Note 5**), Elution Solution (*see* **Note 6**).
2. 200-proof ethanol (ethyl alcohol), ACS grade or better (*see* **Note 7**).
3. β-mercaptoethanol.
4. 7 M ammonium acetate.
5. 2 mL cryovial (for sample).
6. Bench-top centrifuge capable of at least 800 × *g*.
7. Biosafety cabinet.
8. Fume hood with negative airflow (*see* **Note 8**).
9. Large centrifuge capable of maintaining room temperature and centrifuging at least 10,000 × *g* using a rotor able to hold 15 mL conical tubes (*see* **Note 9**).
10. Laboratory heating block set to 95–100 °C.
11. Rocking or rotating platform (*see* **Note 10**).
12. RNase-free low-bind 1.5 mL polypropylene microfuge tubes (*see* **Note 11**).
13. RNase decontamination wipes or spray (*see* **Note 12**).

3 Methods

3.1 Sample Handling

Once the biofluid is collected from the host, flash-freeze 1 mL in a 2 mL cryovial either in liquid nitrogen or in a dry ice/200-proof-ethanol slurry to preserve the RNA profile (*see* **Note 13**). Use of a biosafety cabinet is required when handling biological samples to protect researchers from human pathogen exposure.

3.2 Prepare Kit Solutions

1. Allow mirVana PARIS Kit to come to room temperature (*see* **Note 14**).
2. Add 21 mL 100 % ethanol to miRNA Wash Solution 1 (*see* **Note** 7).
3. Add 40 mL of 100 % ethanol to Wash Solution 2/3 (*see* **Notes** 7 and **15**).
4. Add 375 μL β-mercaptoethanol to 2× Denaturing Solution (*see* **Note 16**).
5. Aliquot 1 mL of nuclease-free molecular biology grade water (*see* **Note 6**) into 1.5 mL microfuge tubes, and place them on heating block set to 95 °C. This preheated water will be used to elute RNA from the column in the final step (*see* **Note 17**).

3.3 Modified mirVana PARIS miRNA Isolation Protocol

1. Add an equal volume of 2× Denaturing Solution to frozen biofluid sample (*see* **Note 18**).
2. Place sample on a rocking or rotating platform at room temperature until fully thawed and mixed (*see* **Note 10**).
3. Incubate at room temperature for 10 min.
4. Add an equal volume of Acid-Phenol:Chloroform (*see* **Note 19**).
5. Vortex for 30 s to mix.
6. Centrifuge at 10,000×*g* for 5 min at room temperature (*see* **Note 20**).
7. Carefully remove the tubes from the centrifuge, and check that there is an upper (aqueous) layer and a lower (organic) layer.
8. Transfer approximately 90 % of the upper aqueous phase of this first extraction to a clean tube and estimate the volume. Take care to leave behind a volume of aqueous liquid so that the meniscus does not touch the interphase (*see* **Note 21**). Set aside.
9. To the leftover organic residuum, add a volume of water equivalent to the aqueous volume that was just transferred to the new tube.
10. Vortex for 30 s to mix.
11. Centrifuge at 10,000×*g* for 5 min at room temperature.
12. Transfer approximately 90 % of the upper aqueous phase of this second extraction to the same tube that contains the first aqueous volume removed from the phenol chloroform (*see* **Note 21**). The remainder of the phenol:chloroform can now be discarded (*see* **Note 5**).
13. Add 1.5× volumes of 100 % ethanol to the total aqueous volume removed from first and second organic extractions (*see* **Note** 7).

14. Invert 10 times to mix, and let solution stand at room temperature for 10 min.
15. Apply solution through column, 700 μL at a time, by centrifugation at not more than 800 × *g* (*see* **Note 22**), discarding flow-through at each pass, and reassemble filter column and reservoir tube (*see* **Note 3**).
16. Apply 700 μL of prepared Wash Solution 1 to the column (*see* **Note 23**), and centrifuge at 800 × *g* for 30 s to pass solution through filter column (*see* **Note 22**). Discard flow-through, and reassemble filter column and reservoir tube (*see* **Note 3**).
17. Apply 500 μL of prepared Wash Solution 2/3 to the column (*see* **Note 24**), and centrifuge at 800 × *g* for 30 s to pass solution through filter column (*see* **Note 22**). Discard flow-through, and reassemble filter column and reservoir tube.
18. Repeat **step 17**.
19. Without applying any other solutions, centrifuge filter column and empty reservoir tube for 30 s to dry residual ethanol.
20. Transfer filter column to fresh tube (*see* **Note 25**).
21. Apply 100 μL of 95 °C (*see* **Note 26**) nuclease-free water (*see* **Note 6**) to the filter column, and incubate at room temperature for 1 min.
22. Centrifuge filter column at 10,000 × *g* for 1 min to elute RNA from the column (*see* **Note 27**).
23. Repeat **steps 21** and **22**.
24. The filter component of the column assembly can be discarded as RNA has been eluted from the filter and is in the flow-through in the collection tube.
25. Centrifuge RNA sample at maximum speed for 1 min to collect residual column fibers.
26. Avoiding the residual fibers from the filter column, transfer the RNA sample to a new microfuge tube. Proceed to ethanol precipitation for small RNA NGS sample preparation (*see* **Note 27**).
27. Add 0.5 volumes 7 M ammonium acetate to a final concentration of 2–2.5 M. Mix well (*see* **Note 28**).
28. Add 4 volumes of 100 % ethanol. Mix well, and place at −20 °C from 4 h to overnight.
29. Centrifuge at 16,000 × *g* for 30 min at 4 °C to precipitate RNA (*see* **Note 29**).
30. Wash pellet twice with 80 % ethanol (*see* **Note 30**).
31. Resuspend RNA pellet in volume of water as downstream protocol dictates.

4 Notes

1. The mirVana PARIS Kit is enough for 40 reactions when using the manufacturer provided protocol and suggested tissues (*see* Ambion mirVana PARIS Kit user guide). With the modified protocol described here, one 40 reaction Kit will purify ~20 mL of biofluid.
2. Wash Solution 2/3 is used for the second and third rinse of the silica-based column containing the immobilized RNA.
3. The filter column and collection tube will be reused at all steps in this modified protocol, with the exception of the last one where the RNA isolation and purification is complete.
4. Cell Disruption Buffer is included in the reagent list; however it will not be used for the current method that was designed for cell-free biofluid samples.
5. The Acid-Phenol:Chloroform is caustic; therefore, care must be taken during the handling and disposal. Personal protective equipment and the use of a fume hood are required.
6. Elution Solution is provided for final elution of the RNA for routine purposes. In the current protocol, nuclease-free molecular biology grade water is used for elution of the RNA.
7. As the ratio of ethanol to aqueous buffer is important to whether or not RNA is dissolved in—or precipitating out of—solution, it is crucial to use 200-proof, ACS grade or better, ethanol in making the alcohol:buffer solutions. Each time dehydrated ethanol is exposed to the environment, water from atmospheric humidity will dissolve in it, subsequently decreasing the ethanol content of the downstream solution. Using a small bottle of 200-proof ethanol, or aliquoting a larger bottle into smaller volumes, will increase the likelihood that the ethanol remains as the stock.
8. For safety reasons, with the exception of the last step, the entire protocol should be performed in a fume hood with negative airflow designed for volatile chemicals.
9. The pH of all buffers and solutions is an important aspect of their molecular function. Since temperature has a significant effect on pH, it should be controlled. All steps described here are done at room temperature unless otherwise stated. However, extended centrifugation may increase the temperature of the sample being centrifuged. Therefore, the centrifuges used in the non-column-based centrifugation steps must be set to the standard ambient temperature of 25 °C. For brief centrifugation steps, such as the ones for passing liquid through microfuge columns, a temperature-controlled centrifuge is not required.

10. It is not important at which speed a standard laboratory rocking or rotating platform is used as long as it allows a thorough mixing of the frozen biofluid in the denaturing buffer.
11. We found that the collection tubes supplied with the mirVana PARIS Kit did not always tightly cap. In addition, the use of low-binding tubes decreases evaporation and residual RNA material left behind in the storage tube. Therefore, once the RNA has been eluted from the column, it should be transferred to a tightly capped nuclease-free low-binding microfuge tube.
12. Clean bench and all equipment that will be used for RNA purification with RNase decontamination spray or wipes according to the manufacturer's recommendation for those products. Overall precaution should be taken to minimize possible exposure of RNA to RNAases.
13. While miRNA has been shown to be relatively stable [12], treating samples the same way each time will ensure that collection bias is minimized, and will preserve the total RNA profile. In frozen samples, RNases are inactive due to the low temperature that does not allow water to be in the liquid form necessary for these proteins to degrade RNA. Samples are thawed in the presence of 2× Denaturing Solution to ensure that RNases are denatured; therefore, they are irreversibly inactivated [11].
14. The mirVana PARIS Kit is shipped at room temperature, and components are stored either at room temperature or at 4 °C according to the manufacturer's specifications. For either the routine use or the current modified protocol, the mirVana Paris Kit components should be allowed to come to room temperature before use.
15. A white precipitate of excess EDTA might form in the Wash Solution 2/3 but it is of no consequence and should be left behind in the bottle when using this solution.
16. The 2× Denaturing Solution forms a precipitate at the recommended storage temperature of 4 °C. Once warmed to room temperature, visually inspect the solution. If a solid white precipitate is present, place the bottle tightly closed at 37 °C and, occasionally, mix until solution is fully reconstituted.
17. Microfuge-tube cap locks or aluminum foil can be used to ensure the tubes stay closed under increased temperature and pressure from the evaporating solution.
18. Estimate the volume of the biological sample. If the sample tube is more than halfway full, which would prevent that an equal volume of 2× Denaturing Solution be added, add only 1/10th volume of 2× Denaturing Solution in the tube, and

mix vigorously until frozen sample is slightly loosened from the tube. Transfer frozen sample and residual solution to a larger tube that has the remaining 2× Denaturing Solution.

19. A small volume of aqueous buffer overlays the organic Acid-Phenol:Chloroform. When using this reagent, be sure that two distinct layers are present. Agitation of this solution should be avoided so that the layers do not mix. If the solution looks cloudy or small bubbles are present, it should be allowed to settle until the two layers are visibly separate. When using this solution, be sure to withdraw Acid-Phenol:Chloroform from beneath the aqueous buffer layer. When the volume of solution gets low, be sure to watch that you are withdrawing the Acid-Phenol:Chloroform and not the overlying buffer.
20. The phenol-chloroform phase separation steps involve centrifuging a relatively large volume. Therefore, it is advisable that the rotor for the temperature-regulated centrifuge (*see* **Note 9**) is confirmed to be compatible with centrifuge tubes that can hold this volume prior to beginning the purification procedure. The tube should be capable of holding 5 times the volume.
21. Depending on the biofluid, a white interphase may or may not be obvious, particularly for the second extraction. Upon careful inspection, the phases should be visible and should not be disrupted when pipetting the upper aqueous volume.
22. The columns from the mirVana PARIS Kit were designed for the manufacturer's protocol. With the modified method, larger volumes than originally intended pass though the column. As RNA will bind to the fibers of the column, it is best to carefully maintain the integrity of the column. Therefore, the maximum centrifugation speed recommended for passing the aqueous extraction/ethanol solution is $800 \times g$.
23. Prepared Wash Solution 1 contains 21 mL 100 % ethanol.
24. Prepared Wash Solution 2/3 contains 40 mL 100 % ethanol.
25. To prevent dried residual material from being introduced into the fresh reservoir tubes, clean the outside of the filter column using a wipe with 70 % ethanol solution but avoid wetting the filter.
26. Preheat an aliquot of nuclease-free molecular biology grade water on a heat block set to 95 °C, and use it to elute RNA from the filter column. To account for evaporation at this temperature, double the volume that will be used should be preheated.
27. If the RNA will be used for any other sequencing aside from small RNA, DNAse treatment of the sample may be necessary.
28. Ethanol precipitation of RNA should always proceed with the salt being added to the RNA sample and thoroughly mixed prior to adding alcohol.

29. Centrifuge the tube with the hinge of the cap out so that the RNA collects under the hinge inside the tube. As the RNA will likely be translucent at this stage, it will be easier to locate and avoid disrupting.
30. Be sure to allow 80 % ethanol to run down the hinge side of the interior of the microfuge tube.

References

1. Baudhuin LM (2013) Quality guidelines for next-generation sequencing. Clin Chem 59:858–859
2. Castel SE, Martienssen RA (2013) RNA interference in the nucleus: roles for small RNAs in transcription, epigenetics and beyond. Nature 14:100–112
3. Fire A, Xu S, Montgomery MK, Kostas SA et al (1998) Potent and specific genetic interference by double-stranded RNA in Caenorhabditis elegans. Nature 391:806–811
4. Zhao Y, Srivastava D (2007) A developmental view of microRNA function. Trends Biochem Sci 4:189–197
5. Yang Q, Hua J, Wang L, Xu B et al (2013) MicroRNA and piRNA profiles in normal human testis detected by next generation sequencing. PLoS One 8:e66809. doi: 10.1371/journal.pone.0066809
6. Schee K, Lorenz S, Worren MM, Gunther CC et al (2013) Deep sequencing the microRNA transcriptome in colorectal cancer. PLoS One 8:e66165. doi:10.1371/journal.pone.0066165
7. Chana G, Bousman CA, Money TT, Gibbons A et al (2013) Biomarker investigations related to pathophysiological pathways in schizophrenia and psychosis. Front Cell Neurosci 7:95
8. Chomczynski P, Sacchi N (1987) Single-step method of RNA isolation by acid guanidinium thiocyanate-phenol-chloroform extraction. Anal Biochem 162:156–159
9. Chomczynski P, Sacchi N (2006) Single-step method of RNA isolation by acid guanidinium thiocyanate-phenol-chloroform extraction: twenty-something years on. Nat Protoc 1: 581–585
10. Burgos KL, Javaherian A, Bomprezzi R, Ghaffari L et al (2013) Identification of extracellular miRNA in human cerebrospinal fluid by next-generation sequencing. RNA 5: 712–722
11. mirBase: the microRNA Database [Internet]. Release 20. Manchester (England): University of Manchester. 2006 [updated 2013 Jun 24; cited 2013 Nov 25]. Available from http://www.mirbase.org/index.shtml
12. Mitchell PS, Parkin RK, Kroh EM, Fritz BR et al (2008) Circulating microRNAs as stable blood-based markers for cancer detection. Proc Natl Acad Sci USA 105:10513–10518

Chapter 9

Sequencing Small RNA: Introduction and Data Analysis Fundamentals

Jai Prakash Mehta

Abstract

Small RNAs are important transcriptional regulators within cells. With the advent of powerful Next Generation Sequencing platforms, sequencing small RNAs seems to be an obvious choice to understand their expression and its downstream effect. Additionally, sequencing provides an opportunity to identify novel and polymorphic miRNA. However, the biggest challenge is the appropriate data analysis pipeline, which is still in phase of active development by various academic groups. This chapter describes basic and advanced steps for small RNA sequencing analysis including quality control, small RNA alignment and quantification, differential expression analysis, novel small RNA identification, target prediction, and downstream analysis. We also provide a list of various resources for small RNA analysis.

Key words small RNA, miRNA, Next generation sequencing

1 Introduction

Small RNA sequencing is a powerful method to quantify the expression of various noncoding small RNAs. The different forms of small RNA are important transcriptional regulators. The most abundant form of small RNA found in cells is microRNA (miRNA). miRNA binds to a target sequence thereby degrading or reducing the expression of target RNA. miRNA primary transcripts (pri-miRNAs) are produced from genomic DNA by RNA polymerase II in a manner similar to other RNAs. The enzyme Drosha cuts the pri-miRNA to a smaller 70 nucleotide stem-loop precursor (pre-miRNA). This pre-miRNA is then cleaved by Dicer to a ~23 nucleotide miRNA duplex. One strand of the final miRNA product, the mature miRNA, forms the microRNA-induced silencing complex (miRISC), which guides binding of the miRNA to the target, thereby degrading or silencing the RNA (Fig. 1).

The advent of next generation sequencing (NGS) platforms has enabled us to sequence small RNA at high throughput, making it feasible to study miRNA expression, identify novel miRNAs, miRNA

M. Lucrecia Alvarez and Mahtab Nourbakhsh (eds.), *RNA Mapping: Methods and Protocols*, Methods in Molecular Biology, vol. 1182, DOI 10.1007/978-1-4939-1062-5_9, © Springer Science+Business Media New York 2014

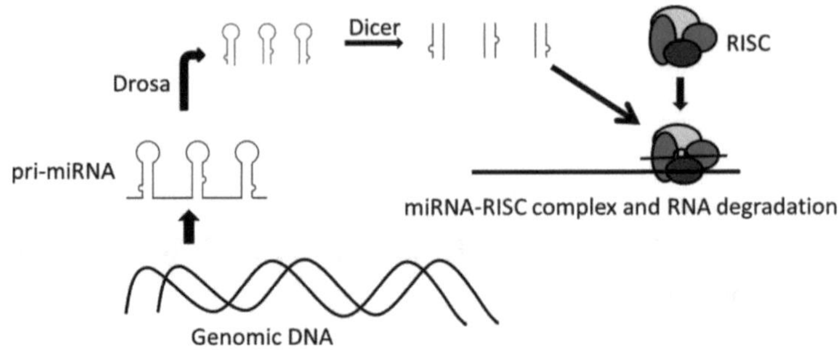

Fig. 1 An illustration of miRNA synthesis and its mechanism of target degradation

mutations, and various miRNA isoforms along with the possibility to study other types of small RNA.

NGS of small RNA (small RNA-Seq) starts with total RNA extraction, followed by fractioning and size selection for 17–25 nucleotides. The fragments are then ligated to adapters, and multiple copies are made using reverse transcription (RT) and PCR. The adapter-ligated PCR products are sequenced using a NGS machine. Since the sample has been size selected, only small RNA are sequenced in this process (Fig. 2).

2 Data Analysis

Sequencing data analysis can be quiet hard and time consuming. One needs to be trained in Linux command line environment because most of the sequencing data analysis still runs on Linux servers. Recently, some web based tools for easy analysis and interpretation of the data have been developed. However, it is practically not possible to transfer the massive raw FASTQ files over internet; thus, developers came up with ingenious ways to compress the files by providing a script included in software such as miRanalyser [1], mirTools [2, 3], and DSAP [4]. Most of the scripts aggregate up similar reads, and provide a count estimate for all miRNA to be uploaded on the web server. A comprehensive list of software and tools as well as their key features is provided in Table 1. Additionally, some important miRNA resources are listed in Table 2.

miRNA sequencing data analysis goes through a number of processes, depending on your questions and experimental design. The most important steps are listed in Fig. 3 and explained below.

2.1 Understanding Your Data

The sequencing machine outputs data as FASTQ format. It is similar to FASTA format but with an added field on quality for each base sequenced. The FASTQ file can be opened in any text editor. The first line indicates the Sequence ID, with adapter sequence at the end (as shown in example) if the samples have been multiplexed;

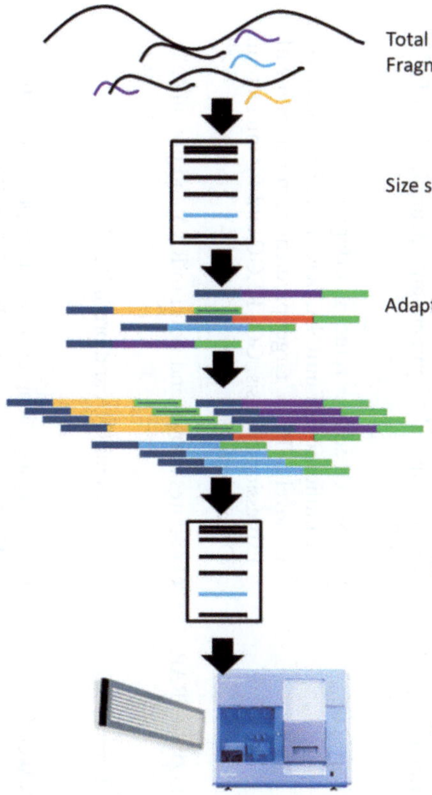

Fig. 2 Basic steps in small RNA sequencing

the second line is the raw sequence output of the machine; the third line contains a "+"; and the fourth line provides information about the quality of each base sequenced in ASCII format, with lower values ASCII indicating low quality, and higher values ASCII, high quality. An example of a single FASTQ entry is depicted below.

@HWI-EAS283:38:63BWGAAXX:1:54:5040:6842 1:N:0:TTAGGC

TATTGCACTCGTCCCGGCCTTGGAATTCTCGGGTG

+

IIIIIIIIIHIIIIEHIIIIHIIHIIIHHHFCEC

Since small RNA sequences are short, and the sequencing is often longer than the miRNA size, the output reads often contain sequences from the adapter, which needs to be removed before performing any further analysis.

2.2 Quality Control Checks

As with any sequencing technique, quality control is essential and should be performed at different steps including the total number of reads, alignment percentage with whole genome, and known small RNA. Other parameters to look for are the number of "N,"

Table 1
List of comprehensive tools for small RNA sequencing data analysis

Name	Year	Availability	Key features
miRanalyser [1]	2009	http://web.bioinformatics.cicbiogune.es/microRNA	Web Server, adapter removal, alignment and quantification using miRBase, RFam and RepBase. Predicting novel miRNA, DE analysis
miRExpress [22]	2009	http://mirexpress.mbc.nctu.edu.tw/	Adapter removal, alignment and quantification of miRNA
mirTools [2]/mirTools2 [3]	2010	http://centre.bioinformatics.zj.cn/mirtools/	Web Server, adapter removal, mapping and quantification of miRNA, ncRNA, piRNA. Prediction of novel miRNA, target prediction, integration with expression analysis, GO, KEGG and PP interaction. DE analysis
miRTRAP [23]	2010	http://flybuzz.berkeley.edu/miRTRAP.html	Computes known and novel miRNA and pre-miRNA
MIReNA [24]	2010	http://www.lgm.upmc.fr/mirena/index.html	Computes miRNA and pre-miRNA
miRNAkey [25]	2010	http://ibis.tau.ac.il/miRNAkey/	GUI, performs adapter removal, mapping, counts and RPKM estimation, calculates DE miRNA
DSAP [4]	2010	http://dsap.cgu.edu.tw/dsap.html	Web Server, GUI, Uploads FASTQ format and reads with counts. Performs adapter removal, poly ATCG/N cleaning, alignment and quantification using Rfam and miRBase database. Detects cross-species presence
miRDeep [11]/miRDeep2 [26]	2012	https://www.mdc-berlin.de/8551903/en/research/research_teams/systems_biology_of_gene_regulatory_elements/projects/miRDeep	Accepts data as FASTQ, SAM and BAM format. Calculates known and novel miRNA. Graphical display of pre-miRNA structure. Target prediction using TargetScan algorithm
miRspring [27]	2013	http://mirspring.victorchang.edu.au/	Standalone JavaScript encoded HTML document for integrated miRNA analysis. Also reports seed-isomiRs

Table 2
Resources for small RNA sequencing data analysis

Name	Availability	Summary
Sequence analysis tools		
FASTQC	http://www.bioinformatics.babraham.ac.uk/projects/fastqc/	Quality check on raw sequences in FASTQ format
FASTX	http://hannonlab.cshl.edu/fastx_toolkit/	Reads trimming, quality filtering, adapter removal, FASTQC to FASTA conversion etc.
BWA [6]	http://bio-bwa.sourceforge.net/	Reads alignment
Bowtie [5]	http://bowtie-bio.sourceforge.net	Ultra-fast reads alignment
SAMTOOLS [28]	http://samtools.sourceforge.net/	Software to analyze aligned files (SAM). SAM to BAM conversion and many more utility tools to handle aligned data
Database		
MiRBase [7–10]	http://www.mirbase.org/	Database of miRNA sequences
Rfam [29]	http://rfam.sanger.ac.uk/	Database of RNA families
Target prediction software		
TargetScan [14–17]	http://www.targetscan.org/	Database of miRNA targets
MiRanda [13]	http://www.microrna.org/microrna	Predicts miRNA targets
PicTar [18]	http://pictar.mdc-berlin.de/	Identification of miRNA targets
MiRDB [19]	http://mirdb.org/miRDB/	miRNA target prediction and functional annotation
PMTED [20]	http://pmted.agrinome.org/	Plant miRNA target expression database
miRNA variant database		
MiRvar [21]	http://genome.igib.res.in/mirlovd	miRNA variant database

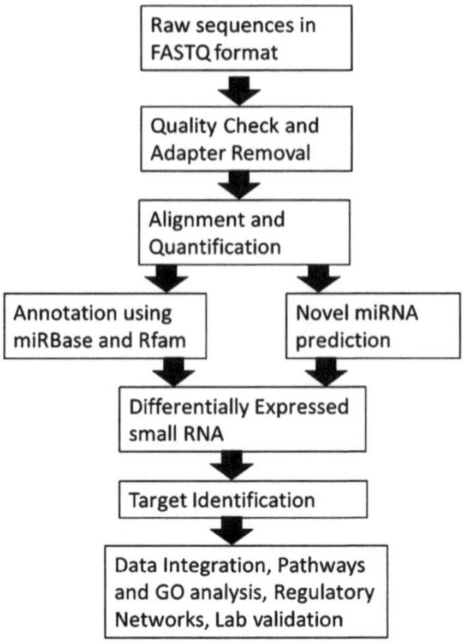

Fig. 3 Work flow of small RNA sequencing data analysis

which stands for "no base was called," and the per-base sequence quality. Reads with low quality or large number of missing base calls (N) may be filtered off. Overrepresentation of certain sequences should be looked for as they might be potential contaminants such as adapter dimers. FASTQC is a handy application to get an overview of the data. The software runs on Windows, Mac, and Linux and can be obtained from http://www.bioinformatics.babraham.ac.uk/projects/fastqc/. An example of FASTQC output is shown in Fig. 4.

FASTX-Toolkit has multiple modules to remove sequences with low quality. The software runs only on Linux and can be downloaded from http://hannonlab.cshl.edu/fastx_toolkit/.

2.3 Adapter Removal

Since the sequencing goes beyond the length of the typical miRNA, adapter contamination in the reads is inevitable. Therefore, it is absolutely important to remove the adapter sequences from any further analysis. Most of the miRNA comprehensive software comes with a utility tool to remove the adapters from the reads. Alternatively, FASTX-Toolkit can be used to remove them. FASTX-Toolkit can be obtained from http://hannonlab.cshl.edu/ fastx_toolkit/.

2.4 Reads Alignment and Quantification

The reads are aligned to a reference sequence, and overlapping reads are used as the expression measure for the respective small RNA. The alignment can be performed using Bowtie [5] or BWA [6], though most miRNA-based tools have these alignment tools embedded in the software. Since different samples are sequenced at varying depth (i.e., the total number of sequence varies among

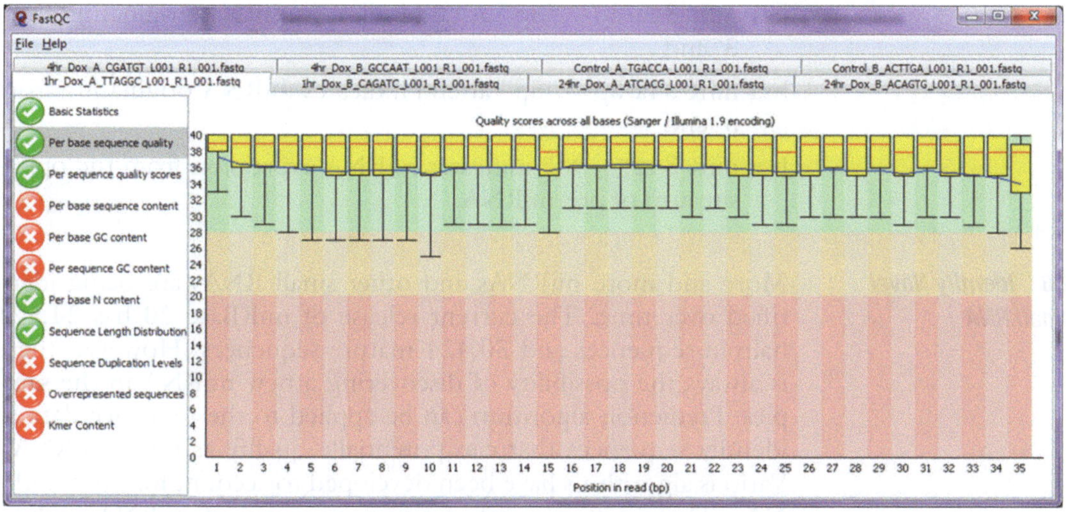

Fig. 4 An example of FASTQC output. The Basic Statistics tab provides the statistics of the library. Per-base sequence quality is an important parameter to consider. Presence of overrepresented sequence is also indicative of a low quality run

samples), normalization is an important step in quantifying the data. A common practice is to calculate expression as "reads per million" or "reads per million mapped." Calculating expression as "reads per million" is more popular as it takes all sequenced reads that passed the quality control measures. Alternatively, "reads per million mapped" is used when there are contaminant sequences of unknown origin.

Another approach used is the alignment with known small RNA sequences. This is the preferred approach when experiments are designed specifically to study miRNAs. miRbase is a database of all known miRNAs from Human as well as other organisms [7–10]. The reads are aligned with known miRNAs, and overlapping reads are used to quantify the expression.

2.5 miRNA Nomenclature

To correctly interpret the results, it is important to understand the nomenclature of miRNA. The name of a miRNA starts with the organism name, e.g., "hsa" for Human, "mmu" for mouse, etc., followed by "miR," and a number, e.g., hsa-miR-4531. Small "r" in mir indicates pre-miRNA. Suffix a, b, c indicates variants of the same miRNA with 1–2 base differences, while "-3p" and "-5p" at end of the name indicate miRNA originating from 3′ and 5′ end, respectively. The opposite arm of a pre-miRNA is indicated by "*" at end. Some miRNA annotation examples are provided below.

hsa-miR-34a → Human mature miRNA

hsa-mir-34a → Small "r" indicated pre-miRNA

hsa-miR-34b → a/b/c at the end indicates closely related family of miRNA with 1 or 2 bases different.

hsa-miR-34a-3p → "3p" at end indicates miRNA originating from 3′ end

hsa-miR-34a-5p → "5p" at end indicates miRNA originating from 5′ end

hsa-miR-34a* → "*" indicates miRNA expression from the opposite arm of pre-miRNA.

2.6 Identify Novel small RNA

More and more miRNAs and other small RNAs are being identified over time. The current release of miRBase 20 has 24,521 hairpin sequences and 30,424 mature sequences. However, there is always the possibility of discovering a new miRNA in the samples. Prediction algorithm can be applied to the sequence data to identify sequences that can potentially qualify as novel miRNAs. Various algorithms have been developed to account for novel miRNAs. These software uses intrinsic property of miRNA such as the presence of Drosha and Dicer cleavage site from the genomic coordinates, RNA folding patterns, etc. Some of the algorithms/software used for such analysis are miRDeep [11], miRanalyzer [1], and mirTools [2, 3].

2.7 Differential Expression

Most of the experiments are designed to identify differentially expressed miRNA among two or more groups of samples. Differentially Expressed (DE) miRNAs can be identified from counts data using software such as DESeq [12], which is a bioconductor package that runs on R, a statistical package (http://www.r-project.org/). miRanalyser [1] is an easy-to-use web based tool that has DESeq incorporated in the pipeline. However, the analysis needs to be done at the same time as alignment and quantification since the web server does not store results for long. An alternative method is to use the "normalized expression" as explained in an earlier section (Subheading 2.4). The "normalized expression" is likely normally distributed; therefore, traditional statistics methods such as "*t*-test" and "ANOVA" can be used.

2.8 Time Series Experiments

For time series experiments, the normalized data can be clustered. If there are replicates, non-differentially expressed genes can be removed using ANOVA. RNA with low variability can be removed for further analysis. Interesting clusters can be further analyzed for its functional role.

2.9 Target Identification

The final aim of any miRNA profiling study is to examine the mechanism by which miRNA affects the cellular function. The driving notion of such analysis is that the miRNA binds to the target mRNA thereby reducing the expression of target genes. It is estimated that 50 % of all known gene have miRNA binding sites, which are usually located in the 3′ UTR region of the target gene. Match at certain region, in particular towards 5′ region, is more important for RNA degradation. However, binding is rarely perfect and RNA degradation is dependent on the binding strength between the miRNA and its target gene. Variations in the miRNA and/or its target sequence can have a strong effect on RNA degradation or silencing. A target gene can have multiple miRNA binding sites, a characteristic linked with higher degree of degradation.

To understand the downstream effect of miRNA, potential targets can be identified using pattern recognition and other advanced algorithms that rely on the presence of a complementary region at the 3′ UTR region of the gene. Many other parameters, such as number and location of mismatches, are accounted when identifying target genes. However these approaches are not perfect, and account for large number of false positives. Therefore, the availability of mRNA or protein expression profiles under similar condition can greatly help in reducing false positives, and contribute to a better understanding of the system. Some of the algorithm/software for target predictions are miRanda [13], TargetScan [14–17], PicTar [18], miRDB [19], and PMTED [20] (Table 2).

2.10 Variant Detection

Since sequencing technology provides information on each single base, it is possible to study the effect of polymorphism among miRNAs. Polymorphism in the miRNA and/or target genes can have profound effect on the binding and downstream effect. MiRvar [21] is a database of polymorphic miRNA, and can be used to integrate with the results (Table 2).

3 miRanalyser, an Easy to Use Software for miRNA Data Analysis

Here we provide a brief description of miRanalyser [1], a simple to use web server for miRNA data analysis. More details can be found on the following website http://bioinfo2.ugr.es/miRanalyzer/miRanalyzer.php.

The software accepts data in the following format:

Format 1

```
SEQUENCE1 COUNTS1
SEQUENCE2 COUNTS2
-
-
SEQUENCEn COUNTSn
```

Format 2

```
>ID1 COUNTS1
SEQUENCE1
>ID2 COUNTS2
SEQUENCE2
-
-
>IDn COUNTSn
SEQUENCEn
```

To convert the FASTQ format, a PERL script is provided by the developers. Install PERL (http://www.perl.org) on the computer. Download the convertor groupReadsV2.pl and execute it with input files in the same folder

e.g., perl groupReadsV2.pl input=inputfile.fq output=inputfile.counts

Typing "perl groupReadsV2.pl" provides all the options.

Once the data has been formatted, upload the data on miRanalyser web server and get back the results.

Apart from quantifying the known miRNA, the software also finds novel miRNA, and is capable of identifying DE miRNA. The differential expression analysis uses DESeq algorithm [12] to identify differentially expressed miRNA. The job ids separated by "," should be provided for each group, and the software reports the list of DE miRNA. A comprehensive list of software for miRNA analysis is listed in Table 1.

References

1. Hackenberg M, Sturm M, Langenberger D et al (2009) miRanalyzer: a microRNA detection and analysis tool for next-generation sequencing experiments. Nucleic Acids Res 37: 68–76
2. Zhu E, Zhao F, Xu G et al (2010) mirTools: microRNA profiling and discovery based on high-throughput sequencing. Nucleic Acids Res 38:392–397
3. Wu J, Liu Q, Wang X et al (2013) mirTools 2.0 for non-coding RNA discovery, profiling and functional annotation based on high-throughput sequencing. RNA Biol 10:1087–1092
4. Huang PJ, Liu YC, Lee CC et al (2010) DSAP: deep-sequencing small RNA analysis pipeline. Nucleic Acids Res 38:385–391
5. Langmead B, Trapnell C, Pop M et al (2009) Ultrafast and memory-efficient alignment of short DNA sequences to the human genome. Genome Biol 10:R25
6. Li H, Durbin R (2009) Fast and accurate short read alignment with burrows-wheeler transform. Bioinformatics 25:1754–1760
7. Griffiths-Jones S, Grocock RJ, van Dongen S, Bateman A et al (2006) miRBase: microRNA sequences, targets and gene nomenclature. Nucleic Acids Res 34:140–144
8. Griffiths-Jones S (2004) The microRNA registry. Nucleic Acids Res 32:D109–D111
9. Griffiths-Jones S, Saini HK, van Dongen S et al (2008) miRBase: tools for microRNA genomics. Nucleic Acids Res 36:D154–D158
10. Kozomara A, Griffiths-Jones S (2011) miRBase: integrating microRNA annotation and deep-sequencing data. Nucleic Acids Res 39: D152–D157
11. An J, Lai J, Lehman ML et al (2013) miRDeep*: an integrated application tool for miRNA identification from RNA sequencing data. Nucleic Acids Res 41:727–737
12. Anders S, Huber W (2010) Differential expression analysis for sequence count data. Genome Biol 11:R106
13. Betel D, Wilson M, Gabow A et al (2008) The microRNA.org resource: targets and expression. Nucleic Acids Res 36:D149–D153
14. Garcia DM, Baek D, Shin C et al (2011) Weak seed-pairing stability and high target-site abundance decrease the proficiency of lsy-6 and other microRNAs. Nat Struct Mol Biol 18: 1139–1146
15. Grimson A, Farh KK, Johnston WK et al (2007) MicroRNA targeting specificity in mammals: determinants beyond seed pairing. Mol Cell 27:91–105
16. Friedman RC, Farh KK, Burge CB et al (2009) Most mammalian mRNAs are conserved targets of microRNAs. Genome Res 19:92–105
17. Lewis BP, Burge CB, Bartel DP (2005) Conserved seed pairing, often flanked by adenosines, indicates that thousands of human genes are microRNA targets. Cell 120:15–20
18. Krek A, Grun D, Poy MN et al (2005) Combinatorial microRNA target predictions. Nat Genet 37:495–500
19. Wang X (2008) miRDB: a microRNA target prediction and functional annotation database with a wiki interface. RNA 14:1012–1017

20. Sun X, Dong B, Yin L et al (2013) PMTED: a plant microRNA target expression database. BMC Bioinformatics 14:174
21. Bhartiya D, Laddha SV, Mukhopadhyay A et al (2011) miRvar: a comprehensive database for genomic variations in microRNAs. Hum Mutat 32:E2226–E2245
22. Wang WC, Lin FM, Chang WC et al (2009) miRExpress: analyzing high-throughput sequencing data for profiling microRNA expression. BMC Bioinformatics 10:328
23. Hendrix D, Levine M, Shi W (2010) miRTRAP, a computational method for the systematic identification of miRNAs from high throughput sequencing data. Genome Biol 11:R39
24. Mathelier A, Carbone A (2010) MIReNA: finding microRNAs with high accuracy and no learning at genome scale and from deep sequencing data. Bioinformatics 26: 2226–2234
25. Ronen R, Gan I, Modai S et al (2010) miRNAkey: a software for microRNA deep sequencing analysis. Bioinformatics 26:2615–2616
26. Mackowiak SD (2011) Identification of novel and known miRNAs in deep-sequencing data with miRDeep2. Curr Protoc Bioinformatics 12, Unit 12.10
27. Humphreys DT, Suter CM (2013) miRspring: a compact standalone research tool for analyzing miRNA-seq data. Nucleic Acids Res 41:147
28. Li H, Handsaker B, Wysoker A et al (2009) The sequence Alignment/Map format and SAMtools. Bioinformatics 25:2078–2079
29. Burge SW, Daub J, Eberhardt R et al (2013) Rfam 11.0: 10 years of RNA families. Nucleic Acids Res 41:D226–D232

Chapter 10

Measuring Expression Levels of Small Regulatory RNA Molecules from Body Fluids and Formalin-Fixed, Paraffin-Embedded Samples

Adrienn Gyongyosi, Otto Docs, Zsolt Czimmerer, Laszlo Orosz, Attila Horvath, Olga Török, Gabor Mehes, Laszlo Nagy, and Balint L. Balint

Abstract

MicroRNAs are involved in the regulation of various pathophysiological processes such as immune regulation and cancer. Next-generation sequencing methods enable us to monitor their presence in various types of samples but we need flexible methods for validating datasets generated by high-throughput methods.

Here we describe the detailed protocols to be used with our MiRNA Primer Design Tool assay design system. The presented methods allow the flexible design of the oligonucleotides needed for the RT-qPCR detection of any variant of small regulatory RNA molecules from virtually any species. This method can be used to measure miRNA levels from formalin-fixed, paraffin-embedded (FFPE) samples and various body fluids. As an example, we show the results of the hsa-miR-515-3p, hsa-miR-325, and hsa-miR-155 quantification using a specific UPL probe (Universal Probe Library) and a stem-loop RT-qPCR assay. The small nucleolar RNA RNU43 is used as endogenous control for normalization of the results. Urine from healthy pregnant women and FFPE samples from patients diagnosed with colorectal cancer and treated with antibody-based anti-EGFR monotherapy were used as samples.

Key words miRNA, RT-qPCR, Stem-loop, Urine, FFPE, Cancer

1 Introduction

MicroRNAs (miRNAs) are short 18–25 nucleotide-long, single stranded, non-coding RNA molecules [1]. MiRNAs act as post-transcriptional regulators of gene expression through induction of translational repression or RNA decay [2]. MiRNAs play an important role in a wide range of normal physiological and pathological processes [3, 4]. Aberrantly expressed miRNAs contribute to the development of various human disorders including cancers, neurodegenerative disorders, and autoimmune diseases [5, 6].

Adrienn Gyongyosi and Otto Docs have contributed equally to this chapter.

M. Lucrecia Alvarez and Mahtab Nourbakhsh (eds.), *RNA Mapping: Methods and Protocols*, Methods in Molecular Biology, vol. 1182, DOI 10.1007/978-1-4939-1062-5_10, © Springer Science+Business Media New York 2014

Unlike mRNAs, miRNAs are highly stable and well-preserved RNA molecules in formalin-fixed' paraffin-embedded (FFPE) samples and different body fluids including plasma, saliva, and urine [7–9]. The high stability as well as their disease-specific expression pattern renders miRNAs suitable for biomarkers in molecular diagnostic applications. Recently, miRNAs have been investigated as potential diagnostic and prognostic markers in patients suffering of cancer and autoimmune diseases by using samples of various origin, like FFPE, plasma, and urine [10–13]. Interestingly, a miRNA expression profile is equally useful for identifying tissue origin of tumors, subtype classifications of different cancers, prediction of clinical outcome, and selection of patients for targeted therapy [14–19].

Stem-loop RT-qPCR is a widely used technique for detection and quantification of mature miRNAs [20]. This method includes two steps: (1) miRNA-specific stem-loop RT primer-based reverse transcription, and (2) quantification of product using conventional TaqMan™ assay with miRNA-specific TaqMan™ probe and forward primer. This method enables us to quantify mature miRNA expression levels from nanograms of total RNA [20]. A simplified setup of stem-loop RT-qPCR uses a general stem-loop primer-specific probe from the Universal Probe Library (UPL, Roche Applied Science) for quantification of RT products without loss of specificity and efficiency [21]. More recently, we developed a novel UPL probe-based stem-loop RT-qPCR assay design software (MiRNA Primer Design Tool, freely available online at: http://genomics.dote.hu:8080/mirnadesigntool/) that enables anyone to design the stem-loop and specific primers needed for the assay. The MiRNA Primer Design Tool was validated in various species and experimental conditions [22]. The method broadens the application of the stem-loop primer-based miRNA quantitation to virtually any species or variant of small regulatory RNA molecules.

Here we present a detailed protocol for a UPL probe-based stem-loop RT-qPCR assay. We used this method to quantify hsa-miR-515-3p, hsa-miR-325, and hsa-miR-155 using the small nucleolar RNA RNU43 as a normalization control. Human urines from healthy pregnant donors as well as FFPE samples from patients diagnosed with colorectal cancer were analyzed using the method described here.

2 Materials

2.1 Urine Samples

Urine samples were collected from pregnant women during a prospective study in the Department of Obstetrics and Gynecology, University of Debrecen, Hungary. Samples were collected according to rules and regulations of the University of Debrecen, Medical and Health Science Center, with the approval of the local ethics committee (file number: DEOEC RKEB/IKEB 3092-2010). All patients were informed about the purpose of sample collection and signed an informed consent form.

2.2 FFPE Samples

FFPE samples were obtained during routine diagnostic procedures for histological and molecular analysis performed in the Department of Pathology, University of Debrecen. All samples had KRAS mutant genotype confirmed by capillary sequencing and were divided into two groups of five patients according to disease-free survival after anti-EGFR therapy (good and poor responders). Samples were collected according to rules and regulations of the University of Debrecen, Medical and Health Science Center, with the approval of the local ethics committee (file number: RKEB/IKEB 3856-2013).

2.3 Reagents and Equipments

1. The High Pure miRNA Isolation Kit is used to prepare small RNA (Roche Applied Science). Before miRNA isolation prepare the following buffers:
   (a) 20 % Binding Buffer: for each sample, mix 80 μl of Binding Buffer (from the High Pure miRNA Isolation Kit) with 320 μl of nuclease-free water (NFW) in a sterile, nuclease-free 15 ml centrifuge tube (*see* **Note 1**).
   (b) Wash Buffer: if the High Pure miRNA Isolation Kit is freshly opened, prior of the first use, add 40 ml absolute ethanol.
   (c) Tissue Lysis Buffer: 100 μl Paraffin Tissue Lysis Buffer, 8 μl 20 % SDS, and 40 μl Proteinase K working solution per sample.
2. Transcriptor First strand cDNA Synthesis Kit (Roche Applied Science).
3. First Strand cDNA Synthesis kit for RT-PCR (AMV) (Roche Applied Science).
4. Sodium dodecyl sulfate (SDS).
5. Xylene (Sigma-Aldrich).
6. Absolute ethanol.
7. Nuclease-free water (NFW, Lonza).
8. Microcentrifuge with centrifugal force of 15,000 ×*g*.
9. Microcentrifuge tubes (1.5 ml).
10. 0.2 ml PCR strip (Axygene).
11. Eppendorf Thermomixer comfort.
12. Microcentrifuge Combi Spin FVL-2400N.
13. Agilent 2100 Bioanalyzer and a Small RNA Chip (Agilent Technologies).
14. Applied Biosystems 2720 Thermal Cycler.
15. Roche LightCycler 480 qPCR.

16. Roche LightCycler 480 compatible qPCR plates.
17. The assays used for the qPCR measurements were designed by using the MiRNA Primer Design Tool at: http://genomics.dote.hu:8080/mirnadesigntool/ (*see* **Note 2**).

3 Methods

3.1 Sample Preparation and Requirements

Collected urine samples should be stored temporarily at 4 °C. Samples are centrifuged at 3,000 × *g* for 10 min to remove cellular debris. The supernatant is transferred to 1.5 ml pre-labeled screw tubes (sterile), and stored at −75 to −80 °C.

Regarding your FFPE samples: use only tumor blocks with malignant cell content higher than 30 %. A dedicated pathologist should evaluate the tumor cell content of the paraffin embedded blocks. For each sample to be measured by qPCR, miRNA isolation should be carried out from five pieces of 4 μm-thick FFPE tumor slices and their adjacent non-tumorous tissues.

3.2 Small RNA/miRNA Preparation Protocol from Urine Samples

Urine samples can be frozen and stored at −70 °C until isolation process. Samples should be thawed slowly and kept on ice (*see* **Notes 3–7**).

1. Under a chemical fume hood, mix 150 μl of urine with 312 μl of 20 % Binding Buffer in a 1.5 ml microcentrifuge tube (*see* **Note 1**). Mix the solutions by vortexing.
2. Pour the mixture into a High Pure filter tube.
3. Place your filter tube into a clean 1.5 ml microcentrifuge tube.
4. Centrifuge at 15,000 × *g* for 1 min at room temperature. Collect the flow-through in a clean 1.5 ml microcentrifuge tube.
5. Add 200 μl Binding Enhancer to the flow-through, and mix well by vortexing. This fraction contains your miRNAs.
6. Add the mixture to a new High Pure filter tube.
7. Centrifuge at 15,000 × *g* for 1 min.
8. Discard the flow-through.
9. Add 500 μl of Wash Buffer and centrifuge columns at 15,000 × *g* for 1 min.
10. Repeat the washing **steps 8** and **9**.
11. Transfer High Pure Filter columns into a new collection tube and centrifuge at 15,000 × *g* for 2 min to completely dry the filter fleece.
12. Place High Pure filter tube in a new 1.5 ml microcentrifuge tube.
13. Add 50 μl of Elution Buffer to the top of the filter and wait for 1 min.

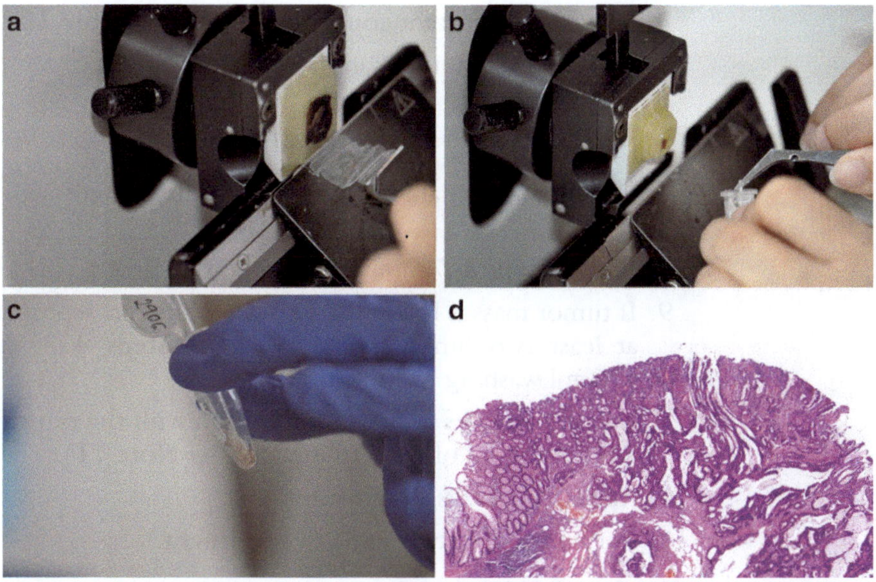

Fig. 1 FFPE sample preparation for deparaffinization. (**a**) FFPE blocks were sectioned with microtome. (**b** and **c**). Individual slices were removed from the microtome and placed into 1.5 ml microcentrifuge tubes. (**d**) Marginal area of a moderately differentiated colorectal adenocarcinoma with a tubular architecture and mucin production showing invasion through the muscular layer of the bowel wall (Hematoxylin and eosin (H&E or HE) staining, 20-fold magnification). The normal colorectal epithelium is visible on the *left side* of the picture. The importance of the histological selection relies on accurate identification and enrichment of tumor tissue in FFPE samples applied for molecular testing. Ez nem ennek az ábrának az aláírása

14. Centrifuge at 15,000×*g* for 1 min to elute the purified miRNAs.
15. Store samples below −20 °C.
16. For miRNA isolation from higher amount of urine, *see* **Note 8**.

3.3 miRNA Preparation from FFPE Samples

Tumor slices from FFPE (formalin-fixed, paraffin-embedded) blocks can be obtained by microtome dissection and should be collected in 1.5 ml microcentrifuge tubes as shown in Fig. 1. Five slices of 4 μm thickness should be combined in one microcentrifuge tube (*see* **Note 9**). For nucleic acid isolation, Xylene deparaffinization is used with subsequent washing by absolute ethanol. The deparaffinization steps should be performed 4–5 times to remove all the paraffin from samples. All steps of deparaffinization are recommended to be done under a chemical fume hood; solutions can be stored at room temperature.

Perform miRNA isolation using the Roche High Pure miRNA Isolation Kit as it follows.

Day 1: deparaffinization

1. Add 1 ml Xylene to each sample and incubate for 5 min at 56 °C in a thermomixer at 800 rpm.
2. Centrifuge the samples for 2 min at 15,000×*g*.

3. Discard supernatant containing xylene (*see* **Note 10**).
4. Add 1 ml absolute ethanol.
5. Vortex for 10 s.
6. Centrifuge for 2 min at 15,000 × *g*.
7. Discard supernatant containing absolute ethanol with traces of xylene.
8. Repeat **steps 4–7** at a minimum of two times.
9. If tumor mass is larger than 30 % of the slice, repeat **steps 1–3** at least two times, then proceed to **steps 4–7** and repeat ethanol washing for four times.
10. Dry the samples at 56 °C for 15 min with the centrifuge tube lid open to rid of residual ethanol (*see* **Note 11**).
11. Add Tissue Lysis Buffer solution.
12. Vortex and incubate at 55 °C overnight.

Day 2: miRNA isolation

13. Centrifuge samples for 5 min, 15,000 × *g* at room temperature.
14. Transfer supernatant to a new microcentrifuge tube.
15. Add 325 μl 20 % Binding Buffer working solution and vortex briefly.
16. Add 120 μl Binding enhancer and vortex briefly.
17. Pipette mixtures into High Pure filter columns.
18. Centrifuge at 15,000 × *g* for 1 min and collect the flow-through. This fraction contains your miRNAs.
19. Add 205 μl Binding Enhancer to flow-through and vortex.
20. Pipette mixtures onto a new High Pure filter column.
21. Centrifuge at 15,000 × *g* for 1 min and discard the flow-through.
22. Add 500 μl Wash Buffer and centrifuge at 15,000 × *g* for 1 min and discard the flow-through.
23. Add 300 μl Wash Buffer and centrifuge at 15,000 × *g* for 1 min and discard the flow-through.
24. Centrifuge at 15,000 × *g* for 2 min to completely dry the filter fleece.
25. Place the High Pure filter tube into a fresh 1.5 ml microcentrifuge tube, and then add 50 μl Elution Buffer and incubate for 1 min at room temperature.
26. Centrifuge for 1 min at 15,000 × *g*.
27. The flow-through contains the isolated miRNA, which can be stored at −20 °C.

3.4 RNA Quality Control and Yield

For detection of miRNA content of the isolated samples and for quality control, use an Agilent Bioanalyzer 2100 System (Agilent Technologies) with a Small RNA Agilent Chip.

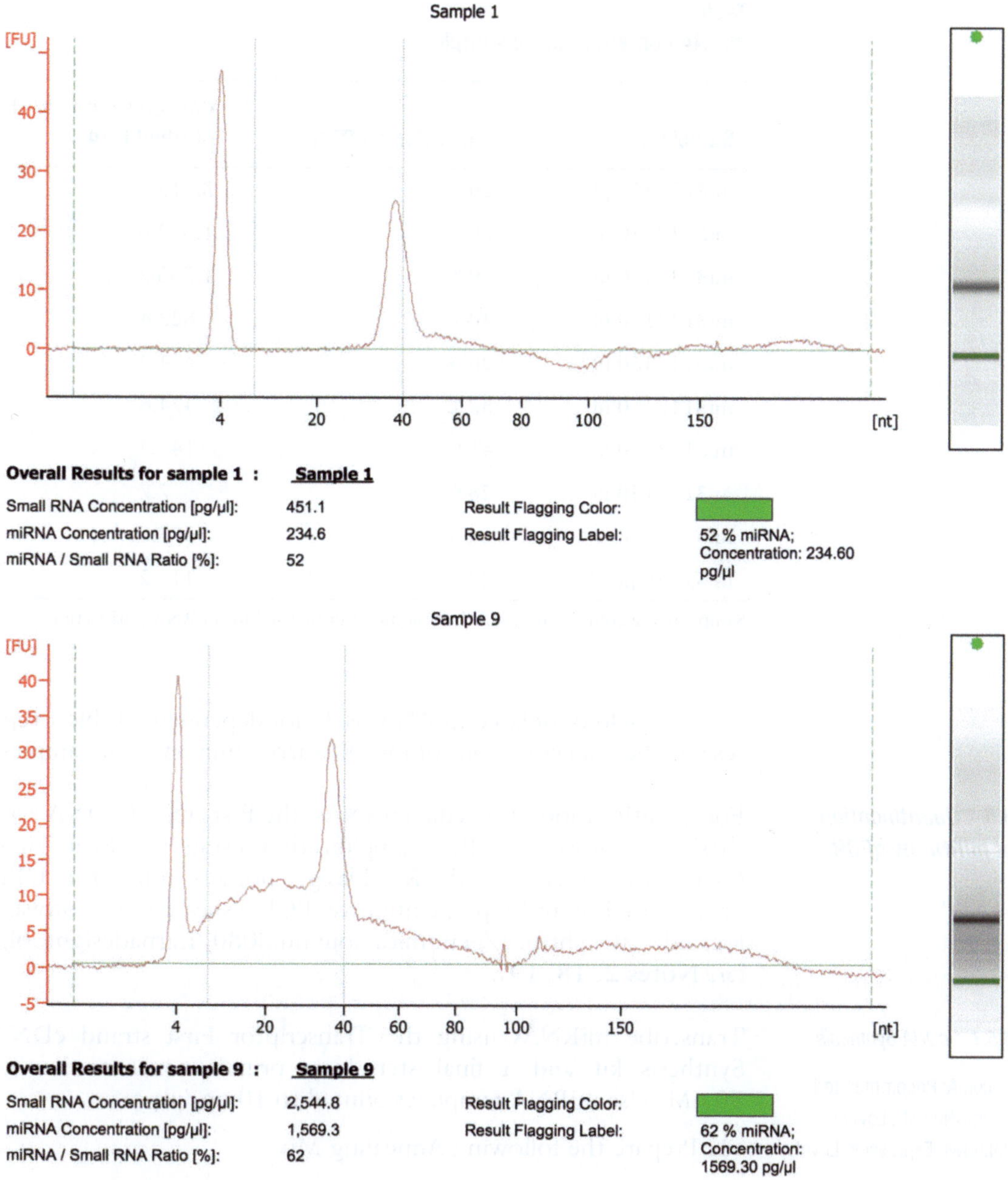

Fig. 2 Assessment of urinary miRNA concentration and quality using the Agilent 2100 bioanalyzer and associated Small RNA Agilent Chip

Perform the run according to the instructions of the manufacturer.

The report will contain the amount of RNA purified and the percentage of miRNA in the purified fraction. Sensitivity of the Agilent microchip kit is 50 pg/µl for the 40 bp fraction. Representative results of the analysis are shown in Fig. 2 (*see* **Notes 12–17**).

Table 1
miRNA content of urine samples

Sample name	miRNA/small RNA	Concentration pg/μl purified RNA
mi312/450 μl	20 %	3,115.8
mi312/150 μl	30 %	1,043.6
mi313/450 μl	39 %	1,553.7
mi313/150 μl	40 %	823.4
mi314/450 μl	28 %	454.2
mi314/150 μl	53 %	474.6
mi315/450 μl	41 %	144.7
mi315/150 μl	28 %	7.7
mi5/450 μl	54 %	573.5
mi5/150 μl	34 %	117.2

Sample name column indicates the volume of urine used for miRNA purification

The yield of isolated miRNAs is donor dependent. Table 1 represents the concentrations of miRNAs from human urine samples.

3.5 Quantification of miRNA by qPCR

For quantification of specific miRNAs, the first step of cDNA synthesis is followed by PCR amplification using the Real Time Quantitative method (qPCR). Design primers using the UPL probe-based stem-loop quantitative PCR assay design software located at: http://genomics.dote.hu:8080/mirnadesigntool/ (*see* **Notes 2, 18, 19**).

3.5.1 cDNA Synthesis

Option A: Recommended for miRNA with Low or Unknown Expression Levels

Transcribe miRNAs using the Transcriptor First strand cDNA Synthesis kit and a final stem-loop primer concentration of 50 nM. Use miRNA templates diluted to 10 ng/μl.

1. Prepare the following Annealing Mix:

miRNA [10 ng/μl]:	1 μl
Stem-loop RT primer [1 μM]:	1 μl
NFW:	11 μl
Total:	13 μl/reaction

If processing more than eight samples, prepare a mastermix with a minimum of 10 % excess (*see* **Note 20**). Mix components in a 0.2 ml PCR strip by pipetting up and down five times and spin briefly in a Combi Spin microcentrifuge.

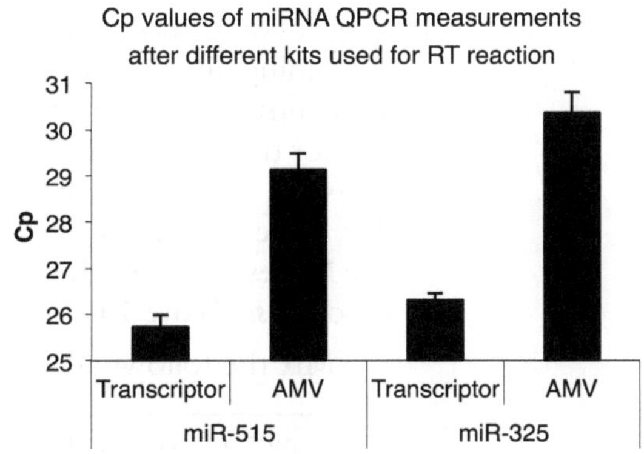

Fig. 3 Comparison of two different RT-qPCR methods. Ten nanograms of miRNA were reverse transcribed by Transcriptor First Strand cDNA Synthesis kit (Transcriptor) or First Strand cDNA Synthesis kit for RT-PCR (AMV) according to Kit protocols. Cp values of the RT-qPCR measurements for hsa-miR515-3p and hsa-miR-325 from urine sample of one representative donor are shown. Error bars represent the SD value of the qPCR triplicates

2. Incubate samples for 10 min at 65 °C in an Applied Biosystems 2720 Thermal Cycler (*see* **Notes 3–7**).
3. For each reaction, prepare the RT Mix as follows:

5× reaction buffer:	4 µl
RNase inhibitor [40 U/µl]:	0.5 µl
10 nM dNTP mix:	2 µl
RT enzyme:	0.5 µl
Total	7 µl/well

4. Add the RT Mix to the Annealing mix in the 0.2 ml PCR strip, vortex, and spin briefly (*see* **Note 21**). Perform the RT reaction on a thermocycler using the following thermal profile:

50 °C	60 min
85 °C	5 min
4 °C	∞

Option B: Recommended for miRNA with Relatively High Expression Levels

A more rapid protocol can be used to reverse transcribe miRNAs with high expression level. This method uses the First Strand cDNA synthesis kit for RT-PCR (AMV) that has shorter hands-on time but lower sensitivity (higher average crossing point (Cp) values); therefore it should be used only for miRNAs with relatively high expression levels. Figure 3 shows the performance of the two different options.

1. If processing more than eight samples, prepare a mastermix containing all components of the reaction listed below except your miRNA sample (*see* **Note 20**). Prepare separate mastermix for the measured miRNA and the normalizer miRNA. It is advisable to measure the normalizer and the target miRNA on the same qPCR plate from the same batch of reagents (*see* **Notes 3–7**). In this setup, a separate annealing step is not needed (*see* **Note 21**).

   Mix the following components per sample:

miRNA [10 ng/μl]:	1 μl
25 mM $MgCl_2$:	4 μl
10× buffer:	2 μl
Stem-loop RT primer [100 μM]:	0.2 μl
RNase inhibitor:	1 μl
RT enzyme:	0.8 μl
NFW:	11 μl
Total:	20 μl/well

2. Perform the RT reaction on a thermocycler using the following thermal profile:

25 °C	10 min
42 °C	60 min
99 °C	5 min
4 °C	∞

3.5.2 qPCR Amplification

1. Dilute reverse transcribed miRNA twofold with NFW (*see* **Notes 3–7**).
2. Prepare qPCR mix in triplicate for each RT reaction. Components for one single qPCR reaction are:

NFW:	10.28 μl
25 mM $MgCl_2$ (Fermentas):	2 μl
10× buffer (Fermentas, $MgCl_2$ free buffer):	1.6 μl
2.5 mM dNTP (Fermentas):	0.8 μl
miRNA-specific forward primer [100 μM]:	0.06 μl
Universal Reverse primer [100 μM]:	0.06 μl
UPL Probe 21:	0.1 μl
Taq polymerase [5 U/μl] (Fermentas):	0.1 μl
Total:	15 μl/well

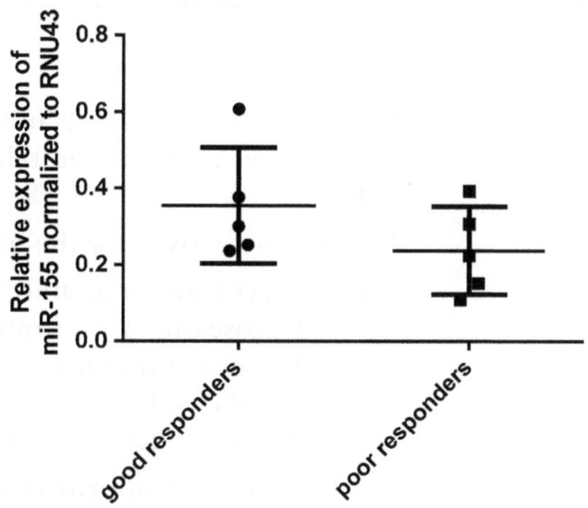

Fig. 4 Relative expression of miR-155 normalized to RNU43 in colorectal adenocarcinoma. miR-155 expression profile of five good and five poor responder patients with short and long disease-free survival period after anti-EGFR monotherapy

3. Mix 5 μl of the two-fold diluted reverse transcribed miRNA with 15 μl qPCR mix/reaction. Perform all qPCR measurements in triplicate (*see* **Notes 20, 22–24**).
4. Run the qPCR amplification in a thermocycler using the following temperature profile:

95 °C	3 min
95 °C	10 s
58 °C	30 s
72 °C	1 s
Repeat **steps 2–4** for 45 cycles	
40 °C	10 min

5. Collect data from fluorescence signal at **step 3** in "Single mode" (*see* **Notes 25–29**).

We quantified miR-155 using this protocol and RNU43 as endogenous control. Our results correlate with previously published results [23] and suggest that in colorectal cancer increased levels of miR-155 might predict a better overall survival (OS) and disease-free survival (DFS) after treatment with anti-EGFR therapy. Results of the measurements are presented in Fig. 4. Due to the low number of samples (five independent samples per group), the presented results are not statistically significant and further investigation is needed.

4 Notes

1. Guanidine-thiocyanate in Binding Buffer may form toxic gases; therefore, perform the dilution step under a chemical fume hood.
2. Preparatory steps for the protocol:
   (a) Select your target miRNA sequence from mirBase (http://mirbase.org/) or from your next-generation sequencing dataset(e.g.,hsa-mir-155:UUAAUGCUAAUCGUGAUAGGGGU). Use the identified sequence to design your assay at: http://genomics.dote.hu:8080/mirnadesigntool/.

      Useful information and details on how to use the software have been collected and stored at our dedicated website: http://genomics.med.unideb.hu/en/research-interest/mirna-research. The MiRNA Primer Design Tool can be used freely under GNU/GPL license and installed on local servers from the following location: https://github.com/lbalintatmedunidebhu/mirnaprimerdesigntool.

   (b) During selection of your target, perform a sequence search in the miRBase database to identify microRNAs with similar sequences. If similar sequences on the 3′ end are present, evaluate your assays for differences in the other regions and consider using alternative methods for detection.
   (c) Order your stem-loop oligo by HPLC purification. All other oligos used can be of standard desalted purification.
3. Always work with samples kept on ice.
4. Work in RNase-free conditions.
5. Always wear gloves.
6. Use only nuclease-free water.
7. Use RNase-free plasticware.
8. A higher amount of body fluid can be used if the miRNA concentration is too low. It is important to keep the same proportion of urine sample and Binding Buffer. For example, add 920 μl Binding Buffer to 450 μl urine, and repeat the binding steps to the column (**steps 5–8**).
9. During microtome dissection of the FFPE blocks, carefully clean tweezers with nuclease-free water and 70 % ethanol to exclude ribonucleic acid transfer and contamination between samples.
10. Xylene should be discarded according to the local rules of waste management. Xylene is an organic solvent and should be treated as a chemically hazardous waste.
11. To remove all the supernatant during deparaffinization, additional centrifugation steps can be performed.

12. The Agilent small RNA kit contains a highly viscous gel and, therefore, it is essential to pipette carefully.
13. If your sample contains significant amounts of tRNA then you will observe a 60 nt band in chromatogram of the Agilent Bioanalyzer run [24].
14. Specific miRNA concentrations can be measured only by an Agilent Bioanalyzer miRNA microchip. Due to the lower sensitivity, Nanodrop or Qubit systems are not appropriate.
15. If starting with a sample isolated from cells or tissues, the initial material should be between 5 ng and 50 ng of total RNA per RT reaction. This value stands for total RNA isolated by Tri reagent. The concentration of total RNA mentioned here was measured by Nanodrop or Qubit systems due to the fact that besides the miRNAs it contains all the other RNA fractions of a cell.
16. When using a specific miRNA isolation protocol, this starting material can be much lower and should be optimized for each sample type.
17. Conventional, non-miRNA optimized column-based RNA purification methods are not suitable for microRNA purification since they are optimized for larger RNA molecules and the low molecular weight fraction of miRNAs are not bound to the column.
18. Universal Reverse Primer and the UPL probe 21 can be used with all your assays designed with the previously presented MiRNA Primer Design Tool.
19. Each assay should be tested on a sample known to contain your miRNA of interest.
20. For the PCR step, if processing more than eight samples, prepare a Mastermix. Always calculate the volumes for your mastermix with a minimum of 10 % excess.
21. Perform a No Amplification Control (NAC) by leaving out the RT (reverse transcription) enzyme. Your NAC control should be at least a 3.33-C_P higher value (meaning lower signal) than your control. Some background noise is acceptable but it should be less than 10 % of your signal (3.33 C_P would mean 10 % signal with an optimal PCR efficiency).
22. You can test efficiency of your qPCR by performing a qPCR on a dilution series of ten-fold dilution steps that cover five orders of magnitude.
23. At the qPCR step, always perform a No Template control (NTC) by replacing your reverse transcribed sample with nuclease-free water.
24. qPCR step should be performed in triplicate (three technical replicates).

25. Normalizers should be carefully selected by measuring their expression levels across several samples used in your experiment. Usually small nucleolar RNAs (snoRNAs) and nuclear RNAs like U6 can be used after careful consideration.
26. Normalization should be performed with the $\Delta\Delta C_T$ method [25].
27. If working with body fluids, you can normalize your sample to a relevant marker that reflects the changes in the concentration of the body fluid, e.g., total protein concentration, density.
28. If working with body fluids, the purification step can be controlled by adding spike-in artificial miRNA molecules to the sample.
29. Synthetic RNA molecules can be used for carefully monitoring the efficiency of your reactions and generating standard curves. If using synthetic RNA molecules as positive controls for your reaction, order them separately from your assay components in order to avoid cross contamination at the manufacturing site. Dissolve them under a chemical hood. Keep them isolated from your assays, reagents and samples, if possible in another freezer. The danger of contaminating your reagents with traces of the amplicon is extremely high.

Acknowledgements

Balint L. Balint is a Szodoray fellow of the University of Debrecen Medical and Health Science Center, Hungary and is supported by the TÁMOP 4.2.4.A/2-11-1-2012-0001 grant implemented through the New Hungary Development Plan co-financed by the European Social Fund and the European Regional Development Fund. Jim Hesson of AcademicEnglishSolutions.com revised the English.

References

1. Bartel DP (2004) MicroRNAs: genomics, biogenesis, mechanism, and function. Cell 116: 281–297
2. Huntzinger E, Izaurralde E (2011) Gene silencing by microRNAs: contributions of translational repression and mRNA decay. Nat Rev Genet 12:99–110
3. Ivey KN, Srivastava D (2010) MicroRNAs as regulators of differentiation and cell fate decisions. Cell Stem Cell 7:36–41
4. Baltimore D, Boldin MP, O'Connell RM et al (2008) MicroRNAs: new regulators of immune cell development and function. Nat Immunol 9:839–845
5. Tili E, Michaille JJ, Costinean S et al (2008) MicroRNAs, the immune system and rheumatic disease. Nat Clin Pract Rheumatol 4: 534–541
6. Du L, Pertsemlidis A (2011) Cancer and neurodegenerative disorders: pathogenic convergence through microRNA regulation. J Mol Cell Biol 3:176–180
7. Weber JA, Baxter DH, Zhang S et al (2010) The microRNA spectrum in 12 body fluids. Clin Chem 56:1733–1741
8. Xi Y, Nakajima G, Gavin E et al (2007) Systematic analysis of microRNA expression of RNA extracted from fresh frozen and formalin-

fixed paraffin-embedded samples. RNA 13: 1668–1674

9. Jung M, Schaefer A, Steiner I et al (2010) Robust microRNA stability in degraded RNA preparations from human tissue and cell samples. Clin Chem 56:998–1006
10. Hui A, How C, Ito E et al (2011) MicroRNAs as diagnostic or prognostic markers in human epithelial malignancies. BMC Cancer 11:500
11. Mengual L, Lozano JJ, Ingelmo-Torres M et al (2013) Using microRNA profiling in urine samples to develop a non-invasive test for bladder cancer. Int J Cancer 133:2631–2641
12. Shen N, Liang D, Tang Y et al (2012) MicroRNAs—novel regulators of systemic lupus erythematosus pathogenesis. Nat Rev Rheumatol 8:701–709
13. Alevizos I, Illei GG (2010) MicroRNAs as biomarkers in rheumatic diseases. Nat Rev Rheumatol 6:391–398
14. Mattie MD, Benz CC, Bowers J et al (2006) Optimized high-throughput microRNA expression profiling provides novel biomarker assessment of clinical prostate and breast cancer biopsies. Mol Cancer 5:24
15. Rosenfeld N, Aharonov R, Meiri E et al (2008) MicroRNAs accurately identify cancer tissue origin. Nat Biotechnol 26:462–469
16. Lu J, Getz G, Miska EA et al (2005) MicroRNA expression profiles classify human cancers. Nature 435:834–838
17. Takamizawa J, Konishi H, Yanagisawa K et al (2004) Reduced expression of the let-7 microRNAs in human lung cancers in association with shortened postoperative survival. Cancer Res 64:3753–3756
18. Hayashita Y, Osada H, Tatematsu Y et al (2005) A polycistronic microRNA cluster, miR-17-92, is overexpressed in human lung cancers and enhances cell proliferation. Cancer Res 65:9628–9632
19. Chen JQ, Russo J (2009) ERalpha-negative and triple negative breast cancer: molecular features and potential therapeutic approaches. Biochim Biophys Acta 1796:162–175
20. Chen C, Ridzon DA, Broomer AJ et al (2005) Real-time quantification of microRNAs by stem-loop RT-PCR. Nucleic Acids Res 33:e179
21. Varkonyi-Gasic E, Wu R, Wood M et al (2007) Protocol: a highly sensitive RT-PCR method for detection and quantification of microRNAs. Plant Methods 3:12
22. Czimmerer Z, Hulvely J, Simandi Z et al (2013) A versatile method to design stem-loop primer-based quantitative PCR assays for detecting small regulatory RNA molecules. PLoS One 8:e55168
23. Shibuya H, Iinuma H, Shimada R et al (2010) Clinicopathological and prognostic value of microRNA-21 and microRNA-155 in colorectal cancer. Oncology 79:313–320
24. Becker C, Hammerle-Fickinger A, Riedmaier I et al (2010) mRNA and microRNA quality control for RT-qPCR analysis. Methods 50:237–243
25. Livak KJ, Schmittgen TD (2001) Analysis of relative gene expression data using real-time quantitative PCR and the 2(-Delta Delta C(T)) method. Methods 25:402–408

Chapter 11

MicroRNA Profiling in Plasma or Serum Using Quantitative RT-PCR

Marina C. Costa, Ana Lúcia Leitão, and Francisco J. Enguita

Abstract

MicroRNAs (miRNAs) are important cellular modulators that regulate gene expression at the posttranscriptional level. Circulating miRNAs present in human plasma or serum have recently become an emerging field in biomedical research, mostly due to its potential applications in the diagnosis and prognosis of several diseases. Although miRNA profiling in biofluids holds great promise, there are several challenges to overcome. Here, we present an experimental procedure for profiling miRNA in plasma or serum with high sensitivity and specificity using qRT-PCR. This method is also suitable for studying miRNAs in other body fluids or clinical samples that also contain low amounts of RNA.

Key words Circulating miRNA, Plasma, Serum, Biomarker

1 Introduction

MicroRNAs (miRNAs) are single-stranded noncoding RNA molecules of ~22 nucleotides that inhibit mRNA translation to protein by interacting with its 3′ untranslated region (UTR). Phylogenetically conserved, miRNAs constitute a layer of epigenetic regulation that is involved in a variety of biological processes, such as proliferation, differentiation, development, and apoptosis [1, 2].

Cells are able to actively secrete miRNAs and other RNA molecules by a mechanism that is not totally understood. Circulating miRNAs are intracellular mediators that can be isolated from plasma, serum, urine, and other body fluids and exhibit remarkable stability to degradation [3]. Interestingly, deregulation of circulating miRNAs has been associated with a variety of human diseases, including cancer [4, 5] and cardiovascular diseases [6, 7]. In addition, miRNA profiles have been shown to be altered in serum and plasma taken from patients suffering from cancer [8–10] and cardiovascular and degenerative diseases [11–13]. These features in combination with their tissue specificity make circulating miRNAs promising candidates for use as noninvasive biomarkers in a wide

M. Lucrecia Alvarez and Mahtab Nourbakhsh (eds.), *RNA Mapping: Methods and Protocols*, Methods in Molecular Biology, vol. 1182, DOI 10.1007/978-1-4939-1062-5_11,

range of disease, toxicology, or injury studies. Moreover, blood samples are easy to obtain and carry little risk to the patient compared to biopsies and surgical procedures.

However, profiling of circulating miRNAs presents several unique challenges that should be addressed. Serum and plasma only contain a small fraction of the RNA found in whole blood. Most of the RNA is found in the buffy coat, which mainly consists of white blood cells and platelets. In order to get accurate circulating miRNA profiles, it is important to ensure that the serum or the plasma fraction is not contaminated with RNA from blood cells. Proper collection and preparation of blood plasma and serum are necessary in order to minimize confounding variation attributable to plasma processing or blood cell contamination in circulating miRNA studies [8, 14, 15]. In addition, when working with serum or plasma it is very important not to work with hemolyzed samples because even small traces of red blood cells in serum or plasma may greatly affect the miRNA profile. Indeed, serum and plasma samples contain a high concentration of enzyme inhibitors that could interfere with the cDNA synthesis (reverse transcription or RT) and qPCR quantification. Serum samples are cleaner than plasma ones since they lack all the proteins and blood structures involved in coagulation. However, plasma contains a higher diversity of miRNAs because some of them are related with platelets and microparticles that are lost during the coagulation process and serum purification. In the case of the miRNA profiling from plasma samples, blood must be treated with anticoagulants such as citrate. Heparin is not recommended since it is able to bind and block nucleic acids resulting in a loss of the majority of the circulating miRNAs.

Data normalization is another major challenge for the analysis of circulating miRNA because RNAs from some commonly used reference genes, such as U6, are not present in serum and plasma. A common strategy for normalization is the use of miRNAs that have been reported to be stably expressed such as miR-191, miR-103, let-7a, and miR-16 [16–20]. These miRNAs may constitute the best possible normalization controls, because they belong to the same family of RNA molecules than the quantified target. However, they often may not ensure constant expression under all experimental conditions [21, 22]. In contrast, totally unrelated spiked-in RNAs from *Caenorhabditis elegans* (e.g., cel-miR-39, cel-miR-54, and cel-miR-238) not only can monitor the efficiency of RNA purification and reverse transcription reaction but also can be used as normalization controls [4, 23]. Thus, stable reference controls can be obtained by adding the same amount of spiked-in RNAs in an equal volume of serum or plasma. Another strategy of data normalization is to identify suitable reference for each study through the systematic evaluation of the expression level of the miRNAs present in the circulation. Instead of using a single or a set of reference control(s), an alternative approach known as "global mean normalization" can be used when the number of analyzed miRNAs is high (typically more than 50). This method employs the average expression level of all miRNAs detected in a sample as normalization factor [24].

Here, we describe a procedure for the detection of circulating miRNA by qRT-PCR profiling, which consists in two major steps: (1) miRNA purification and (2) miRNA profiling using qRT-PCR.

2 Materials

2.1 miRNA Isolation from Serum or Plasma

- Trizol LS™ reagent (Invitrogen).
- RNeasy mini kit (QIAGEN).
- Extra RPE-buffer (QIAGEN).
- Chloroform.
- Absolute ethanol.
- RNase-free pipette tips.
- RNase-free 1.5 and 2 ml tubes.
- Benchtop refrigerated microcentrifuge capable of 13,000×*g* centrifugal force.
- Benchtop refrigerated centrifuge for PCR or ELISA plates.

2.2 miRNA Quantification by qRT-PCR

- Exiqon Universal cDNA synthesis kit.
- Exiqon miRCURY LNA™ primers.
- Exiqon SYBRGreen mastermix.
- Real-time PCR plates.

3 Methods

3.1 MiRNA Isolation from Serum or Plasma

The method described here combines the miRNA extraction from liquid samples using Trizol™ LS reagent (Invitrogen), with the RNeasy Mini Kit, a silica membrane-based method for total RNA purification. Samples are initially mixed with trizol/chloroform and centrifuged. The upper, aqueous phase containing total RNA is extracted, and ethanol is added to provide appropriate binding conditions to an RNeasy Mini Kit purification column for all RNA molecules of 20 nucleotides or longer.

1. Slowly thaw the sample (plasma or serum) on ice. In the meantime, cool the benchtop centrifuge to 4 °C.
2. Place 300 μl of sample in a 1.5 ml tube and centrifuge at 15,000×*g* for 5 min at 4 °C to remove cell debris (*see* **Note 1**).

3. Transfer 250 μl of sample to a new 1.5 ml tube, and add 750 μl Trizol™ LS reagent (do not transfer cell debris or serum coagulate into the purification). Vortex and incubate for 5 min at room temperature. The volume ratio of Trizol™ LS reagent to sample should be 3:1 (*see* **Notes 2** and **3**).
4. Add 200 μl of chloroform to each sample, and mix by vortexing. Incubate for 2 min at room temperature.
5. Centrifuge the sample at 12,000 × *g*, for 15 min, at 4 °C.
6. Transfer the upper aqueous phase to a fresh 2 ml tube, and add 1.5 volumes of ethanol. Mix by pipetting.
7. Transfer 750 μl of the mixture to an RNeasy Mini Spin column (QIAGEN) and centrifuge at room temperature for 30 s at 13,000 × *g*. Discard the flow-through (*see* **Notes 3** and **4**).
8. Repeat the previous step until the entire sample is used (mix by pipetting before every load) (*see* **Note 5**).
9. Wash the column by adding 700 μl of RWT buffer (QIAGEN); centrifuge at room temperature for 1 min at 13,000 × *g*, and discard the flow-through.
10. Wash three times by adding 500 μl of RPE buffer (QIAGEN); centrifuge at room temperature for 1 min at 13,000 × *g*, and discard the flow-through.
11. Transfer the column to a new 2 ml collection tube (QIAGEN), and centrifuge at room temperature for 2 min at 13,000 × *g*.
12. Leave the tubes open under a chemical hood for 1 min to evaporate the rest of ethanol from the washing buffers (*see* **Note 6**).
13. Transfer the column to a 1.5 ml collection tube (QIAGEN), and add 50 μl RNase-free water (make sure that the liquid is centered on the membrane). To increase the RNA recovery yield use pre-warmed water at 65 °C.
14. Incubate for 1 min, and centrifuge at room temperature for 1 min at 13,000 × *g*.
15. Store RNA samples at −80 °C (*see* **Notes 7** and **8**).

3.2 MiRNA Profiling in Plasma or Serum Samples

For miRNA expression profiling in plasma or serum, we routinely use the Exiqon's miRCURY LNA™ microRNA PCR System. This qPCR system combines a universal reverse-transcription reaction with locked nucleic acid (LNA) PCR primers, which results in higher sensitivity and specificity. Other platforms for qPCR profiling of miRNAs are also available in the market such as the TaqMan™ system described in other chapters of this book. The protocol for PCR profiling described here uses the total RNA isolated as described in Subheading 3.1, which undergoes reverse transcription (Subheading 3.2.1) followed by real-time PCR (Subheading 3.2.2).

3.2.1 cDNA Synthesis

1. Add 40 µl nuclease-free water to the tube containing the RNA spike-in (synthetic control template, Exiqon).
2. Mix by vortexing, and spin down. Store in aliquots at −20 °C.
3. Thaw all the Universal cDNA Synthesis kit (Exiqon) reagents on ice, and spin down all the components.
4. For each sample, prepare the reaction mix by adding 4 µl of 5× reaction buffer, 2 µl of the enzyme mix, and 1 µl of the RNA spike-in to 8–13 µl of extracted RNA. Complete with RNase-free water to 20 µl, mix well, and spin down.
5. Incubate for 60 min at 42 °C followed by a heat inactivation of the reverse transcriptase for 5 min at 95 °C (use a thermocycler).
6. Store the cDNA at 4 °C for up to 4 days or at −20 °C for up to 5 weeks.

3.2.2 Real-Time PCR Amplification

1. Dilute the cDNA template obtained in Subheading 3.2.1 to a maximum of 1:12 in nuclease-free water. Recommended cDNA dilutions range from 1:5 to 1:10 depending on the amount of RNA extracted from the samples.
2. In the case of using commercially designed miRNA primers, dilute forward and reverse miRNA-specific primers in the volume of RNase-free water recommended by the manufacturer and place them on ice. In the case of custom-made miRNA primers, dilute each primer to a final concentration of 10 µM.
3. For each sample, prepare the real-time PCR reaction mixture by adding 10 µl of SYBR® Green Master Mix, 2 µl to PCR primer mix (1 µl of forward primer plus 1 µl of reverse primer), and 8 µl diluted cDNA template in a total volume of 20 µl. This reaction volume must be enough to fulfill a single well of a 96-well qPCR plate. In the case of using high-throughput 384-well plates, the reaction volume must be scaled down to 10 µl. In order to ensure statistical significance of the results the preparation of technical triplicates is recommended. Mix the reaction, and seal the qPCR plate with optical film. Spin down in a cooled centrifuge at 1,000 × *g* during 5 min. Reaction plates must be used immediately or kept frozen at −20 °C in the absence of light for a maximum period of 7 days.
4. Perform real-time PCR amplification followed by melt curve analysis. Cycling conditions: 95 °C for 10 min and then 45 cycles of 95 °C for 10 s and 60 °C for 60 s.

3.2.3 Data Analysis

There are two common methods used for qRT-PCR data analysis: absolute and relative quantification. Absolute quantification determines the copy number of a target by using a standard curve with a sample of known concentration. In relative quantification, target molecules are compared to a reference target relative to a reference group (control sample). Usually, endogenous controls are used as reference targets because the gene expression remains relatively constant across tissue and cell types. The most usual method for miRNA profiling in biofluids is the relative quantification.

For miRNA profiling in plasma or serum samples, numerous strategies have been proposed to normalize data from qRT-PCR. To date, there is no consensus on the type of normalization that should be used. Spike-in normalization has the advantage of being always controlled in the amount of the normalization RNA added to the samples. On the other hand, internal reference gene has to be adapted to each situation and is not easy to perform when the amount of isolated RNA is low, because its quantification is less accurate. Finally, the overall normalization strategy is only recommended for quantifying a high number of miRNAs in the same sample, for instance when a genome low-density qPCR array is performed for miRNA screening (*see* **Note 9**).

Data processing can be performed by spreadsheet analysis (Excel or other similar software) or by using specialized software. Some General Public License (GPL) software are available, such as Data Assist system from Applied Biosystems and pyQPCR (http://pyqpcr.sourceforge.net), being a good alternative to commercial software as GenEx (Exiqon), MxPro ET (Agilent), GeneSpring (Agilent), and Statminer (Life Technologies). Another web-based alternative is the RT^2 profiler server by SA-Biosystems (http://pcrdataanalysis.sabiosciences.com/pcr/arrayanalysis.php).

4 Notes

1. Before starting with the miRNA isolation procedure, samples should be centrifuged at 15,000×g for 5 min to pellet any debris and insoluble components present in plasma or serum. Make sure to start with an excess of sample in order to secure sufficient supernatant for the next step. Typically the recommended starting sample volume is 500 μl.
2. In order to perform more efficient phenolic RNA extractions from liquid samples, we recommend the use of Trizol LS, a more concentrated version than regular Trizol, which will increase the yield of RNA recovery from diluted samples. However, there are some alternatives to phenolic-based extractions that are available in the market including the extraction kits for plasma or biofluids commercialized by Exiqon, Qiagen, and Norgen. These kits are based on the precipitation of plasma proteins with sodium acetate and RNA concentration and purification by column chromatography.
3. An important challenge when working with plasma or serum is the high content of RT and/or PCR enzyme inhibitors in this type of sample. Thus, increasing the amount of sample input in

order to obtain a better sensitivity is not possible in samples with high content of enzyme inhibitors. The quantity of inhibitors that remain after RNA extraction may vary greatly not only between different extraction methods but also from sample to sample.

4. Because serum and plasma contain only very small amounts of RNA, there is a high risk of loss of a significant amount of the RNA during purification. For this reason, some protocols also recommend adding a carrier RNA during the extraction procedure. Standard RNA carriers used to improve RNA extraction are the RNA from bacteriophage MS2 (commercialized by Roche) and the tRNA extracted from yeast (commercialized by Ambion, Sigma-Aldrich, and Invitrogen, among others).
5. The efficiency of the RNA extraction protocol may also be monitored by adding RNA spike-in, a synthetic control template from Exiqon. Usually, spike-in RNA molecules are small RNAs totally unrelated with the organism which is being studied and having approximately the same size than a typical miRNA (19–23 nucleotides). The RNA spike-in will be added to the extracted RNA, reverse transcribed with it, and detected with a specific primer set. In addition, the RNA spike-in can be used to monitor the efficiency of the cDNA synthesis step, allowing for assessment of residual components present in the RNA preparation which may act as inhibitors of the cDNA synthesis and PCR enzymes. To monitor the efficiency of the cDNA synthesis step, the RNA spike-in should be added to the cDNA synthesis reaction instead.
6. The additional centrifugation steps and the incubation with the tubes open under the chemical hood promote the total clearance of residual phenol and ethanol used during the purification procedure.
7. Plasma and serum are cell-free samples with very low concentrations of RNA. Thus, quantification of RNA preparations from serum or plasma by routine methods such as absorbance at 260 nm using Nanodrop spectrophotometer or equivalent is often inaccurate. Microfluidics analysis with systems as the Agilent Bioanalyzer offer more precision in RNA quantification, but they have the main disadvantage of being expensive when compared with other techniques.
8. Serum and plasma should be separated carefully to avoid hemolysis. The quantification of miR-451, which is almost exclusively located in red blood cells, can be used to detect traces of hemolysis in plasma or serum.
9. It is recommended to investigate which of the available methods for data normalization will provide the best results for your experimental conditions. Spike-in RNA and stable endogenous

miRNAs (endogenous controls) are recommended for the profiling of a small number of miRNAs in each sample. The global mean normalization method is more accurate for screening panels of a large number of miRNAs (more than 50).

Acknowledgments

M.C.C. was supported by a postdoctoral fellowship from Fundação para a Ciência e Tecnologia, Portugal (Ref. SFRH/BPD/65131/2009).

References

1. Bartel DP (2009) MicroRNAs: target recognition and regulatory functions. Cell 136: 215–233
2. Zhang S, Chen L, Jung EJ et al (2010) Targeting microRNAs with small molecules: from dream to reality. Clin Pharmacol Ther 87:754–758
3. Jung M, Schaefer A, Steiner I et al (2010) Robust microRNA stability in degraded RNA preparations from human tissue and cell samples. Clin Chem 56:998–1006
4. Brase JC, Johannes M, Schlomm T et al (2011) Circulating miRNAs are correlated with tumor progression in prostate cancer. Int J Cancer 128:608–616
5. Yu DC, Li QG, Ding XW et al (2011) Circulating MicroRNAs: potential biomarkers for cancer. Int J Mol Sci 12:2055–2063
6. D'Alessandra Y, Devanna P, Limana F et al (2010) Circulating microRNAs are new and sensitive biomarkers of myocardial infarction. Eur Heart J 31:2765–2773
7. Dimmeler S, Zeiher AM (2010) Circulating microRNAs: novel biomarkers for cardiovascular diseases? Eur Heart J 31:2705–2707
8. Chen X, Hu Z, Wang W et al (2012) Identification of ten serum microRNAs from a genome-wide serum microRNA expression profile as novel noninvasive biomarkers for non-small cell lung cancer diagnosis. Int J Cancer 130:1620–1628
9. Gilad S, Meiri E, Yogev Y et al (2008) Serum microRNAs are promising novel biomarkers. PLoS One 3:e3148
10. Lawrie CH, Gal S, Dunlop HM et al (2008) Detection of elevated levels of tumour-associated microRNAs in serum of patients with diffuse large B-cell lymphoma. Br J Haematol 141: 672–675
11. Fichtlscherer S, De Rosa S, Fox H et al (2010) Circulating microRNAs in patients with coronary artery disease. Circ Res 107: 677–684
12. Komatsu S, Ichikawa D, Takeshita H et al (2011) Circulating microRNAs in plasma of patients with oesophageal squamous cell carcinoma. Br J Cancer 105:104–111
13. Wang GK, Zhu JQ, Zhang JT et al (2010) Circulating microRNA: a novel potential biomarker for early diagnosis of acute myocardial infarction in humans. Eur Heart J 31: 659–666
14. Blondal T, Jensby NS, Baker A et al (2013) Assessing sample and miRNA profile quality in serum and plasma or other biofluids. Methods 59:S1–S6
15. Pritchard CC, Kroh E, Wood B et al (2012) Blood cell origin of circulating microRNAs: a cautionary note for cancer biomarker studies. Cancer Prev Res (Phila) 5:492–497
16. Bargaje R, Hariharan M, Scaria V et al (2010) Consensus miRNA expression profiles derived from interplatform normalization of microarray data. RNA 16:16–25
17. Chang KH, Mestdagh P, Vandesompele J et al (2010) MicroRNA expression profiling to identify and validate reference genes for relative quantification in colorectal cancer. BMC Cancer 10:173
18. Liang Y, Ridzon D, Wong L et al (2007) Characterization of microRNA expression profiles in normal human tissues. BMC Genomics 8:166
19. Peltier HJ, Latham GJ (2008) Normalization of microRNA expression levels in quantitative RT-PCR assays: identification of suitable reference RNA targets in normal and cancerous human solid tissues. RNA 14:844–852

20. Song J, Bai Z, Han W et al (2012) Identification of suitable reference genes for qPCR analysis of serum microRNA in gastric cancer patients. Dig Dis Sci 57:897–904
21. D'Haene B, Mestdagh P, Hellemans J et al (2012) miRNA expression profiling: from reference genes to global mean normalization. Methods Mol Biol 822:261–272
22. Wu F, Zhang S, Dassopoulos T et al (2010) Identification of microRNAs associated with ileal and colonic Crohn's disease. Inflamm Bowel Dis 16:1729–1738
23. Li Y, Kowdley KV (2012) Method for microRNA isolation from clinical serum samples. Anal Biochem 431:69–75
24. Mestdagh P, Van Vlierberghe P, De Weer A et al (2009) A novel and universal method for microRNA RT-qPCR data normalization. Genome Biol 10:R64

Chapter 12

MicroRNA Profiling of Exosomes Isolated from Biofluids and Conditioned Media

Sweta Rani

Abstract

Exosomes are membrane-bound 50–100 nm vesicles released from many cell types including normal and tumorous tissues. Exosomes transport mainly miRNAs, mRNAs, enzymes, cytokines, etc. from the cells of origin to the neighbor cells mediating the communication between them.

The content of exosomes can be explored using RNA profiling after their isolation from medium conditioned by cultured cells or from other biofluids. This chapter includes detailed discussion on isolation, characterization, and miRNA profiling of exosomes. First, exosomes are isolated by filtration and ultracentrifugation, and then characterized using immunoblotting and transmission electron microscope. Finally, we used low density arrays to profile exosomal miRNA.

Key words Exosomes, Extracellular, Cell line, Conditioned medium, Biofluids, miRNA

1 Introduction

Exosomes are membrane-bound, 50–100 nm diameter vesicles formed by the inward budding of cellular compartments called internal vesicles or endosomes [1]. These internal vesicles carry nucleic acids such as miRNAs, and proteins like enzymes and cytokines, and are secreted by merging with the cell membrane when they become exosomes [2]. Exosomes interact with target cells by fusing with their membrane and releasing their content. Alternatively, exosomes can be taken up by the target cells through their endocytic pathways [3].

Exosomes mediate cell-to-cell communication and transfer several RNAs, proteins, enzymes etc. between cells [4]. Exosomal content depends on its cell of origin [5]. Tumor cells-derived exosomes have been associated with accelerating tumor growth [5] and invasiveness [6, 7]. Exosomes have also been reported to confer resistance to drug [8] and transfer phenotypic traits to secondary cells [9]. This chapter focuses on the isolation and profiling of exosomal miRNA.

M. Lucrecia Alvarez and Mahtab Nourbakhsh (eds.), *RNA Mapping: Methods and Protocols*, Methods in Molecular Biology, vol. 1182, DOI 10.1007/978-1-4939-1062-5_12, © Springer Science+Business Media New York 2014

2 Material

2.1 Cell Culture and Conditioned Medium Collection

1. BT474 (American Type Tissue Collection).
2. RPMI-1640.
3. 2 mM L-Glutamine.
4. 5 % fetal bovine serum (FBS).
5. Trypsin.
6. Tissue culture grade vented flasks (e.g., 175 cm^2).
7. Sterile pipettes.
8. Biosafety cabinet for basic mammalian cell culture.
9. Hemocytometer.

2.2 Exosomes Isolation

1. 50 mL centrifuge tubes.
2. Ultracentrifuge with fixed-angle or swinging-bucket rotors.
3. Polyallomer or polycarbonate tubes suitable for the specific ultracentrifuge rotor.
4. Phosphate-buffered saline (PBS) (Sigma-Aldrich).
5. 12 mL syringes.
6. 18 G needle.
7. 0.2 μM filter.

2.3 Immunoblotting

1. 4× Laemmli sample buffer: 277.8 mM Tris–HCl, pH 6.8, 4.4 % LDS, 44.4 % (w/v) glycerol, 0.02 % bromophenol blue (Sigma-Aldrich). Store at −20 °C.
2. Resolving buffer (Bio-Rad): 1.5 M Tris–HCl, pH 8.8, 0.4 % SDS. Store at room temperature.
3. Stacking buffer (Bio-Rad): 0.5 M Tris–HCl, pH 6.8, 0.4 % SDS. Store at room temperature.
4. Acrylamide, 30 % solution (Sigma-Aldrich). Store at 4 °C.
5. *N,N,N′,N′*-Tetramethyl-ethylenediamine (TEMED; Sigma-Aldrich). Store at 4 °C.
6. Ammonium persulfate: 10 % solution in ddH_2O. Store at −20 °C.
7. 10× Running buffer (Tris/glycine/SDS):25 mM Tris, 192 mM glycine, 0.1 % SDS, pH 8.3. Store at room temperature.
8. Prestained molecular weight markers: SeeBlue Plus Protein Standard (Life Technologies). Store at 4 °C.
9. Mini-slab size electrophoresis system (ATTO).
10. 10× Transfer buffer: (Tris/glycine): 25 mM Tris, 192 mM glycine, pH 8.3 (Bio-Rad). Store at room temperature.
11. Immuno-Blot PVDF membrane (Bio-Rad).

12. Extra thick blot paper (Bio-Rad).
13. Tris-buffered saline (TBS, 10×): 100 mM Tris–HCl pH 7.5, 1.5 M NaCl. Store at room temperature.
14. Blocking buffer: Membranes were blocked in 5 % low-fat dry milk (Bio-Rad) or bovine serum albumin (BSA; Sigma-Aldrich).
15. Washing buffer (TBS-T): 1× TBS solution supplemented with 1 % Tween-20. Store at room temperature.
16. Antibody dilution buffer: 1× TBS supplemented with 3 % (w/v) BSA and 1 % Tween-20. Single aliquots frozen at −20 °C.
17. Primary antibodies: mouse anti-TSG101 (Abcam), anti-PDC6I/Alix (Abcam), and anti-CD63 (Abcam); dilute in 3 % nonfat milk/BSA and 0.1 % Tween (Sigma-Aldrich) as recommended by the manufacturer.
18. Secondary antibody: Horseradish peroxidase (HRP) conjugated anti-mouse antibody (Cell Signaling).
19. Semi-dry electroblotting system (Bio-Rad). Other standard electroblotting system can be used.
20. Enhanced chemiluminescent (ECL): Pierce ECL Western Blotting Substrate (Thermo Scientific).
21. Imaging system: Detection was performed with the Chemidoc exposure system (Bio-Rad).

2.4 Isolation of Protein from Exosomes

1. Bradford reagent (Thermo Scientific).
2. Lysis buffer: 10 mM Tris–HCl, pH 7.4; 100 mM NaCl; 1 mM EDTA; 1 mM EGTA; 1 mM NaF; 20 mM $Na_4P_2O_7$; 2 mM Na_3VO_4; 1 % Triton X-100; 10 % glycerol; 0.1 % SDS; 0.5 % deoxycholate (Life Technologies). Store at −20 °C.
3. Protease inhibitors (Sigma-Aldrich), store at 4 °C.

2.5 Isolation of Total RNA from Exosomes

1. TRI Reagent (Sigma-Aldrich) (*see* **Note 1**); store at 4 °C.
2. Chloroform (Sigma-Aldrich) (*see* **Note 1**); store at room temperature.
3. Glycogen (Sigma-Aldrich): 120 μg/mL final concentration used (store at −20 °C for up to 1 year).
4. Isopropanol (Sigma-Aldrich): store at room temperature.
5. Ethanol 75 % using ddH_2O (*see* **Note 2**), store at room temperature.
6. RNase-free water (Life Technologies).
7. NanoDrop spectrophotometer or equivalent.

2.6 TaqMan microRNA Low Density Arrays (TLDA)

1. Megaplex Reverse Transcription (RT) Human Pool A. Store at −20 °C.
2. Megaplex RT Human Pool B. Store at −20 °C.
3. Megaplex PreAmp Primers, Human Pool A (*see* **Note 3**). Store at −20 °C.
4. Megaplex PreAmp Primers, Human Pool B (*see* **Note 3**). Store at −20 °C.
5. TaqMan Human MicroRNA Array A. Store at −20 °C.
6. TaqMan Human MicroRNA Array B. Store at −20 °C.
7. MicroAmp Clear Adhesive Film (Life Technologies) (*see* **Note 4**).
8. ABI Prism 7900 HT Sequence Detection System (Life Technologies).
9. 0.1× TE Buffer, pH 8.0 (Life Technologies). Store at room temperature.
10. Centrifuge (Heraeus).

3 Methods

3.1 Experimental Design

A common experimental design consists of a treated sample and a control. For example, here we have included BT474 age-matched parent cells and drug resistant variant of BT474 cells. Biofluids from the consenting patient should be age and gender matched before proceeding to exosomes isolation (*see* **Note 5**) [10].

3.2 Exosome Isolation from Conditioned Media (CM)

1. Seed BT474 and drug resistant variant, and allow growing for 24 h.
2. Replace with fresh FBS-depleted medium (dFBS) (*see* **Note 6**), and allow cells to grow for 5 days (*see* **Note 7**).
3. Pass the CM through a 0.22 μM filter to ensure that no cells or large cell particles are present. Move on to the next step of exosome isolation (*see* **Note 8**).
4. Fill the polyallomer tube with the filtered CM and centrifuge at 100,000 × *g* for 1 h 15 min at 4 °C. A swinging bucket rotor is preferred to isolate exosomes using an ultracentrifuge. Do not alternate between swinging and fixed-angle bucket as it makes a difference in the exosome isolation (*see* **Note 9**). Take into consideration the pelleting efficiency of the ultracentrifuge (*see* **Note 10**).
5. Remove the supernatant carefully without disturbing the exosome pellet.
6. Carefully resuspend the exosome pellet in about 200 μL of PBS, pool together all tubes containing exosomes from the same CM, and transfer to a fresh ultracentrifuge tube. Fill the tube with PBS.

7. Centrifuge again at 100,000 × *g* for 1 h 15 min at 4 °C.
8. Resuspend the resulted exosomes in 20–100 μL PBS (depending on the pellet size obtained). Exosomes can be stored up to 1 year at −80 °C. To avoid repeated freezing and thawing, store exosomes in small aliquots (*see* **Note 11**).

3.3 Exosome Isolation from Biofluids

1. Dilute viscous biofluids such as plasma, serum, and urine, with an equal volume of PBS. Dilute it further to a 5 mL total volume with PBS.
2. Filter this suspension using a 0.22 μM filter to remove debris and cells.
3. Centrifuge at 100,000 × *g* for 1 h 15 min at 4 °C.
4. Carefully remove the supernatant with a pipette without disturbing the pellet (*see* **Note 12**). Resuspend the exosome pellet in about 200 μL of PBS, transfer it into a suitable ultracentrifuge tube (depending on the rotor used), and fill it with PBS.
5. Centrifuge at 100,000 × *g* for 1 h 15 min at 4 °C.
6. Remove the supernatant carefully and resuspend the washed exosomes pellet in a small volume (typically 100 μL) of PBS.

3.4 Comparison of Exosome Secretion Levels

Exosomes secreted by cell line variants can be compared with parent controls to determine if the manipulation of the cell lines alters their ability to expel exosomes. Comparison studies can also be performed using serum specimens from cancer patients versus healthy controls to determine if the secretion of exosome is altered in different diseases [8].

(A) Using cells

1. The amount of exosomes secreted by variant and control cell lines (for example, BT474 drug resistant variant and parent cells) can be compared by seeding equal number of cells.
2. After harvesting the CM, proceed with exosome isolation and, simultaneously, trypsinize the cells and count them using a hemocytometer.
3. Lyse the exosomes using any lysis buffer and quantify protein using Bradford protein quantification.
4. Calculate μg of exosomes expelled per 10^4 cells to compare exosomes secreted by each cell line variants.

(B) Using biofluids

1. The amount of exosomes from individual patients and healthy controls can be compared by isolating exosomes from fixed volume of biofluids (see Note 13).
2. Resuspend the exosomes in a fixed volume of PBS.

3. Lyse the exosomes using any lysis buffer and quantify protein using Bradford protein quantification.
4. Calculate μg of exosomes isolated per ml or μl of biofluids.

3.5 Exosomes Characterization

There are several different methods to confirm the presence of exosomes in isolations from CM and biofluids. Some of the techniques frequently used to determine size and morphology of exosomes include the following:

- TEM (Transmission electron microscopy): identifies the presence of vesicular structures [9].
- Cryo-EM (cryo-electron microscopy) [11] (*see* **Note 14**).

Immunoblotting can be used to determine the expression of protein markers associated with exosomes. Some of the commonly used markers include tetraspanins (CD9, CD63, CD81, CD82) and cytosolic proteins (Tsg101, Alix). As there is no specific marker for exosomes, it is important to note that immunoblotting alone cannot identify whether detected proteins are from exosomes alone and not from other type of extracellular vesicles of endosomal origin [12]. Nevertheless, immunoblotting is a useful technique in detecting proteins present in purified exosomes.

1. Lyse the exosomes using lysis buffer and protease inhibitors, or by mixing them directly with loading or Laemmli buffer.
2. Boil samples for 5 min at 95 °C, briefly centrifuge, and keep on ice.
3. Run the electrophoresis using the mini-slab size system or any other form of electrophoresis device.
4. Clean the glass plates with 95 % (v/v) ethanol before assembling the front and back glasses in the opposite clamps. Check for any leakage by pouring dH_2O and then draining it off.
5. Depending on the exosomal marker (kDa of protein) used, prepare 7.5–12 % gels as described in Table 1. For example, use a 12 % gel to detect TSG101 (molecular weight 47 kDa), whereas a 7.5 % gel should be use for Alix (molecular weight 96 kDa).
6. Pour the stacking gel solution prepared as described in Table 1. Immediately insert the combs and allow polymerization.
7. Carefully remove the glass plates containing the gel from the holder and assemble the gasket with the electrodes. Fill the gasket and the outer chamber with 1× running buffer. Gently remove the comb and then wash the wells using a thin pipette tip.
8. Load samples and molecular markers in the wells.
9. Start the electrophoresis at a constant voltage up to 130 V to carry samples through the stacking gel, and then increase to

Table 1
Preparation of polyacrylamide gels

Components	Resolving gel (7.5 %)	Resolving gel (12 %)	Stacking gel
Acrylamide	3.8 mL	5.25 mL	0.8 mL
ddH_2O	8.0 mL	6.45 mL	3.6 mL
1.5 M Tris–HCl	3.0 mL	3.0 mL	–
1.25 M Tris–HCl	–	–	0.5 mL
10 % SDS	150 μL	150 μL	50 μL
10 % APS	60 μL	60 μL	17 μL
TEMED	10 μL	10 μL	6 μL

APS Ammonium persulfate; *TEMED* *N*,*N*,*N′*,*N′*-tetramethylethylenediamine

150 V to run through the resolving gel. Turn off the power supply immediately after the tracking dye (bromophenol blue from the Laemmli sample buffer) has run off the gel.

10. Cut a sheet of PVDF paper to a size that is slightly larger than the gel. If desired, a corner of the PVDF paper can be cut to allow the tracking of the gel orientation. Activate the membrane by 1 min incubation in methanol. Transfer the PVDF to a tray filled with transfer buffer. Wet two blot papers in transfer buffer.
11. Carefully disassemble the gel unit, cut, and remove the stacking gel with a blade. Transfer the resolving gel into a tray containing transfer buffer.
12. Semi-dry transferring machine is used here. Assemble the transfer cassette by laying the PVDF membrane on the top of the blot papers. Carefully place the gel on top of the membrane. Place another blot paper on top of the gel. Use a roller to make sure that no air bubbles exist between the gel and the membrane. Close the transfer cassette.
13. Connect the unit to a power supply and begin the transfer at a constant current of 200 mA for 1 h.
14. After the transfer, disconnect the power supply and disassemble the transfer cassette. Remove the blot papers and the gel. If using prestained markers, check that their corresponding bands are clearly visible on the PVDF membrane.
15. Incubate the PVDF membrane in blocking buffer for 1 h at room temperature on a rocker with gentle shaking.
16. After blocking, wash the membrane twice in TBS-T buffer for 10 min and then incubate for 1 h at room temperature or overnight at 4 °C (depending on the manufacturer's instructions) in primary antibody solution with gentle shaking.

17. Remove the primary antibody, and wash the membrane three times in TBS-T buffer for 10 min each with vigorous shaking.
18. Incubate the membrane in freshly prepared secondary antibody solution for 1 h at room temperature with gentle shaking (*see* **Note 15**).
19. Wash the membrane three times, for 10 min each, with TBS-T buffer and a vigorous shaking.
20. Mix the ECL reagents together at a ratio of 1:1 immediately before use and evenly add to the blot for 5 min.
21. Remove excess of ECL.
22. Proceed with image acquisition using a Chemidoc exposure system or any other form of image acquisition technique.

3.6 Purification of RNA from Exosomes Isolated from CM and Biofluids

RNA from exosomes is isolated using TRI Reagent and quantified using NanoDrop spectrophotometer. RNA is easily degraded by RNase (ribonuclease) enzymes; therefore, wipe the work space with RNase-Zap before working with RNA. All Eppendorfs, PCR tubes, pipette tips, etc. should be RNase-free, and disposable gloves should be worn.

1. Thaw the isolated exosomes pellets on ice. Add 1 mL of TRI Reagent, and shake vigorously for 15 s. Allow to sit for 5 min for a complete dissociation of nucleoprotein complexes.
2. Add 200 μL of chloroform, and shake vigorously for 15 s. Allow it to stand for 15 min at room temperature (*see* **Note 16**).
3. Centrifuge at 15,700 × *g* in a microfuge for 15 min at 4 °C.
4. Transfer the colorless upper aqueous phase (containing RNA) into a fresh RNase-free 1.5 mL Eppendorf tube.
5. Add 1.25 μL of glycogen and 500 μL of ice-cold isopropanol, and mix. Incubate at room temperature for 5 min, and store at −20 °C overnight for a maximum RNA precipitation (*see* **Note 17**).
6. Centrifuge the tubes at 13,400 × *g* for 30 min at 4 °C to pellet RNA.
7. Remove the supernatant carefully without disturbing the pellet.
8. Wash the pellet twice by adding 750 μL of 75 % ethanol and vortex. Centrifuge at 5,400 × *g* for 5 min at 4 °C.
9. Air-dry the RNA pellet for 5–10 min or until it becomes transparent. Resuspend in 10 μL of RNase-free water (repeated pipetting can be performed to facilitate RNA dissolution).

3.7 TaqMan microRNA Low Density Arrays (TLDA)

The TLDA (Life Technologies) is described in this chapter as a platform for microRNA profiling. It consists of two arrays: TLDA panel A (377 functionally defined microRNAs) and TLDA panel B (290 microRNAs whose function is not yet completely defined) for

Table 2
RT reaction master mix

RT reaction mix components[a]	Volume per sample (μL)
MegaPlex RT primers (10×)	0.80
dNTPs with dTTP (100 mM)	0.20
MultiScribe Reverse Transcriptase (50 U/μL)	1.50
10× RT Buffer	0.80
$MgCl_2$ (25 mM)	0.90
RNase inhibitor (20 U/μL)	0.10
Nuclease-free water	0.20
Total	4.50

[a]All components are part of the MegaPlex RT kit (Life Technologies)

a total of 667 microRNA assays. Each array includes the mammalian U6 (MammU6) assay as a positive control, as well as an assay unrelated to mammalian species, ath-miR159a, as a negative control. Both controls are repeated four times on each card.

Isolate RNA as described above or use any kit of your choice. RNA can be pooled from patients and controls, or profiling could be performed on individual specimens. For pooling, equal quantity of RNA must be used before cDNA preparation.

Experimental designs should take into account the need for an appropriate calibrator sample, typically an untreated sample (for example parent BT474 cell line). A calibrator sample is required to determine the relative expression (delta delta C_T). Select an endogenous control or reference miRNA/s with similar levels of expression within the experiment or with different treatments. Some of the commonly used endogenous controls are miR-16, miR-103, miR-93, miR-192, and miR-451 [13].

Use isolated RNA (25 ng/μL) obtained as described in Subheading 3.6 to perform reverse transcription. Adjust the concentration of all the RNA to the lowest concentration of RNA obtained.

3.7.1 Reverse Transcription

1. Thaw on ice the Megaplex RT Primers, all the TaqMan MicroRNA Reverse Transcription Kit components, and $MgCl_2$ (provided with the kit).
2. Prepare two master mixes per specimen, one for each TLDA panel (A and B) according to Table 2.
3. In a 96-well plate, add 4.5 μL RT reaction mix.
4. Add 3 μL of RNA (25 ng/μL) into appropriate wells containing the RT reaction mix.

Table 3
Megaplex reverse transcription conditions using the ABI 9700 thermal cycler

Stage	Temp (°C)	Time
Cycle (40 cycles)	16 42 50	2 min 1 min 5 min
Hold	85	5 min
Hold	4	∞

Table 4
PreAmp reaction master mix

PreAmp Reaction Mix Components[a]	Volume per specimens (μL)
TaqMan PreAmp Master Mix (2×)	12.5
Megaplex PreAmp Primers (10×)	2.5
Nuclease-free water	7.5
Total	22.5

[a]All components are part of the kit (TaqMan PreAmp Master Mix Kit; Life Technologies)

5. Incubate on ice for 5 min.
6. Follow the thermal cycling conditions described in Table 3 and perform reverse transcription using the ABI 9700 thermal cycler with maximum ramp speed (*see* **Note 18**).

3.7.2 Pre-amplification Reaction

1. Pre-amplify cDNA product (2.5 μL per specimen) (*see* **Note 19**).
2. Prepare 22.5 μL of pre-amplification reaction mix according to Table 4, and add to the cDNA product in a 96-well optical plate sealed with MicroAmp Clear Adhesive Film (*see* **Note 4**).
3. Short spin the plate for 5 s and incubate on ice for 5 min.
4. Pre-amplify using ABI 9700 thermal cycler with standard ramp speed, and follow the thermal cycling conditions according to Table 5.
5. Dilute the pre-amplified product by adding 75 μL of TE buffer and mix, spin down, and store at −20 °C before TLDA TaqMan Real-Time PCR assay. Diluted pre-amplified product can only be stored up to 1 week.

3.7.3 Real-Time PCR Assay

1. Thaw the diluted pre-amplified product on ice. Mix by inverting six times and centrifuge briefly.
2. Mix by swirling the TaqMan Universal PCR Master Mix.

Table 5
Pre-amplification thermal cycling conditions

Stage	Temp (°C)	Time
Hold	95	10 min
Hold	55	2 min
Hold	72	2 min
12 Cycles	95 60	15 s 4 min
Hold	99.9	10 min
Hold	4	∞

Table 6
TLDA reaction master mix

Component	Volume for one array (μL)
TaqMan Universal PCR Master Mix (No AmpErase UNG 2×)	450
Diluted PreAmp product	9
RNase-free water	441
Total	900

3. Set up TLDA TaqMan Real Time Assay for each specimen as detailed in Table 6. Calculation includes volume lost during pipetting.
4. Mix the reaction by inverting the tube six times, and spin down.
5. Load 100 μL of the PCR reaction mix into each port of the TLDA array.
6. Centrifuge (Heraeus) the TLDA plate with 9 up-and-down ramp rates at 331 ×*g* for 1 min, and load into the ABI Prism 7900 HT Sequence Detection System using the 384-well TLDA default thermal-cycling conditions.

3.7.4 Data Analysis

1. The ABI TaqMan SDS v2.3 software is used to obtain raw C_T values.
2. The raw C_T data (SDS file format) can be exported from the Plate Centric View.
3. For analysis of TLDA data, values for each specimen must be normalized to the endogenous control values.
4. Fold changes in patient's biofluids versus control, or age-matched parent BT474 and drug-resistant variant using the comparative C_T method.

4 Notes

1. These reagents are toxic when inhaled so are advised to use in a fume hood.
2. ddH_2O (double-distilled water) is water pretreated with activated carbon, followed by pre-filtration and anti-scaling. Water is then purified by a reverse osmosis system (Millipore Milli-RO 10 Plus) and has a resistivity of 18.2 MΩ-cm.
3. Megaplex PreAmp primers uniformly amplify all microRNAs prior to quantitation by real-time PCR using TaqMan Array MicroRNA Cards. It significantly enhances the ability to detect low expressed microRNAs enabling the generation of a comprehensive expression profile using as low amount of sample as 1 ng of total RNA.
4. MicroAmp *Clear* Adhesive Film (Life Technologies) is more sticky secure seal across the entire microplate to prevent evaporation and well-to-well contamination. Do not use MicroAmp *Optical* Adhesive Film to seal the plate as this is less sticky and may not withstand high temperature for cDNA preparation. MicroAmp Optical Film Compression Pad can also be used on top of the plate to make the seal more secure.
5. Study should be well designed and include an appropriate number of controls for analysis of power for a wide variety of statistical procedures and clinically relevant results. Biofluids from patient and healthy controls should be procured from the consenting patients.
6. Fetal bovine serum (FBS) carries exosomes from the species of origin. Therefore, to prevent any contamination with exosomes from FBS, cells must be cultured either without FBS or grown in exosome-depleted FBS. Deplete exosomes from FBS by centrifuging the sample at $100{,}000 \times g$ at 4 °C for 16 h.
7. Time point for collecting the CM for exosome isolation depends on the growth rate of the cell lines. Do not allow the cells to grow over confluent as they might secret apoptotic bodies.
8. It is highly recommended to isolate exosomes from fresh CM. However, if this is not possible, store the filtered CM up to 1 week at 4 °C before proceeding to exosome purification.
9. Use of swinging bucket rotors is highly recommended during ultracentrifugation. Due to centrifugal force in fixed-angle rotors, exosomes are deposited against the wall of the centrifuge tube, and gradually going down the wall to the bottom of the tube; this process might damage the exosomes [14].
10. K-factor (i.e., pelleting efficiency of a rotor) should be taken into consideration if changing between rotors. The smaller the K-factor,

the better the pelleting efficiency. K-factor can be obtained from the website of the respective centrifuge companies.

11. Repeated freezing and thawing might affect the functionality of exosomes. Therefore, it should be avoided if studying the phenotypic effects of exosomes in vivo or in vitro.
12. Carefully remove the CM with a pipette, leaving behind small volume of biofluid above the pellet. Do not attempt to pour off the supernatant as the pellet may dislodge leading to contamination.
13. Different urine samples even from the same patient would have different content of water; therefore, miRNAs would be more or less diluted. Before comparing samples from different patients, the results should be normalized using urinary creatinine or Tamm Horsfall protein [15].
14. Cryo-EM, a new technique that helped determine the "cup-shaped" morphology of exosomes, was an artifact related to fixation for TEM [11]. Frozen samples of exosomes are used for cryo-EM analysis, avoiding the effects of both dehydration and chemical fixatives.
15. The secondary antibody solution can be reused if stored appropriately.
16. Chloroform helps to separate the mixture into three phases: a red organic phase containing protein, an interphase with DNA, and an upper colorless aqueous phase with RNA.
17. Glycogen acts as a carrier or co-precipitant in RNA purification. Glycogen is added only to the exosomes or CM samples, and left overnight at −20 °C to ensure maximum RNA precipitation. However, no overnight incubation is necessary for RNA isolating from cells [16].
18. After reverse transcription, cDNA can be stored at −20 °C for at least 1 week.
19. Pre-amplification is recommended when the starting amount of RNA is less than 350 ng. When comparing directly miRNA expression profile of exosomes versus cells, it is advisable to perform a pre-amplification step on cDNA amplified from both exosome and cell RNA.

Acknowledgment

This work was supported by funding by Health Research Board of Ireland; Grant number: HRA_POR/2013/341.

References

1. Thery C (2011) Exosomes: secreted vesicles and intercellular communications. F1000 Biol Rep. doi:10.3410/b3-15
2. Keller S, Sanderson MP, Stoeck A et al (2006) Exosomes: from biogenesis and secretion to biological function. Immunol Lett. doi:10.1016/j.imlet.2006.09.005
3. Akers JC, Gonda D, Kim R et al (2013) Biogenesis of extracellular vesicles (EV): exosomes, microvesicles, retrovirus-like vesicles, and apoptotic bodies. J Neurooncol 113:1–11
4. Valadi H, Ekström K, Bossios A et al (2007) Exosome-mediated transfer of mRNAs and microRNAs is a novel mechanism of genetic exchange between cells. Nat Cell Biol 9:654–659
5. Skog J, Wurdinger T, van Rijn S et al (2008) Glioblastoma microvesicles transport RNA and proteins that promote tumour growth and provide diagnostic biomarkers. Nat Cell Biol 12:1470–1476
6. Ginestra A, La Placa MD, Saladino F et al (1998) The amount and proteolytic content of vesicles shed by human cancer cell lines correlates with their in vitro invasiveness. Anticancer Res 18:3433–3437
7. Clayton A, Mitchell JP, Court J et al (2007) Human tumor-derived exosomes selectively impair lymphocyte responses to interleukin-2. Cancer Res 67:7458–7466
8. Corcoran C, Rani S, O'Brien K et al (2012) Docetaxel-resistance in prostate cancer: evaluating associated phenotypic changes and potential for resistance transfer via exosomes. PLoS One. doi:10.1371/journal.pone.0050999
9. O'Brien K, Rani S, Corcoran C et al (2013) Exosomes from triple-negative breast cancer cells can transfer phenotypic traits representing their cells of origin to secondary cells. Eur J Cancer 49:1845–1859
10. Wang YC, Morrison G, Gillihan R et al (2011) Different mechanisms for resistance to trastuzumab versus lapatinib in HER2-positive breast cancers—role of estrogen receptor and HER2 reactivation. Breast Cancer Res. doi:10.1186/bcr3067
11. György B, Szabó TG, Pásztói M et al (2011) Membrane vesicles, current state-of-the-art: emerging role of extracellular vesicles. Cell Mol Life Sci 68:2667–2688
12. Kesimer M, Scull M, Brighton B et al (2009) Characterization of exosome-like vesicles released from human tracheobronchial ciliated epithelium: a possible role in innate defense. FASEB J 23:1858–1868
13. Song J, Bai Z, Han W et al (2012) Identification of suitable reference genes for qPCR analysis of serum microRNA in gastric cancer patients. Dig Dis Sci 57:897–904
14. Witwer KW, Buzás EI, Bemis LT, et al (2013) Standardization of sample collection, isolation and analysis methods in extracellular vesicle research. J Extracell Vesicles 2. doi:10.3402/jev.v2i0.20360.
15. Fernández-Llama P, Khositseth S, Gonzales PA et al (2010) Tamm-Horsfall protein and urinary exosome isolation. Kidney Int 77:736–742
16. Rani S, O'Driscoll L (2011) Reverse-transcriptase polymerase chain reaction to detect extracellular mRNAs. Methods Mol Biol. doi:10.1007/978-1-61779-289-2_2

Chapter 13

Isolation of Urinary Exosomes for RNA Biomarker Discovery Using a Simple, Fast, and Highly Scalable Method

M. Lucrecia Alvarez

Abstract

Urinary exosomes are nanovesicles (40–100 nm) of endocytic origin that are secreted into the urine when a multivesicular body fuses with the membrane of cells from all nephron segments. Interest in urinary exosomes intensified after the discovery that they contain not only protein and mRNA but also microRNA (miRNA) markers of renal dysfunction and structural injury. Currently, the most widely used protocol for the isolation of urinary exosomes is based on ultracentrifugation, a method that is time consuming, requires expensive equipment, and has low scalability, which limits its applicability in the clinical practice.

In this chapter, a simple, fast, and highly scalable step-by-step method for isolation of urinary exosomes is described. This method starts with a 10-min centrifugation of 10 ml urine, then the supernatant is saved (SN1), and the pellet is treated with dithiothreitol and heat to release and recover those exosomes entrapped by polymeric Tamm–Horsfall protein. The treated pellet is then resuspended and centrifuged, and the supernatant obtained (SN2) is combined with the first supernatant, SN1. Next, 3.3 ml of ExoQuick-TC, a commercial exosome precipitation reagent, is added to the total supernatant (SN1 + SN2), mixed well, and saved for at least 12 h at 4 °C. Finally, a pellet of exosomes is obtained after a 30-min centrifugation of the supernatant/ExoQuick-TC mix. We previously compared this method with five others used to isolate urinary exosomes and found that this is the simplest, fastest, and most effective alternative to ultracentrifugation-based protocols if the goal of the study is RNA profiling. A method for isolation and quantification of miRNAs and mRNAs from urinary exosomes is also described here. In addition, we provide a step-by-step description of exosomal miRNA profiling using universal reverse transcription and SYBR qPCR.

Key words Chronic kidney disease, Exosomes, Urinary exosomes, Diabetic nephropathy, Biomarkers, microRNAs, ExoQuick-TC, miR-1207-5p, miR-192, *PVT1*

1 Introduction

Approximately 17 % of adults living in the USA are affected by chronic kidney disease (CKD), a progressive deterioration of renal function secondary to diabetes mellitus, hypertension, and glomerulonephritis. The two most commonly used markers of CKD are serum creatinine and urinary albumin. However, significant renal

M. Lucrecia Alvarez and Mahtab Nourbakhsh (eds.), *RNA Mapping: Methods and Protocols*, Methods in Molecular Biology, vol. 1182, DOI 10.1007/978-1-4939-1062-5_13, © Springer Science+Business Media New York 2014

disease can exist with minimal or no change in serum creatinine levels [1, 2], while non-albuminuric pathways to renal insufficiency have been described [3, 4]. Therefore, the modest diagnostic value of albumin and creatinine levels underscore the need for more sensitive and specific markers of renal injury, particularly ones that correlate with early fibrosis and disease progression.

Biomarkers are very useful indicators of pathological developments associated to a particular disease, which are commonly used to diagnose diseases and/or to follow the pharmacological response to a therapeutic intervention. Urine is an ideal source of biomarkers because it can be conveniently collected noninvasively in large amounts over lengthy periods of time. The use of urinary biomarkers to diagnose disease is a long-standing practice. For example, the "foam test" to determine whether albumin was present in urine has been used for centuries as an indicator of renal disease [5]. Urine can provide biomarkers for diseases of not only the kidney but also the urinary tract [1, 5] and prostate [6]. These biomarkers can be freely available in the urine or inside small (40–100 nm) membrane-bound vesicles known as exosomes. Exosomes are an abundant source of intracellular renal biomarkers that are secreted upon fusion of multivesicular bodies (MVB or late endosomes) with the plasma membrane. Urinary exosomes contain apical membrane and intracellular fluids normally secreted into the urine by cells from all nephron segments and may carry protein, mRNA, and microRNA (miRNA) markers of renal dysfunction and structural injury [7]. Thus, urinary exosomes provide a full representation of the entire urinary system [8], a clear advantage over renal biopsy, an invasive and expensive procedure which provides only a small sample from one of the two kidneys. The recent discovery that exosomes contain functional miRNAs and mRNAs, which could be translated into proteins in target cells mediating communication between cells [9], together with the high potential of these molecules as biomarkers, sparked a renewed interest in their isolation from exosomes.

Currently, the most widely used method for the isolation of exosomes from biological fluids is based on a two-step differential centrifugation process [10–13]. While this method efficiently isolates exosomes from urine, it is time consuming, requires expensive high-speed ultracentrifugation, and allows processing of low volumes of urine and only a few samples at a time. Therefore, the identification of a faster and simpler method than the ultracentrifugation to isolate urinary exosomes is required in order to be able to process a greater number of urine samples for validation of potential exosome biomarkers as well as to routinely use exosomal biomarkers in the clinical laboratory. To this end, we previously reported the comparison of six different methods to isolate urinary exosomes: three based on ultracentrifugation [12–15], one using ultrafiltration [16], and two that utilize the exosome precipitation reagent ExoQuick-TC [17]. We found that the highest number of exosome particles per ml of urine was obtained using our Modified ExoQuick-TC protocol [18]. This method also yielded higher quantities and better qualities of exosomal miRNAs and mRNAs compared to all the other protocols, including the standard ExoQuick-TC method.

Here we describe step by step our Modified ExoQuick-TC protocol, a simple, fast, and highly scalable method for isolation of urinary exosomes [18]. Compared to the standard ExoQuick-TC protocol provided by the manufacturer (SA Biosystems), our ExoQuick-TC protocol has three main modifications that may contribute to its higher exosome yield: an initial centrifugation of the urine at higher speed (17,000 × *g* vs. 3,000 × *g*), treatment of the pellet with a reducent agent and heat to depolymerize the Tamm–Horsfall protein (THP) and release trapped exosomes [14], and a higher amount of ExoQuick-TC per 10 ml of urine (3.3 ml vs. 2 ml) (Fig. 1). Our modified ExoQuick-TC protocol starts with a 10-min centrifugation of 10 ml urine, then the supernatant is saved (SN1), and the pellet is treated with dithiothreitol and heat to increase exosome yield. The treated pellet is then resuspended and centrifuged, and the supernatant obtained (SN2) is combined with the first supernatant, SN1. Next, 3.3 ml of ExoQuick-TC is added to the total supernatant (SN1 + SN2) and saved for at least 12 h at 4 °C. Finally, a pellet of exosomes is obtained after a 30-min centrifugation of the supernatant/ExoQuick-TC mix.

A method for isolation and quantification of miRNAs and mRNAs from urinary exosomes is also described here. As example, we show the quantification of three noncoding RNAs that have been previously associated with diabetic nephropathy and are present in urinary exosomes: miR-192 [19, 20], miR-1207-5p [21], and its host gene, the long noncoding RNA *PVT1* [21, 22].

2 Materials

2.1 Urine Collection

1. Sterile receptacle for urine collection (at least 10 ml).
2. Protease Inhibitor Cocktail (PIC) (Sigma) for urinary exosome preservation (*see* **Note 1**).

2.2 Isolation of Urinary Exosomes

1. Gloves (*see* **Note 2**).
2. 15- and 50-ml conical tubes for centrifugation at 17,000 × *g*.
3. Centrifuge and adapters capable of spinning 15- and 50-ml tubes at 17,000 × *g*.
4. ExoQuick-TC (SBI, System Biosciences) for exosome precipitation (*see* **Note 3**).
5. DL-Dithiothreitol (DTT) (*see* **Note 4**).
6. Isolation solution (I.S.): 250 mM Sucrose, 10 mM triethanolamine, 1 PIC tablet (Roche), adjust pH to 7.6 (*see* **Note 5**).

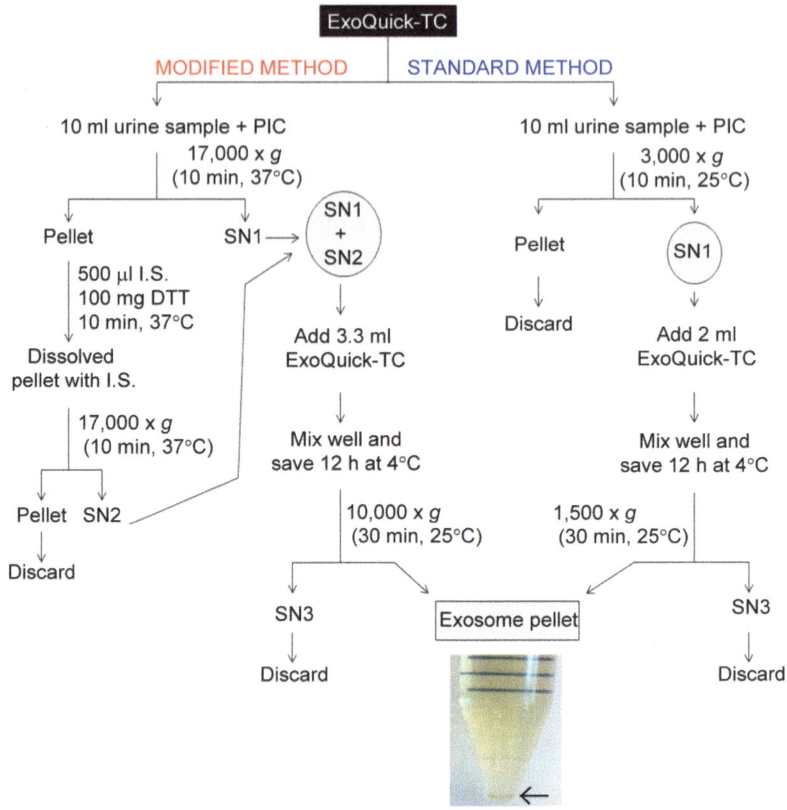

Fig. 1 Isolation of urinary exosomes using the standard ExoQuick-TC method recommended by the manufacturer or our modified ExoQuick-TC method. Urine was initially centrifuged to remove urinary sediment including whole cells, membrane fragments, and other debris. In the modified ExoQuick-TC method, the pellet was dissolved in isolation solution (I.S.) and DL-Dithiothreitol (DTT) and incubated for 10 min at 37 °C to reduce the entrapment of exosomes by Tamm–Horsfall protein. After a second centrifugation, the supernatant (SN2) was combined with the supernatant from the first centrifugation (SN1), treated with ExoQuick-TC, and saved for 12 h at 4 °C. After a final centrifugation, a pellet of urinary exosomes was obtained. Our ExoQuick-TC protocol has three main modifications that may contribute to the higher exosome yield compared to the standard ExoQuick-TC method recommended by the manufacturer (System Biosciences): An initial centrifugation of the urine at higher speed, treatment of the pellet with a reducent agent and heat to depolymerize the Tamm–Horsfall protein and release trapped exosomes (Fernandez-Llama et al. [14]), and a higher amount of ExoQuick-TC per 10 ml of urine

2.3 Separated Purification of Small RNA (miRNAs) and Larger RNA (>200 nt)

1. Gloves (*see* **Note 2**).
2. High-quality, nuclease-free water (*see* **Notes 6** and 7).
3. RNase/DNase-free pipettes, pipette tips, and tubes.
4. RNaseZap for cleaning work surfaces, pipettes, and equipment that must be RNase free (Life Technologies).

Table 1
Life Technologies TaqMan Assays used in real-time quantitative PCR (qPCR)

Gene symbol	Gene name	Assay ID[a]	RefSeq or miR-Base (v20) accession	Assay location	Exon boundary	Amplicon length (bp)
PVT1	Pvt1 oncogene (nonprotein coding)	Hs01069044_m1	NR_003367.1	1374	6–7	83
hsa-miR-1207-5p	microRNA 1207-5p	241060_mat	MIMAT0005871			
hsa-miR-192	microRNA-192	000491	MIMAT0000222			

[a]Assay ID is a unique, alphanumeric string that identifies the assay and encodes basic descriptive information. All the TaqMan assays used were inventoried predesigned TaqMan Gene Expression Assays (Life Technologies, Foster City, CA). Sequences of primers and probe are proprietary of Life Technologies

5. RNeasy Plus Mini kit (Qiagen) (*see* **Notes 8** and **9**).
6. RNeasy MinElute Cleanup kit (Qiagen).
7. β-Mercaptoethanol (β-ME).
8. Ethanol 100 %.
9. 70 % Ethanol: 70 ml Absolute ethanol, 30 ml RNase-free or DEPC-treated water.
10. Microcentrifuge for 1.5 ml tubes.
11. Nanodrop Spectrophotometer to assess RNA purity and concentration (Thermo Fisher) (*see* **Note 10**).
12. Bioanalyzer 2100 to determine RNA quality and integrity (Agilent) (*see* **Note 10**).

2.4 Individual miRNA and mRNA Reverse Transcription and Quantification by TaqMan qPCR

1. Super Script III Reverse Transcriptase (Life Technologies) for reverse transcription of total RNA (except miRNAs).
2. RNaseOUT Recombinant RNase Inhibitor (Life Technologies).
3. 50 ng/μl Random primers.
4. 10 mM dNTP Mix.
5. TaqMan MicroRNA Reverse Transcription kit (Life Technologies).
6. TaqMan Gene Expression Assays specific for the genes that are going to be quantified and the endogenous controls used for normalization (*see* Table 1) [23].
7. TaqMan Gene Expression Master Mix (Life Technologies).
8. 96-Well or 384-well qPCR plates.
9. MicroAmp Optical Adhesive Films (Life Technologies) to seal qPCR plates.
10. Centrifuge with plate adapters.

11. ABI 7900HT thermocycler (Life Technologies) and appropriate blocks for 384- or 96-well PCR plates.
12. CD containing the information about each TaqMan assay (Life Technologies), which will be loaded into the computer when setting up the run.
13. DataAssist software (free at Life Technologies' website) to analyze qPCR results.
14. Urinary Creatinine Assay kit (Cayman): Urinary creatinine can be used to normalize urine samples (input volume) for diuretic dilution effects (optional; *see* Subheading 3.5.1, **step 15**).

2.5 miRNA PCR Profiling

1. Universal cDNA synthesis kit: miRCURY LNA™ Universal RT microRNA PCR System (Exiqon).
2. ExiLENT SYBRR Green master Mix (Exiqon).
3. *Exiqon ready-to-use PCR panels*: Profile 752 miRNAs using only 40 ng total RNA or 372 miRNAs using 20 ng total RNA (no pre-amplification necessary). PCR panels are available for human, rat, and mouse miRNAs. For more information, visit http://www.exiqon.com/mirna-pcr-panels

3 Methods

3.1 Isolation of Urinary Exosomes Using a Modified Exosome Precipitation Method

1. Collect first-void urine samples in sterile receptacles containing 1 ml of PIC (Sigma) per 50 ml urine to reduce protein degradation [11, 13] (*see* **Note 1**).
2. Centrifuge 10 ml of urine/PIC mix at 17,000 ×*g* for 10 min at 37 °C (*see* **Note 11**).
3. Transfer the supernatant, named "SN1," to a fresh tube and save it.
4. Resuspend the pellet in 500 µl of I.S., vortex for 30 s, and add 100 mg of DTT (200 mg/ml final concentration) (*see* **Note 12**).
5. Incubate sample at 37 °C for 5–10 min. During the incubation, vortex the sample every 1–2 min until the pellet is dissolved (*see* **Note 12**).
6. Add 1 ml of I.S. to the dissolved pellet, mix, and centrifuge it at 17,000 ×*g* for 10 min at 37 °C.
7. Combine the supernatant obtained after the centrifugation in the previous step (named "SN2") with SN1 (the supernatant saved in **step 3**). Discard the pellet or resuspend it in Laemmli buffer for future analysis using sodium dodecyl sulfate-polyacrylamide gel electrophoresis (SDS-PAGE).
8. Transfer supernatant to a sterile vessel, measure the total volume, and add 3.3 ml of ExoQuick-TC exosome precipitation

solution per 10 ml of supernatant (SN1 + SN2). Mix well by inverting or flicking the tube.

9. Refrigerate the supernatant/ExoQuick-TC mix overnight (at least 12 h).
10. Centrifuge supernatant/ExoQuick-TC at 10,000 × *g* for 30 min at room temperature or 4 °C.
11. After centrifugation, the exosomes may appear as a beige or a white pellet at the bottom of the tube (Fig. 1).
12. Spin down residual ExoQuick-TC solution by centrifugation at 1,500 × *g* for 5 min. Remove all traces of fluid by aspiration taking great care not to disturb the exosome pellet.
13. Resuspend the pellet in 300 μl buffer RLT Plus supplemented with 1 % β-ME (*see* **Notes 13–15**).

3.2 Separated Purification of Small RNA (miRNAs) and Larger RNA (>200 nt)

To obtain a higher yield of both miRNAs and larger RNAs from the same exosome pellet, it is recommended to purify small RNA (miRNA, 5S RNA, and tRNA) and larger RNA (>200 nt) separately using both RNeasy Plus Mini kit and RNeasy Min Elute Clean up kit (Qiagen) (Fig. 2, **Notes 8** and **9**). Alternatively, total RNA (including small RNAs) can be extracted using miRNeasy Mini kit following the manufacturer's instructions. Other total RNA extraction kits from other suppliers (*see* **Note 16**) can also be used as long as they are able to produce high-quality RNA according to the criteria specified in **Note 10**.

1. Spray RNaseZap onto work surfaces, pipettes, equipment, and the gloves that you are wearing before starting with the RNA extraction. Even trace quantities of RNase can lead to degradation during RNA purification protocols, lower yields from in vitro transcription reactions, and variable results in the qPCR.
2. Rinse the RNaseZap off with nuclease-free water.
3. Disrupt and homogenize exosome pellet (*see* **Note 17**) in 300 μl buffer RLT plus (from **step 13**, Subheading 3.1).
4. To eliminate traces of DNA, transfer the sample to a gDNA eliminator spin column placed in a 2 ml collection tube.
5. Centrifuge for 30 s at >8,000 × *g*. Discard the column, and save the flow-through (*see* **Note 18**).
6. Add 1 volume (about 350 μl) of 70 % ethanol.
7. Add up to 700 μl of the "aqueous phase/ethanol" mixture to a Qiagen RNeasy Mini Column. Centrifuge at 8,000 × *g* for 1 min.
8. Readminister the flow-through to the column to bind RNA not captured in the first pass. Centrifuge mini column again at 8,000 × *g* for 1 min. *Important*: Save both the column and the

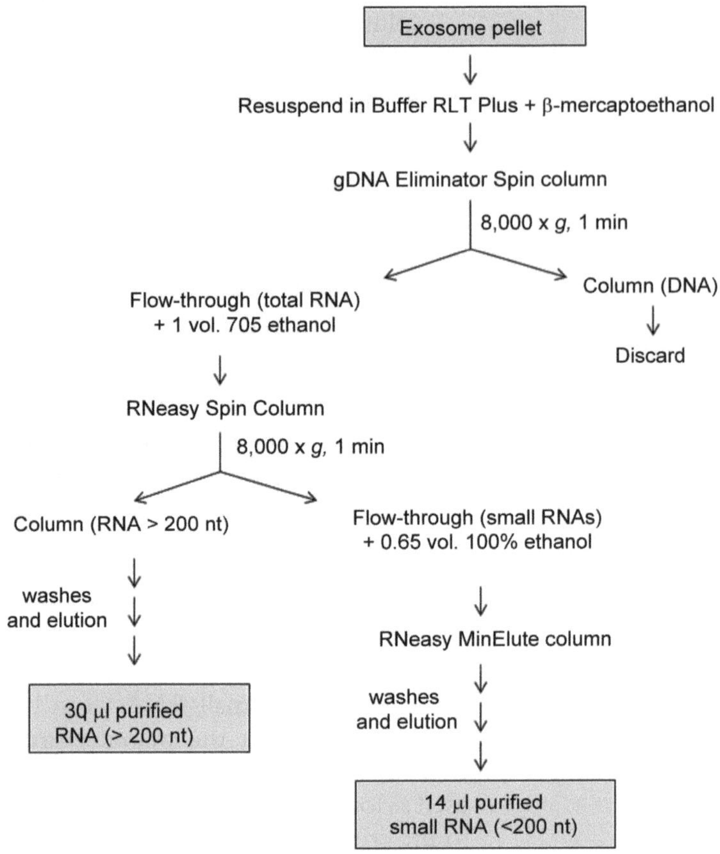

Fig. 2 Procedure for separated purification of small RNA (miRNAs) and larger RNA (>200 nt) from the same exosome pellet. RNeasy Plus Mini kit and RNeasy Min Elute Cleanup kit (Qiagen) are used to purify small RNA (miRNA, 5S RNA, and tRNA) and larger RNA (>200 nt) separately from the same exosome pellet

flow-through for RNA (>200 nt) and miRNA extractions, respectively.

9. *Flow-through*: Add 0.65 volumes of 100 % ethanol, and continue with **step 19** after RNA (>200 nt) has been eluted from the column (**steps 10–18**).
10. *Column*: Add 700 µl buffer RW1 to the RNeasy spin column, and centrifuge for 1 min at ≥8,000 × *g* to wash the spin column membrane. Discard the flow-through.
11. Add 500 µl buffer RPE to the RNeasy spin column, and centrifuge for 1 min at ≥8,000 × *g* to wash the spin column membrane. Discard the flow-through.
12. Add 500 µl buffer RPE to the RNeasy spin column, and centrifuge for 1 min at ≥8,000 × *g* to wash the spin column membrane. Discard the flow-through.
13. Repeat **step 12**, but centrifuge for 2 min (*see* **Note 19**).

14. Place the RNeasy spin column in a new 1.5 ml collection tube.
15. Add 30–50 μl RNase-free water directly to the spin column membrane, and incubate for 5 min at room temperature.
16. Centrifuge for 1 min at ≥8,000 × *g* to elute the RNA (>200 nt).
17. To obtain higher RNA concentrations, repeat **steps 15** and **16** using the eluate from **step 16** (if high RNA concentration is required). Reuse the collection tube from **step 16** (*see* **Note 20**).
18. Save the RNA (>200 nt) at −80 °C, and continue processing the flow-through to purify miRNAs (*see* next step).
19. Apply the flow-through from **step 9** to an RNeasy MinElute spin column placed in a 2 ml collection tube.
20. Close the lid gently, and centrifuge for 1 min at 8,000 × *g*. Discard the flow-through.
21. Place the column in a new 2 ml collection tube, and add 500 μl buffer RPE to the spin column (*see* **Note 21**). Close the lid gently, and centrifuge for 1 min at 8,000 × *g* to wash the spin column membrane. Discard the flow-through.
22. Add 500 μl of 80 % ethanol to the RNeasy MinElute spin column. Close the lid gently, and centrifuge for 2 min at 8,000 × *g* to wash the spin column membrane. Discard the flow-through and collection tube (*see* **Note 22**).
23. Place the RNeasy MinElute spin column in a new 2 ml collection tube and centrifuge at full speed for 5 min with the lid opened to make sure that all the ethanol evaporates (*see* **Note 23**). Discard the flow-through and collection tube.
24. Place the RNeasy MinElute spin column in a new 1.5 ml collection tube.
25. Add 14 μl of RNase-free water directly to the center of the spin column membrane, and incubate for 5 min at room temperature.
26. Centrifuge for 1 min at full speed to elute the RNA (*see* **Notes 24** and **25**).
27. Save the purified small RNA at −80 °C.
28. Determine RNA quantity and quality using the Nanodrop Spectrophotometer (Thermo Fisher) and Bioanalyzer (*see* **Notes 10** and **26**).

Exosomal RNA yield obtained using this protocol: Yield varies depending on how diluted or concentrated is the urine sample, but one should expect a minimum of 15 ± 6 ng miRNA per ml of urine and 5.1 ± 2.2 ng RNA (>200 nt) per ml of urine. Figure 3a, b shows the TaqMan qPCR amplification of miR-1207-5p and its host gene the long noncoding RNA *PVT1*, respectively. These two noncoding RNAs have been previously associated with diabetic

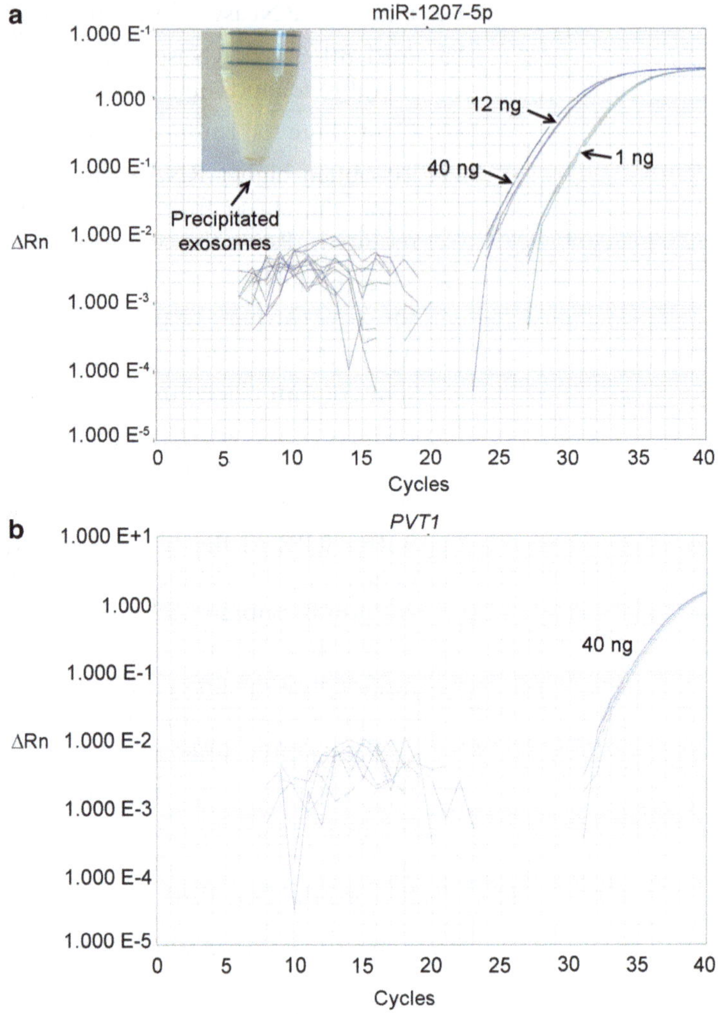

Fig. 3 TaqMan qPCR amplification of small and long noncoding RNA isolated from urinary exosomes. Amplification of miR-1207-5p (**a**) and its host gene, the long noncoding RNA *PVT1* (**b**), isolated from urinary exosomes. These two noncoding RNAs have been previously associated with diabetic nephropathy and suggested as targets and biomarkers for this disease [18, 21]. Different amounts of reverse-transcribed RNA (1, 12, or 40 ng) were used as a template in TaqMan qPCR amplification of miR-1207-5p (**a**). Rn is the fluorescence of the reported dye divided by the fluorescence of a passive reference dye (i.e., reporter signal normalized to the fluorescence signal of ROX). ΔRn is the Rn minus the baseline (increment of fluorescent signal at each time point). Amplification plot shows the log (ΔRn) graphed vs. cycle. All the amplifications were done in triplicate (three technical replicates)

nephropathy [18, 21, 22] and are normally secreted in urinary exosomes. Figure 3a shows TaqMan qPCR amplification of miR-1207-5p from 1, 12, and 40 ng of reverse-transcribed RNA. The initial amount of cDNA template required for qPCR amplification depends on the level of expression of the target gene, but, generally, 1–5 ng is enough for detection of individual exosomal miRNAs using the TaqMan system. In fact, with as little as 40 ng total RNA and without the need for pre-amplification, up to 752 miRNAs can be profiled using Exiqon miRCURY LNA SYBR Green Master Mix (*see* Subheading 3.6).

3.3 Reverse Transcription of Exosomal mRNA

1. Briefly spin down the SuperScript III Reverse Transcriptase, RNaseOUT Recombinant RNase Inhibitor, random primers, and dNTP mix to collect all the components at the bottom of the tube. Use SuperScript III Reverse Transcriptase (Life Technologies) for reverse transcription of total RNA (except miRNAs).
2. Per each sample, assemble the RT reaction on ice in a 0.2 ml RNase-free tube (amounts are for a 20 μl final reaction volume):
   - 10 pg to 5 μg total RNA.
   - 1 μl Random hexamers 50 ng/μl.
   - 1 μl dNTP mix 10 mM (*see* **Note 27**).
   - Up to 13 μl nuclease-free water.
3. Incubate the tube at 65 °C for 5 min and then place on ice for at least 1 min.
4. Prepare the following cDNA synthesis mix per tube (calculate a 10 % more):
   - 4 μl 5× First-strand buffer.
   - 1 μl DTT 0.1 M.
   - 1 μl RNaseOUT Recombinant RNase Inhibitor 40 U/μl.
   - 1 μl of SuperScript III RT (200 U/μl).
5. Mix by pipetting gently up and down, and add 7 μl of cDNA synthesis mix per tube.
6. Incubate the tube at 25 °C for 5 min followed by 60 min at 55 °C.
7. Inactivate the reaction by heating at 70 °C for 15 min (*see* **Note 28**).

3.4 Reverse Transcription of Individual Exosomal miRNAs

TaqMan MicroRNA Assays employ a target-specific stem–loop reverse transcription primer to address the challenge of the short length of mature miRNA. The primer extends the 3′ end of the target to produce a template that can be used in standard TaqMan real-time PCR [24]. The stem–loop structure in the tail of the primer allows the specific detection of the mature, biologically active miRNA. Use this approach to obtain the highest sensitivity and specificity detection of up to six of your favorite miRNAs. For example, we used it to quantify miR-192 and miR-1207-5p, two miRNAs associated with diabetic nephropathy [19–22]. Alternatively, follow protocol in Subheading 3.6 to quantify more than six miRNAs of interest or to do miRNA profiling.

1. Before use, thaw the specific 5× RT primer on ice (*see* **Note 29**). Resuspend the solution completely by gentle vortexing, and then briefly centrifuge the tube.
2. Allow the TaqMan MicroRNA Reverse Transcription Kit components to thaw on ice.
3. In a polypropylene tube, prepare the RT master mix on ice by scaling the volumes below to the desired number of RT reactions. Applied Biosystems recommends adding 10–20 % overage to account for pipetting losses:
   - 0.15 μl dNTPs 100 mM.
   - 1 μl Multiscribe Reverse Transcriptase, 50 U/μl.
   - 1.5 μl 10× Reverse transcriptase buffer.
   - 0.19 μl RNase Inhibitor, 20 U/μl
   - 4.16 μl Nuclease-free water.
4. Mix thoroughly by pipetting up and down few times.
5. Per each sample, combine 7 μl RT master mix (from **step 3**), 3 μl 5× RT primer, and 5 μl miRNA fraction/nuclease-free water (*see* **Note 30**) into a 0.2 ml polypropylene reaction tube. Mix gently, and centrifuge it to bring the solution to the bottom of the tube.
6. Incubate the tube on ice until you are ready to load it in the thermal cycler.
7. Use the following parameter values to program the thermal cycler:
   - 30 min at 16 °C.
   - 30 min at 42 °C.
   - 5 min at 85 °C.
   - ∞ at 4 °C.
8. Load the RT reaction tubes into the thermal cycler, and start the reverse transcription run.
9. Following the reverse transcription run, if you do not proceed immediately to PCR amplification, store the RT reaction at −20 °C.

3.5 Quantification of a Few Individual Exosomal miRNAs and mRNAs by qPCR

3.5.1 Individual miRNA Quantification by TaqMan qPCR

1. Thaw on ice the cDNA samples prepared as described in Subheading 3.4 and the TaqMan® Assay (20×) corresponding to the specific miRNA (Table 1; *see* **Note 31**). Resuspend completely by gently vortexing, and then centrifuge briefly.
2. Mix the TaqMan Gene Expression Master Mix reagent by gently swirling the bottle.
3. Calculate the number of reactions (20 μl per well) that you need for each assay. It is recommended to do at least three technical replicates of each reaction (Table 2). Be sure to include on each plate "no-template controls" (NTCs) for each assay on the plate.

Table 2
Amount of reagents required for miRNA quantification by TaqMan qPCR

Component	Single reaction for 1 well (μl)	Three replicates for 3 wells (μl)[a]
TaqMan small RNA Assay (20×)	1	3.2
Product from RT reaction[b]	1.4	4.5
TaqMan Universal Master Mix (20×)	10	32.0
Nuclease-free water	7.6	24.3
Total volume	20	64

[a]Replicate volumes in the table include 10 % excess for volume loss from pipetting
[b]The amount of RT product recommended in the table is the maximum amount of that can be added to a reaction (the RT primer must be diluted a minimum of 1:15 in the final qPCR reaction)

Table 3
Reaction mix for miRNA quantification by TaqMan qPCR

Component	For 1 sample (μl)[a]	For 15 samples (μl)[b]
TaqMan small RNA Assay (20×)	3.2	48
TaqMan Universal Master Mix (20×)	32	480
Nuclease-free water	24.3	364.8
Total volume	59.5	892.8

[a]Volumes are for three technical replicates (3 wells) of each sample plus about 10 % excess for volume loss from pipetting
[b]Scale up the amount of reagents for your number of samples

4. Prepare the qPCR reaction mix: Use Table 3 to calculate the amount of reagents that you will need for your total number of samples. As an example, Table 3 shows the amount of reagents required to prepare a qPCR reaction mix for 1 and 15 samples.
5. Cap the qPCR reaction mix tube, vortex it, and centrifuge it briefly.
6. Prepare one labeled 1.5 ml microcentrifuge tube per sample, and add 59.5 μl of qPCR reaction mix per tube.
7. Add 4.5 μl of cDNA per tube for a final volume of 64 μl (Table 2).
8. Cap the qPCR reaction mix tube, vortex it, and centrifuge it for 5 min.
9. Transfer 20 μl of the qPCR reaction mix into each of the three wells with technical replicates of the same sample on a 96- or a 384-well plate.
10. Seal the plate with the appropriate cover.

11. Centrifuge the plate briefly.
12. Load the plate into the thermal cycler instrument, and run the qPCR using the following conditions:
    - 10 min at 95 °C.
    - 40 cycles: 15 s at 95 °C (denature) and 60 s at 60 °C (anneal/extend).
13. Start the run.
14. After the run:
    (a) View the amplification plots.
    (b) Set the baseline and threshold values (*see* **Notes 32** and **33**).
15. In order to compensate for any diuretic dilution effects, data should be normalized for creatinine concentration [25]. Urinary creatinine can be measured using the Urinary Creatinine Assay kit (Cayman), which is based on the Jaffe's reaction where creatinine produces quantitatively an orange color with picric acid in alkaline medium. Normalize the results by first converting Log2 data (Cts) into a linear scale assuming amplification efficiency of 2 (100 %) for every assay and all samples [$2^{(40-\text{Ct})}$]. Correct the data by dividing the linear Ct values with the amount of urinary creatinine per subject, and convert the results back to Log2 Cts (1–40 values). Alternatively, THP has been suggested as a reliable normalization variable to allow quantitative analysis in urine, which can be measured in Coomassie-stained gels [14].
16. *Analysis of the results*:
    - If stable endogenous controls are known, use them for normalization in relative quantification of gene expression (*see* **Notes 34** and **35**). Analyze the results using the comparative CT (ΔΔCT) method and the DataAssist software (freely available at Life Technologies' website).
    - If stable endogenous controls are unknown, empirically determine the best reference miRNAs by doing an initial pilot study and use DataAssist software to identify the most stable genes under the used conditions. Alternatively, use the NormFinder [26] and/or GeNorm [23] algorithms to identify the most stable genes (*see* **Notes 36** and **37**).
    - To calculate the relative expression of different targets in the same sample based on PCR efficiency estimate the initial amount of a particular exosomal miRNA in a urine sample using the formula $\text{Ct} = -3.32 \times \text{Log Input cDNA} + 20$. Assume 100 % efficiency in the TaqMan Assay (slope −3.32) and use the Ct results from the qPCR.

Figure 4 shows the relative quantification of exosomal miR-1207-5p and miR-192 by TaqMan qPCR. Higher yield of exosomal miRNA was obtained using our modified ExoQuick-TC protocol compared to the standard protocol.

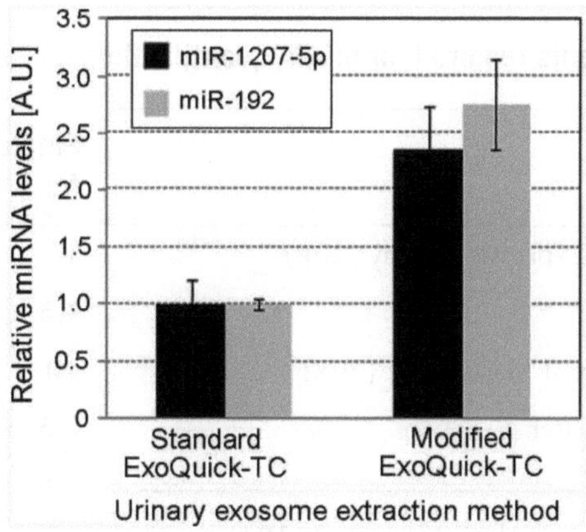

Fig. 4 Relative levels of miRNAs 1207-5p and 192 in urinary exosomes isolated using the standard or the modified ExoQuick-TC method. Urine was processed as explained in Subheading 3.1, and exosomes were isolated using the standard or the modified ExoQuick-TC protocols as described in Subheading 3 and Fig. 1. microRNAs 1207-5p and 192 levels were quantified by TaqMan qPCR and expressed as relative to the levels obtained using the standard ExoQuick-TC method, which was arbitrarily considered as 1. Data are mean ± S.D. of four biological replicates (four urine samples)

3.5.2 Individual mRNA Quantification by TaqMan qPCR

1. Thaw on ice the cDNA samples prepared as described in Subheading 3.3 and the TaqMan Assay (20×) corresponding to the specific mRNA (Table 1; *see* **Notes 31**, **38**, and **39**). Resuspend the TaqMan assay completely by gently vortexing, and then centrifuge briefly.
2. Mix the TaqMan Gene Expression Master Mix reagent by gently swirling the bottle.
3. Calculate the number of reactions (20 μl per well) that you need for each assay. It is recommended to do at least three technical replicates of each reaction (Table 4). Be sure to include on each plate endogenous control assays and "NTCs" for each assay on the plate.
4. Prepare the qPCR reaction mix: Use Table 5 to calculate the amount of reagents that you will need for your total number of samples. As an example, Table 5 shows the amount of reagents required to prepare a qPCR reaction mix for 1 and 15 samples.
5. Cap the qPCR reaction mix tube, vortex it, and centrifuge it briefly.
6. Prepare one labeled 1.5 ml microcentrifuge tube per sample, and add 59.2 μl of qPCR reaction mix per tube.
7. Add 4.8 μl of cDNA per tube for a final volume of 64 μl (Table 4).
8. Follow **steps 8–16** from Subheading 3.5.1.

Table 4
Amount of reagents required for mRNA quantification by TaqMan qPCR

Component	Single reaction for 1 well (μl)	Three replicates for 3 wells (μl)[a]
TaqMan Gene Expression Assay (20×)	1	3.2
cDNA template[b]	1.5	4.8
TaqMan Universal Master Mix (20×)	10	32.0
Nuclease-free water	7.5	24.0
Total volume	20	64.0

[a]Replicate volumes in the table include 10 % excess for volume loss from pipetting
[b]Applied Biosystems recommends that no more than 20 % of the PCR be composed of the reverse transcription reaction (no more than 4 μl of cDNA template per 20 μl reaction)

Table 5
Reaction mix for mRNA quantification by TaqMan qPCR

Component	For 1 sample (μl)[a]	For 15 samples (μl)[b]
TaqMan Gene Expression Assay (20×)	3.2	48.0
TaqMan Universal Master Mix (20×)	32.0	480.0
Nuclease-free water	24.0	360.0
Total volume	59.2	888.0

[a]Volumes are for three technical replicates (3 wells) of each sample plus about 10 % excess for volume loss from pipetting
[b]Scale up the amount of reagents for your number of samples

3.6 Universal Reverse Transcription and SYBR qPCR Quantification of Exosomal miRNAs for Profiling

A universal RT step allows hundreds of miRNA targets to be quantified from the same cDNA synthesis. This approach saves sample, reduces technical variation, and decreases time and effort in the laboratory. However, miRNA amplification using DNA-based primers has restricted sensitivity, especially for AT-rich targets, as well as difficulties with single-nucleotide mismatches [27]. The incorporation of modified nucleotides, e.g., locked nucleic acids [28] or 2′-O-(2-methoxyethyl) derivatives [29], have been used to overcome these problems and normalize melting temperatures.

Here we provide a description of the miRCURY LNA Universal RT microRNA PCR system, a miRNA-specific LNA-based system designed for sensitive and accurate detection of miRNA by qPCR using SYBR Green (Fig. 5). This method allows profiling of up to 752 miRNAs using only 40 ng total RNA. It is based on universal reverse transcription followed by qPCR amplification with specific LNA-enhanced primers [27]. It offers the exceptional sensitivity

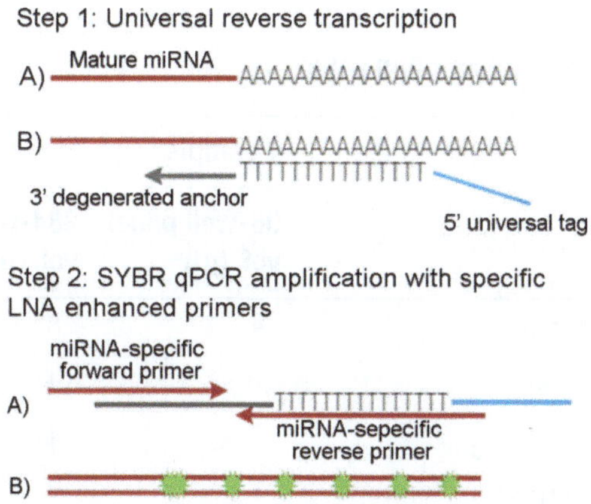

Fig. 5 Outline of the LNA Universal RT SYBR qPCR system for miRNA profiling. A universal RT step allows hundreds of miRNA targets to be quantified from the same cDNA synthesis. A poly-A tail is added to the mature miRNA template (*step 1A*), and cDNA is synthesized using a poly-T primer with a 5′ universal tag and a 3′ degenerated anchor (*step 1B*). The cDNA template is then amplified using miRNA-specific LNA-enhanced forward and reverse primers (*step 2A*) and the fluorescent dye SYBR Green (*step 2B*), which binds to all double-stranded DNA, so the result is an increase in fluorescence intensity proportional to the amount of PCR product produced

and extremely low background required for accurate quantification of very low miRNA levels as well as highly specific assays that allow discrimination between closely related miRNA sequences.

3.6.1 Universal Reverse Transcription

1. Gently thaw the 5× reaction buffer (*see* **Note 40**) and nuclease-free water from the Universal cDNA synthesis kit II (Exiqon; *see* **Note 41**), mix by vortexing, and immediately place on ice.
2. Resuspend the RNA spike-in UniSp6 (*see* **Note 42**) in 80 μl nuclease-free water, mix by vortexing, and spin down. Leave on ice for 20–30 min to fully dissolve RNA spike-in. Store in aliquots at −20 °C.
3. Prior to the RT reaction, add 1 μl synthetic spike-in per RNA sample.
4. Immediately before use, remove the enzyme mix from the freezer, mix by flicking the tubes, spin down all reagents, and place them on ice.
5. Prepare the required amount of RT working solution (*see* Table 6), and place it on ice.
6. Dispense RT working solution into nuclease-free tubes, and add the template RNA in each tube (*see* Table 6).
7. Mix the reaction by very gentle vortexing or pipetting and spin down.

Table 6
Universal reverse transcription setup

Reagent	1 sample		6 samples + 10 % excess vol.	
	96-well panel vol. (μl)[a]	384-well panel vol. (μl)[a]	6× 96-well panels	6× 384-well panel
5× Reaction buffer	4	8	26.4	52.8
Nuclease-free water	7	14	46.2	92.4
Enzyme mix	2	4	13.2	26.4
Synthetic RNA spike-ins (optional) or nuclease-free water	1	2	6.6	13.2
Template exosomal miRNA	6	12	[b]	[c]
Total volume	20	40	92.4	184.8

[a]The 96-well protocol provides sufficient cDNA for one full panel whether located in two 96-well plates or distributed within 192 wells of a 384-well plate. The 384-well protocol is enough for one ready-to-use 384-well plate (panel I or II)
[b]Distribute 14 μl of RT mix per tube (a total of 6), and add 6 μl of exosomal miRNA sample in each tube
[c]Distribute 28 μl of RT mix per tube (a total of 6), and add 12 μl of exosomal miRNA sample in each tube

8. Incubate for 60 min at 42 °C followed by 5 min at 95 °C to heat-inactivate the reverse transcriptase. Immediately, cool down to 4 °C, and save at −20 °C.

3.6.2 SYBR qPCR Amplification with Specific LNA-Enhanced Primers

1. Place cDNA (from **step 8**, Subheading 3.6.1), nuclease-free water, and ExiLENT SYBR Green PCR Master Mix (*see* **Note 43**) on ice and thaw for 15–20 min. Mix the 2× PCR Master Mix by pipetting up and down immediately before using it. Mix the rest of the reagents by vortexing and spin down.
2. Combine 2,000 μl 2× PCR Master Mix and 1,960 μl nuclease-free water. Mix gently, and spin down (*see* **Notes 44–46**).
3. Add 40 μl cDNA to the diluted PCR Master Mix prepared in **step 2**, and mix.
4. Before removing the plate seal from the Human miRNome PCR Panel (Exiqon), briefly spin down the plate(s) in a plate centrifuge.
5. Add 10 μl PCR Master Mix:cDNA mix prepared in **step 3** to each well.
6. Seal the plate with optical sealing as recommended by the instrument manufacturer.
7. Spin plate briefly in a plate centrifuge (1,500 × g for 1 min) to collect the sample in the bottom (*see* **Note 47**).

8. Perform qPCR amplification followed by melting curve analysis (*see* **Note 48**):
   - Polymerase activation/denaturation: 95 °C, 10 min.
   - Amplification: 40 cycles at 95 °C, 10 s, and 60 °C, 1 min. Set the ramp rate to 1.6 °C/s (*see* **Note 49**).
9. Perform the initial data analysis using first the software supplied with the qPCR instrument to obtain row Cts (*see* **Note 50**) followed by a specific software for qPCR data analysis such as Exiqon GenEx (Exiqon) or the freely available data analysis at Qiagen's website (http://www.sabiosciences.com/pcrarray-dataanalysis.php). If the data is going to be analyzed using Exiqon GenEx software (Exiqon), the experiment must be set up as an absolute quantification (AQ), not relative quantification (RQ).

4 Notes

1. Ideally, urine samples should be processed fresh immediately following collection. Alternatively, samples should be frozen at –80 °C but not –20 °C. It has been reported that freezing at –20 °C causes a major loss in exosomes compared to fresh urine. In contrast, recovery after freezing at –80 °C is almost complete. It is essential to do an extensive vortexing after thawing the urine samples to increase exosome recovery in frozen urine [11].
2. It is essential to wear gloves throughout all the procedures described in this chapter.
3. Make sure to use ExoQuick-TC, which is specific for isolation of exosomes from urine and cell culture media, and not the plain ExoQuick because the exosome yield will be lower.
4. The addition of DTT to the pellet disrupts the polymeric network of the THP and release the exosomes entrapped in the network increasing the yield [14].
5. Adjust pH to 7.6 with 1 N NaOH (approx. 220 μl).
6. It is essential to use nuclease-free (RNase-free) water. The most commonly used method for eliminating RNase contamination from water, buffers, and other solutions is treatment with diethylpyrocarbonate (DEPC), which destroys RNase enzymatic activity. However, residual DEPC can negatively affect the activity of the reverse transcriptase; therefore, DEPC-treated water should not be used.
7. We recommend using nuclease-free water that is not DEPC treated as well as RT-PCR-grade water offered by some companies such as Life Technologies, Qiagen, and Abnova.

8. miRNAs and mRNAs are extracted separately in two different fractions from the same sample using both the RNeasy Plus Mini kit combined with RNeasy MinElute Cleanup kit (Qiagen). Alternatively, the RNeasy Plus or the miRNeasy kits (Qiagen) allow the extraction of total RNA including miRNAs.
9. For simultaneous purification of exosomal proteins, small RNAs (<200 nt), and high-molecular-weight (HMW) RNAs (>200 nt) into three different fractions from the same sample, use combined All Prep DNA/RNA/Protein Mini and RNeasy MinElute Cleanup kits (Qiagen).
10. To obtain reliable results from the quantification by TaqMan qPCR, all RNA samples should have high quality according to the following criteria:
    - *RNA concentration and purity by UV spectrophotometry*: Absorbance (*A*) 260:230 ratio should be greater than 1.7 (purity from organic compounds), *A*260:280 at least 1.8 (purity from proteins), and concentration by *A*260 greater than 40 μg/ml total RNA. The readings are affected by pH, and, therefore, be sure to perform the dilutions for spectrophotometry in RNase-free Tris pH 8.0 buffer.
    - *Ribosomal RNA band integrity*: A fraction of each RNA sample should be analyzed on a denaturing agarose gel or on a BioAnalyzer using an RNA 6000 Nano LabChip (Agilent). You should observe sharp bands or peaks without smearing or shoulder (signs of RNA degradation) corresponding to both the 18S and 28S ribosomal RNA. An RNA integrity number (RIN) of 7 or higher obtained using the BioAnalyzer is recommended.
11. The temperature, speed, and time of the urine/PIC mix centrifugation were chosen according to Fernandez-Llama et al. [14]. Low temperature favors THP polymerization and precipitation, which entraps exosomes and reduces yield. Therefore, all the centrifugations are performed at 37 °C to reduce exosome loss.
12. It has been reported that these are the best conditions to release exosomes trapped in the pellets and increase yield [14].
13. If the miRNAeasy kit is going to be used to extract total RNA, the exosome pellet should be resuspended in 300 μl of Qiazol. If the All Prep DNA/RNA/Protein Mini and RNeasy MinElute Cleanup kits (Qiagen) are going to be used (*see* **Note 9**), then resuspend the exosome pellet in 350 μl of buffer RLT (from the All Prep DNA/RNA/Protein Mini kit) supplemented with 3.5 μl of β-ME. Vortex or pipet up and down to mix. Store at −70 °C until you proceed with the RNA extractions.

14. The β-ME should be added to the RLT plus buffer just before using it to help the guanidine isothiocyanate to destroy RNAases.
15. Buffer RLT may form a precipitate during storage. If necessary, warm it up (up to 37 °C) and then place it at room temperature (15–25 °C).
16. Alternatively, a new RNA extraction kit named "miRCURY RNA isolation kit-biofluids" (Exiqon) is optimized for handling low-yield samples and therefore should be well suited for biofluids. According to McAlexander et al. [30], this kit outperforms other RNA isolation methods used to obtain miRNAs from biofluids and requires only 200 μl of urine sample.
17. Incomplete loosening of the exosome pellet may lead to inefficient lysis and reduced RNA yields.
18. Make sure that no liquid remains on the column membrane after centrifugation.
19. Residual ethanol may interfere with downstream reactions. The long centrifugation dries the spin column membrane, ensuring that no ethanol is carried over during RNA elution. In addition, carefully remove the RNeasy spin column from the collection tube after centrifugation so that the column does not contact the flow-through to avoid ethanol carryover.
20. If using the eluate from **step 15**, the RNA yield will be 15–30 % less than that obtained using a second volume of RNase-free water, but the final RNA concentration will be higher.
21. Buffer RPE is supplied as a concentrate. Before using for the first time, add four volumes of ethanol (96–100 %) as indicated on the bottle to obtain a working solution.
22. After centrifugation, carefully remove the RNeasy MinElute spin column from the collection tube so that the column does not contact the flow-through to avoid carryover of ethanol.
23. Place the spin columns into the centrifuge with at least one empty position between columns to avoid the damage of the lid. Orient the lids so that they point in a direction opposite to the rotation of the rotor (e.g., if the rotor rotates clockwise, orient the lids counterclockwise).
24. A minimum volume of 10 μl RNase-free water can be used for elution if a higher RNA concentration is required, but the yield will be reduced by approximately 20 %. Do not elute with less than 10 μl RNase-free water as the spin column membrane will not be sufficiently hydrated.
25. It is very important to pipet RNase-free water into the center of the RNeasy MinElute spin column membrane to ensure that the membrane is completely covered to avoid low RNA yield.

26. Nanodrop does not allow discrimination between different small RNA fractions (miRNA, 5S RNA, and tRNA). The Agilent Small RNA Assay is one of the few analyzers available that can measure miRNA concentration. This chip can quantify miRNAs in absolute amounts (pg/μl) and as a relative percentage of small RNA.

27. dNTP mix: Consists of 10 mM each dATP, dGTP, dCTP, and dTTP at neutral pH.

28. Optional extra step: Amplification of PCR targets >1 kb may require the removal of RNA complementary to the cDNA by adding 1 μl (2 U) of *E. coli* RNase H and incubating at 37 °C for 20 min.

29. Each TaqMan miRNA Assay comes with two tubes per miRNA:
    - RT primer 1 × 150 μl (5×): enough for 50 reactions.
    - TaqMan Assay 1 × 150 μl (20×): enough for 150 reactions.

30. Use the same amount of RNA template for all your samples and complete with nuclease-free water up to 5 μl. The recommended amount of total RNA template is 1–10 ng. If a small RNA fraction was purified from urinary exosomes and the concentration of miRNAs is unknown, use at least 2 μl of RNA template and complete to 5 μl with nuclease-free water. However, the amount of RNA to be used may also vary depending on the miRNA levels in the urinary exosomes.

31. Keep all TaqMan Small RNA Assays protected from light, in the freezer, until you are ready to use them. Excessive exposure to light may affect the fluorescent probes.

32. *Baseline*: Initial cycles of PCR during which there is little change in fluorescence signal (usually cycles 3–15). A reporter fluorescent signal is accumulating but is beneath the limits of detection of the instrument. *Threshold*: It is an arbitrary level of fluorescence chosen on the basis of the baseline variability. A signal that is detected above the threshold is considered a real signal that can be used to define the threshold cycle (Ct) for a sample. *Ct* is defined as the fractional PCR cycle number at which the reporter fluorescence is greater than the threshold.

33. To set the baseline and threshold values using the SDS software (Life Technologies), follow the instructions described in another chapter of this book entitled "SYBR Green and TaqMan Quantitative PCR Arrays: Expression Profile of Genes Relevant to a Pathway or Disease State."

34. Normalization is intended to remove technical variation in data which is not related to the biological changes under investigation. Proper normalization is critical for the correct analysis and interpretation of results from qPCR experiments.

The most commonly used methods for normalization are the following:

(a) To identify and use stably expressed reference genes.

(b) To use the global mean expression value of all commonly expressed miRNAs in a given sample as normalization factor (global normalization method): This method can only be used in a screening with full panels containing several hundred miRNAs [31]. Those miRNAs with an expression similar to the global mean expression value can be selected as reference genes for normalization in follow-up studies where only a limited number of miRNA molecules are profiled [31, 32].

35. Biofluid samples and purified exosomes typically do not contain the larger small RNA species (e.g., 5S, U6, and snoRNAs) that are sometimes used for normalization in other samples. Another difference is that often the number of miRNAs detected in biofluid samples is rather low.

36. Selection of invariant miRNAs can be identified by algorithms specifically developed for reference gene evaluation and selection based on (a) reference gene ranking and stepwise elimination of the least stable gene (GeNorm; [23]), (b) repeated pairwise correlation and regression analysis (BestKeeper software; [33]), and (c) statistical linear mixed-effects modelling (NormFinder; [26]) of the respective experimental data.

37. Characteristics of a good endogenous control candidate for normalization of miRNA quantification:
    - Similar expression level to the miRNA of interest.
    - Invariant expression across all samples of the study (i.e., not affected by the experimental conditions or the disease).
    - Similar small size as the miRNA (i.e., similar stability, extraction, and quantification efficiency).

38. One tube for each 20× TaqMan assay contains:
    - Two unlabeled primers (1× final concentration is 900 nM per primer; 20× stock concentration is 18 μM per primer).
    - One 6-FAM™ dye-labeled TaqMan® MGB probe (1× final concentration is 250 nM; 20× stock concentration is 5 μM).

39. Store 20× TaqMan assays at −15 to −25 °C, and keep them protected from light. To minimize freeze–thaw cycles, consider dividing the solutions into smaller aliquots.

40. The 5× buffer already includes universal reverse transcription primer.

41. This kit contains all reagents required for first-strand cDNA synthesis for 8–64 reactions based on a standard reaction volume of 10–80 μl reactions.

42. The UniSp6 RNA Spike-in template is provided with the Universal cDNA synthesis kit II. Additionally four RNA spike-in templates can be obtained with the separate RNA Spike-in kit. The primary purpose of the RNA spike-ins is to provide controls for the quality of the RNA isolation, the cDNA synthesis reaction, and the PCR. RNA isolations may vary in yield, purity, and integrity, and some sample types, such as serum and plasma, may contain compounds that inhibit the cDNA synthesis or the PCR even though the RNA has been purified using the best standard procedures.
43. The ExiLENT SYBR® Green master mix contains all reagents required for PCR amplification of miRNAs. In addition, a positive control assay named "UniSp6 RNA Spike-in control primer set" is provided with this kit for amplification of the synthetic UniSp6 RNA Spike-in included in the Universal cDNA synthesis kit II.
44. The PCR Master Mix does not include the passive reference dye ROX, and it should be added in the cDNA dilution if advised by instrument manufacturer. The reference dye is used to normalize signals from individual PCR wells in order to enable comparison of qPCR amplification signals across an entire PCR plate. The amount of ROX required is instrument dependent, and it is important to refer to the manufacturer's recommendations when deciding how much ROX to use.
45. The protocol can be interrupted at this stage.
46. The undiluted cDNA may be kept at –20 °C for up to 5 weeks (optional store at 4 °C for up to 4 days). It is recommended that synthesized cDNA is stored in "low-nucleic acid-binding" tubes or plates.
47. The experiment can be paused at this point. Store the reactions protected from light at 4 °C for up to 24 h.
48. Melting curve analysis is an analysis step built into the software of qPCR instruments and is recommended to verify specificity and identity of the amplification reaction. Please follow the instructions provided by the supplier to do a melting curve analysis at the end of the qPCR.
49. If the ramp rate of cooling is too rapid, performance may be compromised. Set the ramp rate of cooling to 1.6 °C/s. This is equivalent to 100 % under standard cycling conditions on the ABI 7500, 7900, and Viia7 instruments.
50. If you are using an ABI instrument, please note that it is not recommended to use auto Ct settings. For a guide on how to set manual baseline and threshold, *see* Chapter 27.

Acknowledgements

The author is grateful to the Roney Family Foundation for the support of this study, to Madieh (Tala) Khosroheidari and Dr. Rupesh Kanchi Ravi for their valuable help in performing part of the experiments, and to Paul Arnold for the critical review of this manuscript.

References

1. Hewitt SM, Dear J, Star RA (2004) Discovery of protein biomarkers for renal diseases. J Am Soc Nephrol 15:1677–1689
2. Tomlanovich S, Golbetz H, Perlroth M et al (1986) Limitations of creatinine in quantifying the severity of cyclosporine-induced chronic nephropathy. Am J Kidney Dis 8:332–337
3. Tsalamandris C, Allen TJ, Gilbert RE (1994) Progressive decline in renal function in diabetic patients with and without albuminuria. Diabetes 43:649–655
4. Macisaac RJ, Jerums G (2011) Diabetic kidney disease with and without albuminuria. Curr Opin Nephrol Hypertens 20:246–257
5. Pisitkun T, Johnstone R, Knepper MA (2006) Discovery of urinary biomarkers. Mol Cell Proteomics 5:1760–1771
6. Nilsson J, Skog J, Nordstrand A et al (2009) Prostate cancer-derived urine exosomes: a novel approach to biomarkers for prostate cancer. Br J Cancer 100:1603–1607
7. Van Balkom BW, Pisitkun T, Verhaar MC et al (2011) Exosomes and the kidney: prospects for diagnosis and therapy of renal diseases. Kidney Int 80:1138–1145
8. Miranda KC, Bond DT, McKee M et al (2010) Nucleic acids within urinary exosomes/microvesicles are potential biomarkers for renal disease. Kidney Int 78:191–199
9. Valadi H, Ekstrom K, Bossios A et al (2007) Exosome-mediated transfer of mRNAs and microRNAs is a novel mechanism of genetic exchange between cells. Nat Cell Biol 9:654–659
10. Pisitkun T, Shen RF, Knepper MA (2004) Identification and proteomic profiling of exosomes in human urine. Proc Natl Acad Sci U S A 101:13368–13373
11. Zhou H, Yuen PS, Pisitkun T et al (2006) Collection, storage, preservation, and normalization of human urinary exosomes for biomarker discovery. Kidney Int 69:1471–1476
12. Gonzales PA, Pisitkun T, Hoffert JD et al (2009) Large-scale proteomics and phosphoproteomics of urinary exosomes. J Am Soc Nephrol 20:363–379
13. Gonzales PA, Zhou H, Pisitkun T et al (2010) Isolation and purification of exosomes in urine. Methods Mol Biol 641:89–99
14. Fernandez-Llama P, Khositseth S, Gonzales PA et al (2010) Tamm-Horsfall protein and urinary exosome isolation. Kidney Int 77: 736–747
15. Thery C, Clayton A, Amigorena S et al (2006) Isolation and characterization of exosomes from cell culture supernatants and biological fluids. Curr Protoc Cell Biol 3:22.1–22.29
16. Cheruvanky A, Zhou H, Pisitkun T et al (2007) Rapid isolation of urinary exosomal biomarkers using a nanomembrane ultrafiltration concentrator. Am J Physiol Renal Physiol 292:F1657–F1661
17. Taylor DD, Zacharias W, Gercel-Taylor C (2011) Exosome isolation for proteomic analyses and RNA profiling. Methods Mol Biol 728:235–246
18. Alvarez ML, Khosroheidari M, Kanchi Ravi R et al (2012) Comparison of protein, microRNA, and mRNA yields using different methods of urinary exosome isolation for the discovery of kidney disease biomarkers. Kidney Int 82: 1024–1032
19. Kato M, Zhang J, Wang M et al (2007) MicroRNA-192 in diabetic kidney glomeruli and its function in TGF-beta-induced collagen expression via inhibition of E-box repressors. Proc Natl Acad Sci U S A 104:3432–3437
20. Putta S, Lanting L, Sun G et al (2012) Inhibiting microRNA-192 ameliorates renal fibrosis in diabetic nephropathy. J Am Soc Nephrol 23:458–469
21. Alvarez ML, Khosroheidari M, Eddy E et al (2013) Role of microRNA 1207-5p and its host gene, the long non-coding RNA *PVT1*, as mediators of extracellular matrix accumulation in the kidney: implications for diabetic nephropathy. PLoS One 8:e77468

22. Alvarez ML, DiStefano JK (2011) Functional characterization of the plasmacytoma variant translocation 1 gene (*PVT1*) in diabetic nephropathy. PLoS One 6:e18671
23. Vandesompele J, De Preter K, Pattyn F et al (2002) Accurate normalization of real-time quantitative RT-PCR data by geometric averaging of multiple internal control genes. Genome Biol 3, research0034
24. Chen C, Ridzon DA, Broomer AJ et al (2005) Real-time quantification of microRNAs by stem-loop RT-PCR. Nucleic Acids Res 33: e179
25. Gautier JC, Riefke B, Walter J et al (2010) Evaluation of novel biomarkers of nephrotoxicity in two strains of rat treated with Cisplatin. Toxicol Pathol 38:943–956
26. Andersen CL, Jensen JL, Orntoft TF (2004) Normalization of real-time quantitative reverse transcription-PCR data: a model-based variance estimation approach to identify genes suited for normalization, applied to bladder and colon cancer data sets. Cancer Res 64: 5245–5520
27. Andreasen D, Fog JU, Biggs W et al (2010) Improved microRNA quantification in total RNA from clinical samples. Methods 50: S6–S9
28. Castoldi M, Schmidt S, Benes V et al (2008) miChip: an array-based method for microRNA expression profiling using locked nucleic acid capture probes. Nat Protoc 3:321–329
29. Beuvink I, Kolb FA, Budach W et al (2007) A novel microarray approach reveals new tissue-specific signatures of known and predicted mammalian microRNAs. Nucleic Acids Res 35:e52
30. McAlexander MA, Phillips MJ, Witwer KW (2013) Comparison of methods for miRNA extraction from plasma and quantitative recovery of RNA from cerebrospinal fluid. Front Genet 4:83
31. Mestdagh P, Van Vlierberghe P, De Weer A et al (2009) A novel and universal method for microRNA RT-qPCR data normalization. Genome Biol 10:R64
32. Chang KH, Mestdagh P, Vandesompele J et al (2010) MicroRNA expression profiling to identify and validate reference genes for relative quantification in colorectal cancer. BMC Cancer 10:173
33. Pfaffl MW, Tichopad A, Prgomet C et al (2004) Determination of stable housekeeping genes, differentially regulated target genes and sample integrity: BestKeeper–Excel-based tool using pair-wise correlations. Biotechnol Lett 26: 509–515

Part II

Functional RNA Mapping

Chapter 14

Identification of Actively Translated mRNAs

Marc R. Reboll and Mahtab Nourbakhsh

Abstract

Ribosomes are large protein–RNA complexes involved in translation of mRNA nucleotide sequences into proteins. Multiple ribosomes, polyribosomes (polysomes), bind to a single mRNA in order to initiate translation and protein synthesis. In order to distinguish actively translated RNAs, total polysomes can be isolated from cell lysates and purified by centrifugation through sucrose density gradients. The polysome fraction represents all actively translated cellular RNAs that can be specifically detected using common RNA detection assays, e.g., RT-PCR. The quantity of the target RNA in polysomes fraction indicates its translation state. This chapter provides a protocol for the isolation and fractionation of polysomes from mammalian cell lines.

Key words Ribosome, Polysomes, Ribosomal subunits, Sucrose density gradient, Optical density

1 Introduction

Identification of transcribed genomic sequences subsequently raises the question of their biological functions. In general, cellular RNA transcripts are classified in four groups based on their structures and biological functions. These are messenger RNAs (mRNAs) encoding amino acid sequence of a polypeptide (protein), transfer RNAs (tRNAs) binding amino acids to ribosomes during translation, ribosomal RNAs (rRNAs) building ribosomes with ribosomal proteins, and small nuclear RNAs (snRNAs) which play an important role in RNA processing [1].

By encoding proteins mRNAs are directly involved in diverse cellular functions whereas other types of RNAs are involved in regulatory mechanism of gene expression. The translation of mRNA to protein synthesis consists in the recruitment of ribosomes. In eukaryotes, translation apparatus is a large and complex machinery comprising at least 11 factors. This complexity provides a variety of regulatory mechanism which allows for a selective translation of mRNAs depending on physiological conditions. Either actively translated or not, mRNAs are permanently

M. Lucrecia Alvarez and Mahtab Nourbakhsh (eds.), *RNA Mapping: Methods and Protocols*, Methods in Molecular Biology, vol. 1182, DOI 10.1007/978-1-4939-1062-5_14, © Springer Science+Business Media New York 2014

Table 1
Troubleshooting guide

Problem	Recommendations
The interface between organic and aqueous phases after phenol extraction is thick and cannot be separated from aqueous phase	This occurs when the amount of protein is very high in the sample. Try to pipet out all the aqueous phase with interface and repeat the phenol/chloroform extraction again
The amount of the extracted RNA is too low	Phenol used in this protocol is buffered to prevent oxidized products in the phenol from damaging the nucleic acids. Check the pH of Phenol or use freshly buffered phenol
The polysome profile does not reveal distinct peaks	Polysome profiles can significantly differ between different cell types. Control experiments using different stimuli, like virus or bacterial LPS, can help to identify the most relevant fractions. These agents strongly alter the polysome profile in most cell lines and can reveal if the observed profile is specific for your cell line. Alternatively, you can start to use the fractions at highest sucrose concentrations first. In this case, it is helpful to analyze abundant RNAs as control
No RNA pellet can be observed	The precipitation of RNA does not necessarily lead to a pellet. Moreover, visible RNA pellet indicated a high salt concentration or contaminant proteins. Continue to analyze the RNA anyway. If the concentration of the target RNA is too low, consider more sensitive RNA methods such as RT-PCR

associated with few or multiple ribosomes. A cluster of ribosomes, bound to an mRNA, are known as polyribosomes or polysomes also known as ergosomes [2]. The high molecular weight of polysome provides a unique physical feature which allows for their selective isolation and purification from cellular content. Polysomes can be size-fractionated by sucrose density gradient centrifugation. Identification of a specific mRNA in polysomes fraction implies a highly active translation state of the mRNA.

In this chapter, we provide the complete experimental procedure for sucrose density gradient centrifugation and fractionation of polymosomes from mammalian cells [3]. This method needs to be combined with an RNA detection assay, such as northern blot, RNase protection assay, or RT-PCR in order to identify the target mRNA in polysome fractions (Table 1).

2 Materials

2.1 Inhibition of Cellular RNA Transcription

1. Protease inhibitors (SigmaP8340-5ML).
2. DTT (Sigma 43815).
3. Cycloheximide (CHX) (Sigma C7698).

4. Heparin (Sigma H3149).
5. Sucrose (Sigma S1888).

2.2 Polysomes Isolation and Fractionation

1. Phenol/chloroform/isoamyl alcohol (Invitrogen 15593-031).
2. Sodium Acetate (Fisher Scientific S209-500).
3. 1× PBS.
4. Trypsin-EDTA 1×, 0.05 % Trypsin/0.53 mM EDTA (Cellgro 25-052-CV).
5. Tris–Base (Fisher Sci BP152-1).
6. KCl (Fisher Sci BP-366-500).
7. MgCl2 (Sigma M-2393).
8. Triton X-100 (SigmaT8787-250ML).
9. DNAase/RNAase-free Ethanol (SigmaE7023-500ML).
10. DNAase/RNAase-free water (Bioexpress UPW-1000).
11. Ultracentrifuge tubes, 14×89 mm (Bechman344059).
12. Polysome Extraction Buffer (PEB with Triton-X100: Triton-X100, 20 mM Tris–HCl (pH 7.5), 50 mM KCl, 10 mM MgCl2, 1 mM DTT, 100 μg/mL CHX, 200 μg/ml Heparin).
13. 50 % Sucrose solution (25 g sucrose in PEB without Triton-X100 to 50 ml).

2.3 Equipments

1. BR-188 Density Gradient Fractionation System (Brandel, MD, USA).
2. Beckman Optima L-70 Ultracentrifuge (Beckman, CA, USA).
3. 7500 Real-Time PCR System (Applied Biosystems, NY, USA).
4. Boekel Scientific Orbitron Rotator I, 115 V (BoekelScientific, NY, USA).

3 Methods

3.1 Polysomes Isolation

1. Prepare 2–5 × 10^7 cells of a desired cell line (2–3 150 cm^2 cell culture flasks).
2. To interrupt cellular translation, incubate cells in complete medium supplemented with a final concentration of 100 μg/mL CHX for 10 min at 37 °C.
3. Discard medium and rinse cells twice with 10 ml 1× PBS containing100 μg/mL CHX at 4 °C.
4. Add 7 ml of trypsin into each flask if necessary to harvest adherent cells and incubate at 37 °C for 3 min (*see* **Note 1**).
5. Collect cells in 10 ml PBS containing 100 μg/mL CHX (*see* **Note 2**).
6. Centrifuge the cells for 10 min at <1,000 × *g*.

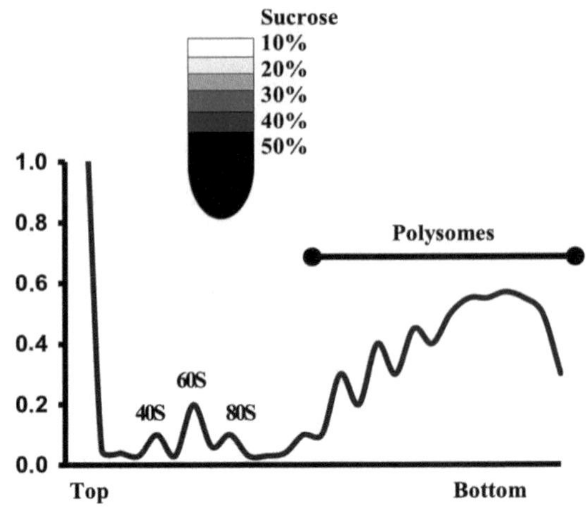

Fig. 1 Polysome profiling. Different layers from a continuous gradient after centrifugation are shown at the figure *top*. The highest sucrose concentration is placed at the bottom of centrifuge tube followed by 40, 30, 20, and 10 % sucrose solution. The diagram indicates a typical UV absorbance profile at 254 nm using cytoplasmic extracts from HeLa cells sedimented through a 10–50 % sucrose gradient. Measured optical density OD_{254} (*y*-axis) was plotted against gradient samples from *top* to *bottom* (*x*-axis)

7. Discard the supernatant and resuspend cell pellet in 1 ml of 1× PBS, and transfer them into 1.7 ml microcentrifuge tube.
8. Lyse cells with 1 ml PEB Buffer containing 1 %Triton-X100, vortex for 15 s, and keep on ice for 30 min.
9. Centrifuge at ≥10,000 × *g* for 30 min at 4 °C and collect supernatants that are ready to be loaded on sucrose gradients.
10. Dilute 50 % sucrose solution into the solutions of 40, 30, 20, and 10 % in PEB buffer (*see* **Note 3**).
11. Make 10 ml gradient by adding 2.0 ml of each solution (50 % on bottom, 10 % on top) and load into ultracentrifuge tube (Fig. 1).
12. Add 1 ml of cell extract to the top of each gradient. Cover each tube with parafilm, put it in a SW41 rotor, and spin samples at 38,000 rpm (250,000 × *g*) for 120 min at 4 °C.
13. Collect 13 fractions from the top to the bottom of the tube using BR-188 Density Gradient Fractionation System (*see* **Note 4**).
14. The OD_{254} nm profile identifies the fractions which are enriched with polysomes.
15. Keep all fractions at −80 °C; it's good for extraction of RNA in 1 month.
16. For RNA extraction, thaw all samples on ice.
17. Mix the water saturation Phenol/chloroform/isoamyl alcohol (25:24:1) completely (*see* **Note 5**).

18. For each sample, put 500 μl of the mixed Phenol/chloroform/isoamyl alcohol into a new 2 ml-microcentrifuge tube (*see* **Note 6**).
19. Add 0.5 ml of each sample into a new tube with Phenol/chloroform/isoamyl alcohol.
20. Vortex for 30 s and then keep rotating for 5 min.
21. Centrifuge at 21,000×*g* for 10 min at 4 °C, at the same time prepare new 2 ml-microcentrifuge tube for each sample, add 3 M Sodium Acetate (pH5.2) 50 μl/tube.
22. Very carefully pipet out all the aqueous phase without disturbing interphase and put it into the prepared fresh 2 ml-microcentrifuge tube. During organic extraction, protein contaminants are denatured and partition either with the organic phase or at the interface between organic and aqueous phases, while nucleic acids remain in the aqueous phase (*see* **Note 7**).
23. Then add 1.5 ml (3 volumes) of 100 % Ethanol, vortex for 30 s.
24. Keep the samples at −80 °C for 3 h.
25. Centrifuge at 14,000 rpm (21,000×*g*) for 30 min at 4 °C.
26. Wash pellet with cold 70 % RNase-free ethanol by rotating for 5 min.
27. Centrifuge pellet at 21,000×*g* for 5 min at 4 °C.
28. Discard supernatants and remain the pellet in the tubes.
29. Put tubes on ice to let all the liquid evaporated.
30. Add 30–50 μl of DNase/RNase-free H_2O.

4 Notes

1. Use a recommended harvesting protocol for your cell line. Eventually skip this step for non-adherent cell lines.
2. Skip this step when using non-adherent cell lines.
3. We recommend using the master solution of 50 % sucrose to prepare 40–10 % solutions. It is crucial to have a constant gradient. The final concentration is secondary.
4. Machine setting needs to be adjusted before. We observed the optimal setting by 75 s fraction collection time and 0.75 ml/min pump speed.
5. Be aware that phenol can cause severe chemical burns on skin and will damage clothing. Wear gloves, safety glasses, and a laboratory coat when working with phenol.
6. Keep shaking with hand when pipetting to avoid water separating out.
7. If disturbed the interphase by accident, repeat centrifugation step again.

References

1. Kowalczyk MS, Higgs DR, Gingeras TH (2012) Molecular biology: RNA discrimination. Nature 482:310–311
2. Lodish H (1999) 4.5. Stepwise formation of proteins on ribosomes. In: Molecular cell biology. Scientific American Books, New York. ISBN 0-7167-3136-3
3. Gu L, Zhang H, He J, Li J, Huang M, Zhou M (2012) MDM2 regulates MYCN mRNA stabilization and translation in human neuroblastoma cells. Oncogene 31:1342–1353

Chapter 15

Mapping of Internal Ribosome Entry Sites (IRES)

Sarah Mehrtens and Marc R. Reboll

Abstract

The dicistronic luciferase reporter gene system is the most common method to isolate and characterize internal ribosome entry sites (IRES). It is based on the expression of a dicistronic RNA comprising two independent reporter genes in 3′ and 5′cistrons, and the putative IRES inserted into intercistronic region. The most convenient aspect of using *Renilla* and firefly luciferase genes is that both gene products can be detected in a single assay using Dual-Luciferase® Reporter Assay System from Promega. The *Renilla* luciferase coding sequence is often inserted into the 5′cistron and serves as internal control. It is translated cap dependently, as it is close to the cap structure at the 5′ end. The 3′cistron located far downstream to the cap structure can only be translated by a cap-independent mechanism when the intercistronic sequence is capable of ribosome binding and re-initiation of translation. Expression level of the 3′cistron is usually normalized to the expression of 5′cistron to estimate the relative IRES activity of intercistronic sequences.

Key words Internal ribosome entry site, Translation, Dicistronic reporter gene plasmid, Luciferase

1 Introduction

The initiation of translation is an important and strictly regulated step in protein expression. Most cellular mRNAs are translated by a cap-dependent translation mechanism. The cap structure at the 5′ end of mRNAs is crucial for binding of initiation factors and ribosome assembly [1]. An alternative cap-independent translation initiation is feasible by an internal ribosome entry site (IRES). The IRES sequences shape stable RNA structures which serve as docking stations for initiation factors, IRES trans-acting factors (ITAFs) and ribosomes. The cap structure is not essential for IRES-dependent translation and the ribosomes can bind closely upstream of the start codon [2]. IRES elements were first identified in the 5′UTRs of viral RNA especially in picorna viruses [3]. However, a number of IRES elements were found in cellular mRNAs later which allow for translation initiation under cellular stress conditions like virus infection, UV and heat stress, or lack of nutrients [4].

M. Lucrecia Alvarez and Mahtab Nourbakhsh (eds.), *RNA Mapping: Methods and Protocols*, Methods in Molecular Biology, vol. 1182, DOI 10.1007/978-1-4939-1062-5_15,

The secondary RNA structure is one of the most important features of IRES. It was shown that IRES elements are composed of several modular structures; each structure contributes to the full activity of the IRES element. Nevertheless, there is often a core element which is indispensable for the IRES activity [5]. In order to identify and characterize an IRES element it is necessary to monitor its ability to initiate translation. The gold standard to detect IRES activity is a dicistronic reporter gene plasmid [6]. The RNA transcribed from this plasmid comprises two reporter genes. The 5′ reporter gene is translated by a cap-dependent mechanism. The 3′ reporter gene can only be translated if the sequence between both reporter genes acts as an IRES. Therefore, the expression of the first reporter gene serves as internal control whereas the expression of the second reporter indicates the IRES activity. These bicistronic reporter plasmids together with cloning strategies to modify or partially delete the intercistronic sequence offer a very versatile experimental tool to define and characterize minimal IRES sequences. Despite a variety of available reporter genes, this chapter will focus on the dicistronic *Renilla* and firefly luciferase reporter gene system which turned to be the most convenient approach for IRES studies.

2 Materials

2.1 Transfection of Reporter Gene Plasmids

1. A dicistronic reporter gene plasmid containing the coding sequences of the firefly luciferase and the *Renilla* luciferase, the known or putative IRES has to be inserted between these two reporter genes (*see* **Note 1**).
2. Transfection reagents 2× HEBS Buffer (0.28 M NaCl, 0.05 M HEPES, 1.5 mM Na_2HPO_4, pH 7.1, store at −20 °C) and 2.5 M $CaCl_2$ solution (store at −20 °C) (*see* **Note 2**).
3. An appropriate cell line for transfection and corresponding media (*see* **Note 3**).
4. 6-well plates (Nunc).

2.2 Cell Lysis

1. *Renilla* luciferase assay lysis buffer, 5× (part of the Renilla Luciferase Assay System from Promega, #E2810).
2. 1× PBS (137 mM NaCl, 2.7 mM KCl, 10 mM Na_2HPO_4, 2 mM KH_2PO_4).

2.3 Luciferase Measurement

1. Cell lysate.
2. Renilla luciferase assay buffer (from Renilla luciferase assay system, Promega, #E2810).
3. Firefly luciferase assay buffer (20 mM Tricin, 0.2 mg/ml coenzyme A, 1.07 mM $(MgCO_3)\ Mg(OH)_2 \cdot 5H_2O$, 2.67 mM

MgSO$_4$, 33.3 mM DTT, 5.3 mM ATP, 472 μM D-luciferin, pH 7.8) (*see* **Note 4**).

4. Luminometer (*see* **Note 5**).

3 Methods

To measure the activity of known or putative IRES, the corresponding DNA sequence is inserted between the two open reading frames of a dicistronic reporter gene plasmid. This plasmid is transfected in suitable cell lines or primary cells. It is transcribed in these cells and results in a dicistronic RNA. The first cistron of this RNA is translated in a cap-dependent manner while the translation of the second cistron is cap independent. The expressed reporter proteins accumulate in the cells which can be lysated after 48 h. In the lysates the amount of the reporter proteins is measured.

The most widely used pair of reporter genes in dicistronic plasmids are the *Renilla* and firefly luciferases. These proteins catalyze enzymatic reactions which emit light [7]. The enzymatic activity of the luciferases is determined by mixing the cell lysates with corresponding luciferase assay buffers. These buffers contain all substrates and co-factors needed for the enzymatic reactions in excess. Therefore, the measured luminescence is just dependent on the amount of luciferase enzyme in the cell lysates and provides information about the cap-dependent and cap-independent translation efficiencies. The *Renilla* and firefly luciferases have completely different substrates and offer the possibility to measure both luciferase reactions in the same cell lysate. Moreover, bioluminescence is very sensitive. In contrast to fluorescence no excitation light is required and background signals are very low [8]. Using this method to measure IRES activity needs several control experiments which should be carefully addressed [9]. One point of concern is the integrity of the dicistronic RNA. If the integrity is not ensured, fragments of this RNA—especially fragments encoding the IRES-dependent reporter gene—might be translated individually and cause misleading reporter gene activity. To confirm the integrity of the dicistronic RNA, Northern Blot experiments can be performed (*see* Table 1). After transfection of the dicistronic reporter gene plasmid, purified mRNA from the transfected cells is subjected to Northern Blot analysis. Using specific probes against *Renilla* and firefly luciferase sequences should result in both cases in a single band. Different band sizes for each probe might indicate a fragmented dicistronic RNA. Another possibility to control the integrity of the dicistronic RNA is a siRNA-mediated knockdown of one of the reporter genes. So, a siRNA knockdown of the *Renilla* luciferase should also result in a decrease of firefly activity.

Table 1
Troubleshooting guide

Problem	Recommendations
The luminescence signals from both luciferases are very low	Low signals for both luciferases indicate bad transfection efficiency. You might either increase the amount of dicistronic reporter gene plasmid for transfection or test another transfection reagent
The luminescence signals of the firefly luciferase are very low when it acts as first or second cistron	If you use the firefly luciferase assay buffer described in this protocol, carefully adjust the pH to 7.8. Lower pH values result in low light unit values. Additionally, you might also increase the amount of transfected reporter gene plasmid
The luminescence signals of the firefly luciferase are very low only when it acts as second cistron (first cistron light units are satisfying)	If the intercistronic sequence is an unknown IRES this might indicate a low IRES activity of the sequence itself. When a known IRES with strong activity is used, you should check the dicistronic reporter gene plasmid for linker sequences between the IRES 3′ end and the start codon of the firefly luciferase. Some IRES doesn't tolerate linker sequences. Therefore, cutting these out might increase the translation efficiency
In Northern Blot analysis using firefly luciferase probes, smaller fragments occur	Smaller fragments in the Northern Blot analysis show that not only the intact dicistronic RNA is transcribed from the reporter gene plasmid. We recommend excising all linker sequences from the intercistronic region of the plasmid

A second control experiment addresses the concern the sequence between the reporter genes might act as a cryptic promoter. To rule this out, a promoter deletion of the dicistronic reporter gene plasmid should lead to the complete loss of both luciferase activities.

3.1 Transfection of Reporter Gene Plasmids

1. Detach HeLa cells in culture flasks using trypsin/EDTA and suspend them in DMEM media containing 5 % FCS. For transfection in 6-well plates seed 400,000 HeLa cells per well in 3 ml media.
2. Transfect the cells 1 day after seeding. Before you start the transfection change media to DMEM containing 10 % FCS.
3. Dilute the $CaCl_2$ solution 1:10 with water to obtain a 250 mM solution.
4. For a single well (of a 6-well plate) dilute 3 μg of the reporter gene plasmid in 120 μl of 250 mM $CaCl_2$ solution in a tube. Carefully add 120 μl of 2× HEBS and mix. The transfection mixture should become turbid; incubate for 2 min at room temperature (*see* **Note 6**).
5. Shortly mix by flipping the tube and add the whole transfection mix (240 μl) drop wise to the cells.
6. After 24 h replace the media by fresh DMEM containing 10 % FCS and incubate the cells for another 24 h.

3.2 Cell Lysis

1. Dilute *Renilla* luciferase assay lysis 5× buffer 1:5 with water.
2. 2 days after transfection wash the cells with 3 ml of 1× PBS per well.
3. Remove all PBS from the cells and add 500 μl of 1× *Renilla* luciferase assay lysis buffer.
4. Scrape the cells using a rubber policeman and transfer the lysate in a 1.5 ml tube.
5. Incubate the lysate at room temperature for 15 min.
6. You may clear the lysate by centrifugation at 13,000 × *g* for 10 min at 4 °C. This step is not crucial; also the lysate containing cell debris will work in luciferase measurements.
7. Measure the lysates directly after lysis or store them at −80 °C.

3.3 Luciferase Measurement

1. Thaw the firefly luciferase assay buffer at room temperature (*see* **Note 7**).
2. Prepare the *Renilla* luciferase assay buffer by diluting 100× *Renilla* Luciferase Substrate 1:100 in *Renilla* assay buffer.
3. Thaw the cell lysates on ice.
4. Using the mentioned luciferase assay buffers measure the firefly and the Renilla luciferase activities separately (*see* **Note 8**).
5. For the measurement in a single sample luminometer, add 20 μl of cell lysate in a 5 ml polypropylene tube (e.g., from Sarstedt). If the luminometer has an injector for the luciferase assay buffers, program an injection volume of 100 μl. Otherwise add 100 μl of the luciferase assay buffers manually and immediately start the measurement of the luminescence. A common protocol for the luminescence measurement is a delay of 2 s after the injection and a measurement period of 10 s. Follow this protocol once for each luciferase assay buffer.
6. If you use a multi-well plate luminometer, start with the same volume of the lysates. Prepare the whole plate, add one of the luciferase assay buffers, and then measure the luminescence of all wells (*see* **Note 9**). Repeat the measurement by using another 20 μl of the lysate in a new plate, add the other luciferase assay buffer, and measure again.
7. In the luciferase measurement you obtain the light units generated by the firefly and the *Renilla* luciferase. To calculate the activity of the known or putative IRES between both reporter genes, refer the light units from the IRES-dependent translated reporter gene (often firefly luciferase) to the light units from the cap-dependent translated reporter gene (often *Renilla* luciferase). The ratio reflects the activity of the IRES.
8. By very low luciferase activity follow the instructions in Table 1.

4 Notes

1. Luciferase reporter gene vectors are commercially available from, e.g., Promega.
2. Next to calcium phosphate precipitation also other transfection protocols can be used like transfection with polyethylenimine (PEI) or Fugene (Roche).
3. We used HeLa cells which are easy to handle and show good transfection rates. Principally, all kinds of cell lines or primary cells can be used for transfection of reporter gene plasmids. The efficiency of the transfection is strongly cell dependent and needs to be optimized.
4. The firefly luciferase assay buffer is very stable; store it at −20 °C in aliquots. It can be thawed several times. In contrast, we only used freshly prepared *Renilla* luciferase assay buffer from Promega. Protect both buffers from light.
5. Next to luminometers which measure single samples, you might also use luminescence multi-well plate readers. In general single measurements are more sensitive than measurements in multi-well plates which offer a higher throughput of samples but lower sensitivity. Also the problem of cross talk between adjacent wells might occur. To reduce this effect you should use solid white multi-well plates.
6. Due to variations in transfection efficiency we recommend to transfect in triplicates. Therefore, prepare a 3.5-fold master mix using 420 μl 250 mM $CaCl_2$ solution, 10.5 μg reporter gene plasmid, and 420 μl 2× HEBS. Add 240 μl of this mix to each of the 3 wells.
7. The firefly luciferase assay buffer needs a long time to thaw, put it on room temperature several hours before you start the luciferase measurement.
8. You might use different injectors of the luminometer for the firefly and the *Renilla* luciferase assay buffer. When you use the same injector for both buffers, thoroughly flush the injection system with water after using the first buffer.
9. If the plate-reading luminometer has no injection system, add 100 μl of the luciferase assay buffers using a multichannel pipette. Add the assay buffer immediately before the measurement.

References

1. Hershey JWB, Merrick WC (2000) The pathway and mechanism of initiation of protein synthesis. In: Sonenberg N, Hershey JWB, Mathews MB (eds) Translational Control of Gene Expression. Cold Spring Harbor Laboratory Press, Cold Spring Harbor, NY, pp 42–52
2. Stoneley M, Willis AE (2004) Cellular internal ribosome entry segments: structures, trans-acting

factors and regulation of gene expression. Oncogene 23:3200–3207

3. Lopez-Lastra M, Rivas A, Barria MI (2005) Protein synthesis in eukaryotes: the growing biological relevance of cap-independent translation initiation. Biol Res 38:121–146
4. Bonnal S, Boutonnet C, Prado-Lourenco L, Vagner S (2003) IRESdb: the internal ribosome entry site database. Nucleic Acids Res 31: 427–428
5. Jopling CL, Spriggs KA, Mitchell SA, Stoneley M, Willis AE (2004) L-Myc protein synthesis is initiated by internal ribosome entry. RNA 10:287–298
6. Sachs AB (2000) Cell cycle-dependent translation initiation: IRES elements prevail. Cell 101:243–245
7. Wood KV (1998) The chemistry of bioluminescent reporter assays. Promega Notes 65:14–20
8. Thorne N, Inglese J, Auld DS (2010) Illuminating insights into firefly luciferase and other bioluminescent reporters used in chemical biology. Chem Biol 17:646–657
9. Van Eden ME, Byrd MP, Sherrill KW, Lloyd RE (2004) Demonstrating internal ribosome entry sites in eukaryotic mRNAs using stringent RNA test procedures. RNA 10:720–730

Chapter 16

Mapping of Protein Binding RNA Elements

Marc R. Reboll

Abstract

The RNA electrophoretic mobility shift assay is a very versatile method to study a broad spectrum of RNA–protein interactions. The technique is based on the separation of protein RNA mixtures through a native acrylamide gel. Compared to unbound RNA, RNA–protein complexes migrate slower through the gel resulting in a mass shift. The RNA EMSA can be used to identify unknown RNA–protein complexes, to map the RNA binding site of single proteins or determine the specificity of RNA–protein complexes using specific antibodies resulting in retarded migration through the gel.

Key words Electrophoretic mobility shift assay, RNA–protein interaction, Supershift, RNA binding site

1 Introduction

Cellular protein expression is regulated on different levels. The control of RNA translation and turnover are key mechanism in adjusting protein expression to altering cellular conditions like stimulation with inflammatory mediators and growth factors or cellular stress [1]. These regulatory mechanisms are mediated by RNA elements in the untranslated regions of mRNAs. The coding sequence of mRNAs is flanked by a 5′ and a 3′untranslated region (UTR). Regulatory elements in the 5′UTR are for instance open reading frames or RNA sequences resulting in stable secondary RNA structures. Both types of these elements decrease the efficiency of a cap dependent translation. Their impact on translation efficiency can be modulated by binding of regulatory proteins [2]. A specific type of regulatory elements in 5′UTRs is the internal ribosome entry sites known as IRES. They allow for a cap independent translation initiation and their activity is mediated by RNA binding proteins called IRES trans-acting factors (ITAFs) [3].

Most important regulatory RNA elements in the 3′UTR are the AU-rich elements (ARE). These elements control the stability and therefore the half-life of the RNA. A group of RNA binding

M. Lucrecia Alvarez and Mahtab Nourbakhsh (eds.), *RNA Mapping: Methods and Protocols*, Methods in Molecular Biology, vol. 1182, DOI 10.1007/978-1-4939-1062-5_16, © Springer Science+Business Media New York 2014

proteins, the ARE binding proteins (ARE-BP), interact with these motifs and are able to increase or decrease the RNA stability [4].

The biological function of the majority of regulatory RNA elements depends directly or indirectly on interaction with proteins. Therefore, it is crucial to identify and characterize the RNA–protein interactions to identify the underlying mechanisms. The RNA EMSA is a highly versatile method to study RNA–protein interactions. However, the initial step for this assay requires the production of sufficient number of labeled RNA of interest through specific transcription in vitro. Therefore, the RNA sequence of interest needs to be inserted downstream to a synthetic promoter in commercially available vectors which serve as DNA template for transcription machinery in vitro.

2 Materials

2.1 In Vitro Transcription (See Note 1)

1. A DNA template plasmid encoding the RNA sequence of interest linearized with a restriction enzyme and purified using phenol chloroform precipitation (*see* **Note 2**).
2. T7 RNA Polymerase and corresponding 5× reaction buffer.
3. 100 mM DTT solution.
4. Nucleotide-Mix containing 10 mM of each nucleotide except the nucleotide used for radioactive labeling, its concentration is reduced to 0.25 mM.
5. Radioactively labeled nucleotide [α-^{32}P] rNTP.
6. Water, RNase-free.
7. RNase Inhibitor (e.g., RNasin from Promega).
8. DNase (e.g., RQ1 DNase from Promega).
9. NucTrap Probe Purification Columns and STE Buffer (100 mM NaCl, 20 mM Tris/HCl pH 7.5, 10 mM EDTA) (Stratagene Agilent) or Qiagen RNA columns to separate transcribed and labeled RNA from nucleotides and truncated transcription products.
10. TBE (Tris Borate EDTA) Buffer, 10× (900 mM Tris, 900 mM Boric acid, 400 mM EDTA, pH 8.0).
11. Acrylamide/Bisacrylamide solution (37.5:1) 30 % (Roth).
12. Urea (Sigma).
13. 10 % APS (Sigma) solution in water.
14. *N*,*N*,*N*′,*N*′-tetramethyl-ethylenediamine (TEMED) (Roth).
15. Formamide loading buffer (80 % formamide, 1 mg/ml Bromophenol Blue, 9 mM Tris, 9 mM boric acid, 40 mM EDTA, pH 8.0).
16. X-ray films (Kodak).

2.2 RNA EMSA

1. 5× Shift-Buffer (100 mM HEPES pH 7.9, 500 mM KCl, 5 mM $CaCl_2$, 10 mM $MgCl_2$, 10 % (w/v) Ficoll), freshly add PMSF (2.5 mM), DTT (2.5 mM), Pepstatin A (2.5 μg/ml), Leupeptin (2.5 μg/ml) (*see* **Note 3**).
2. t-RNA (2 μg/μl).
3. Protein extracts (Cytosolic or nuclear) or purified protein.
4. Labeled RNA from in vitro transcription.
5. 5× Shift-Loading Buffer (20 % Glycerol, 0.1 % (w/v) Bromophenol Blue).
6. TBE (Tris Borate EDTA) Buffer, 10× (900 mM Tris, 900 mM Boric acid, 400 mM EDTA, pH 8.0).
7. Acrylamide/Bisacrylamide solution (37.5:1) 30 % (Roth).
8. 10 % ammonium persulfate (APS) (Sigma) Solution in water.
9. *N*,*N*,*N*′,*N*′-tetramethyl-ethylenediamine (TEMED) (Roth).
10. X-ray films (Kodak).

3 Methods

The RNA EMSA is a powerful in vitro method to identify and characterize RNA–protein interactions [5]. For an RNA EMSA a crude cell lysate, a purified fraction, or single protein is incubated with RNA. Potentially forming RNA–protein complexes are then separated on a native polyacrylamide gel. To detect these RNA–protein complexes, the RNA needs to be labeled. The most efficient way is to use ^{32}P radioactively labeled ribonucleotides by transcription in vitro. Using a corresponding DNA template plasmid the RNA is transcribed in vitro and one of the four nucleotides (ATP, UTP, CTP, GTP) is exchanged by a ^{32}P labeled nucleotide. RNA EMSA is a very sensitive method using small protein and RNA amounts. It is a very variable system and allows many different experimental settings. For instance, new RNA–protein interactions can be revealed using protein lysates. A specific protein in a complex can be detected in a supershift assay. Therefore, a specific antibody is added to the protein lysate incubating with the RNA. During the gel run, the RNA–protein complex with the antibody migrates slower through the gel than the complex without the antibody. This results in an additional mass shift which is called supershift [6].

Another application for the RNA EMSA technique is the characterization of specific RNA binding sites. By using wild type and mutated or partly deleted RNA sequences together with a single protein the RNA binding site for this protein can be easily mapped.

Moreover, the specificity of an RNA–protein interaction can be shown by using unlabeled specific and unspecific competitor RNAs. Increasing amounts of unlabeled specific RNA should reduce the binding affinity of the RNA–protein complex, while an unspecific RNA should not affect the binding behavior [7].

Next to these experimental possibilities offered by the RNA EMSA, there are also limitations. In our hands we found that RNAs larger than 600 nt are difficult to use. So large RNAs might be dissected in smaller fragments and analyzed separately. The formation of an RNA–protein complex is very sensitive to salt concentrations in the reaction buffers. This point must be carefully addressed when establishing an RNA EMSA.

Due to the native running conditions in RNA EMSA the retarded RNA–protein complex provides you no information about the size of the protein(s) bound to the RNA. Here, the proteins can be cross-linked to the RNA and a denaturing SDS polyacrylamide gel is used instead [8]. This technique offers also the possibility to blot the EMSA gel. On the other hand, this approach gives you no information about the binding specificity and the binding strength of the RNA–protein complex since the protein is covalently linked to the RNA.

3.1 In Vitro Transcription

Generally consider the common guidelines for working with RNA. Work with autoclaved tips and tubes, wear gloves, and prepare your RNA working place with RNase inhibiting agents. Carry out all procedures on ice. Carefully obey the rules for working with radioactive substances (here especially working with beta radiators). In this example reaction, rUTP was used as radioactive labeled nucleotide.

1. For an in vitro transcription reaction mix: 4 μl 5× T7 RNA polymerase reaction buffer, 2 μl 100 mM DTT solution, 1 μg of linearized DNA template, 3 μl nucleotide mix (10 mM each rATP, rGTP, and rCTP), 1 μl rUTP (not labeled, 0.25 mM), 50 μCi [α-^{32}P] rUTP, 1 μl RNase Inhibitor (40 U), 1 μl T7 RNA polymerase (20 U), add water up to a final volume of 20 μl.
2. Incubate for 1 h at 37 °C.
3. Add 1 U RQ1 DNase to degrade the DNA template and incubate for another 15 min at 37 °C.
4. To separate non-incorporated nucleotides and truncated fragments from the complete labeled RNA, use NucTrap Probe Purification Columns (Stratagene) or Qiagen RNA columns. Using NucTrap Probe Purification Columns, first add 50 μl of STE Buffer to the transcription reactions. Wash the columns three times with 70 μl of STE Buffer, then add the transcription reaction (70 μl) on the column and use a syringe to press it through the column. Discard the flow through. Elute the RNA by adding 70 μl STE Buffer on the column, press it through the column, and keep the eluate. Repeat this step two times to a final number of three elutions.

In case of using Qiagen RNA columns, fill up the transcription reaction to a volume of 100 µl with water and add the same volume of 70 % ethanol. Follow the manufactures instructions and elute the RNA with 30 µl of water three times.

5. Determine the radioactive activity of the labeled RNA using a scintillation counter giving you activity values in counts per minute (cpm).
6. Store the RNA in a suitable, shielded jar at −20 or −80 °C.
7. To control the quality of the labeling reaction and the purification step, you may perform a denaturing 8 M urea gel. For a 40 ml gel solution solve 19.2 g urea in 4 ml of 10× TBE Buffer and 29.4 ml water 8 (*see* **Note 4**). Subsequently add 6.6 ml of 30 % Acrylamide/Bisacrylamide solution (37.5:1), 320 µl 10 % APS and 42 µl TEMED, gently mix the gel solution, and cast the gel avoiding air bubbles. Place the comb and let the gel polymerize.
8. Load 10,000 cpm of your labeled RNA per lane. Fill up with water to a final volume of 5 µl.

   Add the same volume of formamide loading buffer to the RNA, incubate samples 5 min at 95 °C, shortly spin them down and load them on the urea gel. After a pre-run for 1 h at 150 V, run the gel at 100 V for 2 h after loading the samples. Depending on the size of the labeled RNA, time and voltage have to be adjusted.
9. Uncast the gel after the run by taking away one of the glass plates and put a piece of Whatman paper (3 mm) on the gel. The gel will stick to this paper, carefully free it from the second glass plate. Place a piece of plastic foil on the gel and dry it at 80 °C for 1 h on a vacuum dryer.
10. Expose an X-ray film on the dried gel, incubate it for 16–20 h (*see* **Note 5**), and develop the film (*see* Table 1).

3.2 RNA EMSA

1. Prepare a native 6 % polyacrylamide gel in 0.5× TBE. For a proper separation of RNA–protein complexes from unbound RNA we recommend a gel length of 20 cm. For a 40 ml gel solution mix 8 ml of 30 % Acrylamide/Bisacrylamide solution (37.5:1) with 2 ml of 10× TBE Buffer and 30 ml deionized water. Add 266 µl 10 % APS and 66 µl TEMED, gently mix the gel solution, and cast the gel avoiding air bubbles. Place the comb and let the gel polymerize (*see* **Note 6**).
2. Prepare the different samples using: 4 µl 5× Shift Buffer, 2 µl t-RNA (2 µg/µl), 25 µg of protein extract or 1 µg of purified protein, 500,000 cpm labeled RNA and add water up to a final volume of 20 µl (*see* **Notes** 7 and **8**).
3. Incubate all samples for 10 min at 30 °C, then add 30 U RNase T1 (*see* **Note 9**) and incubate another 30 min at 37 °C.

Table 1
Troubleshooting guide

Problem	Recommendations
No radioactively labeled RNA can be detected on the denaturing control gel	Make sure that the in vitro transcription kit fits to the vector harboring the template sequence. For the T7 RNA polymerase you need an intact T7 promoter upstream of your sequence of interest
The denaturing control gel does not show clear bands, just smears	The smears on the gel indicate degraded RNA. To a certain extent you will find smears next to the full length RNA. If there is only a continuous smear, make sure that your template is intact. You might also increase the amount of RNase inhibitor in your reaction and/or check all solutions you work with for an RNase contamination (exchange them one by one)
No complexes form in the RNA EMSA experiment	If the in vitro transcribed RNA is decent, the ion strength of the shift reaction might be too high. Reduce the salt concentrations (KCl) in the shift buffer
The addition of a supershift antibody weakens the signal of the RNA–protein complex	The antibody seems to bind the part of the protein needed for the RNA binding. If possible, try an antibody directed to alternative parts of the protein
In the RNA EMSA the whole lane is a black smear	The T1 RNase treatment seems to fail. Increase the amount of enzyme, extend the incubation time, or exchange the enzyme
Complexes are formed in RNA EMSA but the bands are smeared	RNA–protein bands in these EMSA experiments are often less clear and distinct as bands known from DNA EMSA. Nevertheless, strong smears might indicate a separation of a formed RNA–protein complex. Reduce gel running times and/or salt concentrations in the shift buffer
A high amount of protein in the RNA EMSA results in a weaker complex than a low amount of protein	This might be possible if the protein used for the EMSA is dissolved in a buffer containing high salt concentrations. You might dialyze the protein to exchange the buffer or lower the salt concentrations
The complexes in RNA EMSA are stuck in the wells of the gel	The protein might be aggregated. Aggregation of the protein can be prevented by adding detergents like Tween or Triton-X to the protein storage buffer

4. Spin down all samples and add 5 μl of 5× shift loading buffer.
5. Prepare the gel for the run, use 0.5× TBE as running buffer. Pre-run the gel without samples for 2 h at 100 V. Load the samples on the acrylamide gel and run the gel for 4 h at 120 V. Depending on the size of the RNA–protein complexes, time and voltage have to be adjusted.
6. Uncast the gel after the run by taking away one of the glass plates and put a piece of Whatman paper (3 mm) on the gel. The gel will stick to this paper, carefully free it from the second glass plate. Place a piece of plastic foil on the gel and dry it at 80 °C for 1 h on a vacuum dryer.

7. Expose an X-ray film on the dried gel, incubate it for 16–20 h (*see* **Note 5**) and develop the film. Follow the instructions in Table 1 to optimize the results.

4 Notes

1. Kits, which include all components for in vitro transcription reactions (except the radioactive nucleotide), are commercially available, e.g., from Promega #P1440.
2. We found that RNAs up to 600 nt worked well in the RNA EMSA experiments described here. For linearizing the DNA plasmid we used 30 μg DNA and incubated over night at 37 °C with 40 U of a single cutter restriction enzyme.
3. The salt concentrations of the 5× shift buffer might be critical for the formation of RNA protein complexes. If no RNA protein complexes form you might decrease the KCl concentration down to 50 mM.
4. To properly solve the urea constantly stir the solution and warm up at about 40 °C.
5. The signal can be increased by using exposure cassettes with intensifying screen set and incubating them at −80 °C for the exposure time. Allow the cassette to come to room temperature before you develop the film.
6. The gel can also be casted the day before the experiment. Wrap it in wet paper towels and store it at 4 °C.
7. Always use one sample without protein as control. If you use a purified, recombinant protein with a tag, consider adding the tag alone as control.
8. The binding of a single protein can be detected by using an antibody and perform a supershift. Therefore, the mass of the RNA–protein complex is increased by the antibody binding the protein of interest. For a supershift, add the antibody (the amount needs to be determined for every assay) 15 min after adding RNAse T1 and incubate 20 more minutes at 37 °C. For a polyclonal antibody start with 1 μl per reaction. A serum without specific antibody is an important control.
9. The RNase T1 digests all RNA except the fragments which are protected by bound proteins.

References

1. Holcik M, Sonenberg N (2005) Translational control in stress and apoptosis. Nat Rev Mol Cell Biol 6:318–327
2. Mignone F, Gissi C, Liuni S, Pesole G (2002) Untranslated regions of mRNAs. Genome Biol 3:Reviews0004, Reviews0004.10

3. Stoneley M, Willis AE (2004) Cellular internal ribosome entry segments: structures, trans-acting factors and regulation of gene expression. Oncogene 23:3200–3207
4. Barreau C, Paillard L, Osborne HB (2005) AU-rich elements and associated factors: are there unifying principles? Nucleic Acids Res 33: 7138–7150
5. Hellman LM, Fried MG (2007) Electrophoretic mobility shift assay (EMSA) for detecting protein-nucleic acid interactions. Nat Protoc 2(8):1849–1861
6. Song CS, Echchgadda I, Seo Y, Oh T, Kim S, Kim S, Cho S, Shi L, Chatterjee B (2006) An essential role of the CAAT/enhancer binding protein-α in the vitamin D-induced expression of the human steroid/bile acid-sulfotransferase (SULT2A1). Mol Endocrinol 20:795–808
7. Reboll MR, Oumard A, Gazdag AC, Renger I, Ritter B, Schwarzer M, Hauser H, Wood M, Yamada M, Resch K, Nourbakhsh M (2007) NRF IRES activity is mediated by RNA binding protein JKTBP1 and a 14-nt RNA element. RNA 13:1328–1340
8. Hamada T, Kurachi S, Kurachi K (2010) Heterogeneous nuclear ribonucleoprotein A3 is the liver nuclear protein binding to age related increase element RNA of the factor IX gene. PLoS One 5:1–12

Chapter 17

Purification of RNA-Binding Proteins

Birgit Ritter and Marc R. Reboll

Abstract

The majority of regulatory RNA sequences exert their function through interaction with proteins. Therefore, the identification of RNA-binding proteins is the key step in understanding the role of many RNA motifs. Here, we describe a straightforward method to identify RNA-binding proteins. In this approach, RNAs are immobilized on beads and incubated with protein lysates. After removing unbound fraction of proteins, the bound proteins are eluted by successive increasing of salt concentration. The lower the salt concentration, the lower is the binding affinity of a protein to the RNA. According to this principle, each elution fraction can contain a number of proteins which bind with similar affinity to the RNA. After gel electrophoretic separation and staining, each single protein band can be identified, isolated, and analyzed by mass spectrometry.

Key words RNA–protein interaction, RNA-binding proteins, Regulatory RNA sequences, RNA immobilization

1 Introduction

Interactions of proteins with nucleic acids—DNA and RNA—are fundamental for many regulatory processes in gene expression. Proteins accompany RNAs throughout their whole cellular life. After transcription, pre-mRNAs evolve into mRNAs by capping, splicing, and polyadenylation. All these steps are mediated by direct interactions with regulatory proteins [1]. Proteins also regulate the export of mature mRNAs from the nucleus into the cytoplasm by binding to regulatory sequences in 5′ or 3′ untranslated regions (UTRs) [2]. In addition, the translation of mRNA to protein requires the binding of initiation factors and ribosomal proteins to the cap structure or 5′ UTR as known by cap-independent translation [3]. Therefore, identification of RNA-interacting proteins is a crucial step in understanding the molecular mechanisms underlying the posttranscriptional gene regulation.

In this chapter, we describe a rapid and straightforward strategy for purification of proteins binding to a specific RNA sequence from crude cell extracts. Therefore, the RNA sequence of interest

M. Lucrecia Alvarez and Mahtab Nourbakhsh (eds.), *RNA Mapping: Methods and Protocols*, Methods in Molecular Biology, vol. 1182, DOI 10.1007/978-1-4939-1062-5_17, © Springer Science+Business Media New York 2014

Table 1
Troubleshooting guide

Problem	Recommendations
No RNA can be detected on the control agarose gel	Make sure that the in vitro transcription kit fits to the vector harboring the template sequence. For the T7 RNA polymerase you need an intact T7 promoter upstream of your sequence of interest
The agarose gel does not show clear bands, just smears	The smears on the gel indicate degraded RNA. Make sure that your template is intact by checking the size of your linearized plasmid on an agarose gel. You might also increase the amount of RNase inhibitor in your reaction and/or check all solutions you work with for an RNase contamination (exchange them one by one)
The RNA coupling to the oligo(dT) beads always fails	Maybe the RNA forms stable secondary structures and the poly(A) sequence needed for coupling is not accessible anymore. You might extend the time and increase the temperature at which the RNA is heated before it is added to the beads. If this does not help, you might insert additional bases in the template plasmid between your sequence of interest and the poly(A) sequence serving as a linker
No protein bands in the elutions loaded on the SDS gel can be detected	Control the flow through and the wash fractions. If a lot of proteins are visible there, you might decrease the stringency of your washing steps. You might also need to decrease the salt concentrations in the binding buffer. The immobilization of the RNA to the beads maybe did not work as well. Take an aliquot of the beads after coupling the RNA. Add the same volume of 2× RNA loading buffer, heat, and load on a denaturing agarose gel. If no RNA is detectable, try new beads or reduce salt concentrations in the binding buffer
Same protein bands occur in the reactions using specific RNA and control RNA	This might indicate that specific proteins still bind firmly to the RNA. Increase the salt concentrations in the elution buffer. Check the proteins eluted after treatment with 10 mM Tris–HCl pH 7.5 as well
Protein bands are very weak after Coomassie staining	You might increase the amount of protein lysate if possible. Alternatively, you might use a sliver staining instead of Coomassie. It is more sensitive but needs additional working steps during mass spectrometry analysis

needs to be transcribed in vitro and linked to magnetic beads. While various linking strategies can be used, the most convenient strategy involves commercially available magnetic beads with poly-(T) linkers initially designed for isolation of poly(A) RNAs. Accordingly, the target RNA sequence of interest requires a poly(A) tail for hybridization to poly-(T) beads. This is simply achieved by insertion of a poly-(T) track in the DNA template. Thus, the final transcript consists of target sequence and a poly(A) tail. Note that a possible disadvantage of this strategy might be co-purification of poly(A)-binding proteins. Therefore, the length of poly(A) tail should correspond to the length of poly(A) tail linked to the beads. Under certain circumstances, the use of control experiments or pre-purification steps using inverted or mutant target sequences is recommended (Table 1).

2 Materials

2.1 In Vitro Transcription (See Note 1)

1. DNA template, linearized with a single cutter restriction enzyme (*see* **Note 2**); this template has to contain an oligo dT sequence for coupling of the in vitro-transcribed RNA (containing a corresponding poly(A) sequence) to oligo dT beads.
2. T7 RNA polymerase and corresponding 5× reaction buffer.
3. 100 mM DTT solution.
4. Nucleotide mix containing 10 mM of each nucleotide.
5. RNase inhibitor (e.g., RNasin from Promega).
6. RNase-free water.
7. DNase (e.g., RQ1 DNase from Promega).
8. Agarose (Fermentas).
9. Formaldehyde (Merck).
10. 10×3-(N-morpholino) propanesulfonic acid (MOPS) (200 mM MOPS, 80 mM sodium acetate, 10 mM EDTA pH = 8.0).
11. 2× RNA-loading buffer (40 % (v/v) formaldehyde, 40 μg/ml ethidium bromide, 120 mM MOPS, 50 mM sodium acetate, 6 mM EDTA pH 8.0, 15 % (v/v) glycerol, 0.15 % (w/v) bromophenol blue).
12. RNA ladder (Fermentas).

2.2 Protein Purification

1. RNA from in vitro transcription.
2. Protein lysates (*see* **Note 3**).
3. Dynabeads Oligo (dT) (DynalBiotech).
4. Magnetic stand (DynalBiotech).
5. 2× Binding buffer (40 mM HEPES pH 7.5, 300 mM KCl).
6. Wash buffer (20 mM HEPES pH 7.5, 150 mM KCl, 2 mM $MGCl_2$, 0.5 mM DTT, 0.5 mM PMSF, 5 % glycerol, 0.1 % NP-40).
7. Elution buffers with increasing concentrations of KCl (0.5 M KCl, 1 M KCl, 2 M KCl).
8. 4× SDS Loading buffer (Roth).

2.3 SDS Gel Electrophoresis and Coomassie Staining

1. Acrylamide/bisacrylamide (37.5:1) solution 30 %.
2. 1.5 M Tris–HCl pH 8.6.
3. 0.5 M Tris–HCl pH 6.8.
4. 50 % Glycerol solution in water.
5. 10 % SDS in water.
6. 10 % Ammonium persulfate (APS) (Sigma) solution in water.
7. *N,N,N′,N′*-tetramethyl-ethylenediamine (TEMED) (Roth).

8. Isobutanol.
9. 1× SDS running buffer.
10. Optional: Coomassie brilliant blue solution (0.1 % (w/v) Coomassie brilliant blue powder, 4 % methanol, 8 % acetic acid).
11. Destaining solution (10 % acetic acid).

3 Methods

The method described here is a simple and straightforward approach to identify RNA-binding proteins for a known regulatory RNA sequence. Therefore, the RNA is used as a bait to catch interacting proteins. The assay is based on the immobilization of the RNA on a solid phase. Here, we used oligo(dT) beads for immobilization. The RNA is synthesized in an in vitro transcription reaction. A prerequisite for using oligo(dT) beads is a poly(A) sequence in the RNA (at least 25 bases). This sequence can be generated by introducing a poly-(T) sequence in the vector used for in vitro transcription (e.g., by site-directed mutagenesis). There are other techniques to immobilize the RNA on beads. For example, the RNA can be coupled to an RNA aptamer called StreptoTag which properly binds to streptavidin beads [6, 7]. The immobilized RNA is then incubated with protein lysates followed by several wash steps. Only proteins which specifically interact with the RNA sequence remain bound at the RNA. Elutions with buffers containing increasing salt concentrations gradually release the proteins from the RNA. Nevertheless, proteins eluted with high salt concentrations might be unspecific binders. To reveal and exclude those proteins a control sequence should be used in parallel. One possibility to generate a control sequence is to simply invert the specific RNA sequence. After eluting the proteins from the beads, the different elutions can then be analyzed on an SDS acrylamide gel. The proteins can be visualized using a Coomassie brilliant blue staining and identified by excising the bands from the gel and analyzing them with mass spectrometry. The protocol can be modified by using UV cross-linking to covalently link the proteins to the RNA [8]. This procedure allows highly stringent wash steps but disguises the binding strength of the identified proteins.

3.1 *In Vitro Transcription*

Carry out all procedures on ice. Generally consider some common guidelines for working with RNA. Work with autoclaved tips and tubes, wear gloves, and prepare your RNA working place with RNase-inhibiting agents like "RNase away" (Roth).

1. For an in vitro transcription reaction mix:
   4 μl 5× T7 RNA polymerase reaction buffer, 2 μl 100 mM DTT solution, 1 μg of linearized DNA template, 4 μl nucleotide mix (10 mM each NTP), 1 μl RNase inhibitor (40 U), 1 μl

T7 RNA polymerase (20 U), add water up to a final volume of 20 µl.

2. Incubate for 1 h at 37 °C.
3. Add 1 U RQ1 DNase to degrade the DNA template, and incubate for another 15 min at 37 °C.
4. If necessary, store the RNA at −80 °C.
5. Cast a 1 % denaturing agarose gel (1× MOPS, 6.7 % formaldehyde). For a 100 ml gel dissolve 1 g agarose in 72 ml of water by cooking. Subsequently add 10 ml 10× MOPS buffer and 18 ml formaldehyde. Place a comb in the gel solution, and let the gel polymerize. Use 1× MOPS as running buffer.
6. Take an aliquot of the in vitro transcription reaction and add the same volume of 2× RNA loading buffer. Incubate the sample at 90 °C for 5 min, shortly spin down, and incubate on ice until gel loading.
7. Load the sample and a size standard on the gel, and run the gel at 55 V for 1 h. Detect the bands with UV light (gel documentation).

3.2 Protein Purification

Perform all steps on ice. Avoid vortexing or pipetting the oligo(dT) beads, and mix by snapping the tubes. Here, we used magnetic Dynabeads Oligo (dT) together with a magnetic stand. This system offers the possibility to work without centrifugation steps. Placing the tubes with the beads on the magnetic stand will collect all beads in a small area at the wall of the tube and makes it easy to take off the supernatants.

1. Transfer 100 µl Dynabeads in a tube, and equilibrate the beads by washing with 1× binding buffer twice. After putting the tubes on the magnetic stand, allow the beads to assemble at the back of the tube by magnetic force, and then take off the liquid. Take the tubes from the magnetic stand, add fresh buffer, and mix by snapping the tube.
2. Add RNase-free water to the in vitro-transcribed RNA up to a volume of 50 µl, heat the RNA for 2 min at 65 °C, and put on ice again. Add 50 µl of 2× binding buffer.
3. Transfer the RNA solution to the washed oligo(dT) beads, and incubate at room temperature for 5 min with gentle shaking.
4. After binding the RNA to the oligo(dT) beads, take off the binding buffer and add protein lysate in 1× binding buffer (*see* **Note 4**).
5. Incubate the protein lysate with the beads for 15 min at 30 °C with gentle shaking.
6. Wash the beads with wash buffer three times, and keep the wash fractions to analyze them on a polyacrylamide gel.

7. Elute the proteins bound to the RNA with 50 µl (*see* **Note 5**) elution buffers containing increasing concentrations of KCl (0.5 M, 1 M, 2 M). A final elution can be performed using 10 mM Tris–HCl pH 7.5 (*see* **Note 6**).
8. Prepare all wash steps and elutions for analyzing on a 10 % SDS polyacrylamide gel by adding 4× SDS loading buffer, and heat them for 5 min at 95 °C. Shortly spin them down, and load the samples on the gel.

3.3 SDS Gel Electrophoresis and Coomassie Staining

To properly separate the proteins eluted from the RNA on the gel, we recommend using a 10–15 cm running gel.

1. For 20 ml of a 10 % running gel add 6.6 ml Acryl/Bisacrylamide solution (30 %), 6.15 ml water, 5 ml 1.5 M Tris–HCl pH 8.6, 1.8 ml 50 % glycerol, and 200 µl 10 % SDS solution. Gently mix, and then add 33 µl TEMED and 133 µl 10 % APS. Cast the running gel, and overlay the gel with isobutanol to get a sharp separation of running and stacking gel.
2. After polymerization of the running gel (about 30 min) discard the isobutanol and prepare the stacking gel solution. For 6 ml of stacking gel solution add 0.9 ml Acryl/Bisacrylamide solution (30 %), 3.54 ml water, 1.5 ml 0.5 M Tris–HCl pH 6.8, and 60 µl 10 % SDS. Mix gently, and finally add 24 µl TEMED and 48 µl 10 % APS. Cast the stacking gel, and immediately place the comb in the gel. Avoid air bubbles in the pockets. The gel needs about 15 min for polymerization.
3. Place the gel in the running chamber, add 1× SDS running buffer, and pull the comb out of the stacking gel. You might flush the pockets with 1× SDS running buffer with a syringe.
4. Load protein marker and samples on the gel (*see* **Note 7**), and start the gel run with 20 mA per gel. After reaching the running gel, the amperage can be increased to 30 mA per gel.
5. After the gel run, you can follow several procedures to identify the purified proteins. If the protein is known, you can perform western blotting and detect the protein on a membrane using specific antibodies. To identify unknown binders of a certain RNA sequence we recommend staining the gel with Coomassie brilliant blue (*see* **Note 8**) and excise the bands for mass spectrometry analysis (*see* **Note 9**).

4 Notes

1. Kits, which include all components for in vitro transcription reactions, are commercially available, e.g., from Promega # P1440.
2. For linearizing the DNA plasmid we used 30 µg DNA and incubated overnight at 37 °C with 40 U of a single cutter restriction enzyme.

3. You might use nuclear, cytosolic, or whole-cell lysates in common lysis buffers like RIPA buffer.
4. Mix equal volumes of protein lysates and 2× binding buffer to obtain protein lysates in 1× binding buffer.
5. The volume of the elutions can be varied depending on the loading volume of the gel pockets.
6. Using 10 mM Tris–HCl pH 7.5 for elution, the RNA is eluted from the oligo(dT) beads together with proteins binding very tightly to the RNA.
7. Load the different wash steps and all elutions on the gel; they can give you important information for troubleshooting. The last wash step should just contain a low amount of protein. If not, you might increase the number of wash steps and/or the concentration of KCl in the wash buffer.
8. For Coomassie staining incubate the gel with Coomassie brilliant blue solution for 1 h and destain the gel with destaining solution for 2 h.
9. Mass spectrometry analysis is offered in many research institutes as in-house service as well as by companies (e.g., Eurogentec).

References

1. Jurica MS, Moore MJ (2003) Pre-mRNA splicing: awash in a sea of proteins. Mol Cell 12:5–14
2. Shyu AB, Wilkinson MF (2000) The double lives of shuttling mRNA binding proteins. Cell 102:135–138
3. Hershey JWB, Merrick WC (2000) The pathway and mechanism of initiation of protein synthesis. In: Sonenberg N, Hershey JWB, Mathews MB (eds) Translational Control of Gene Expression. Cold Spring Harbor Laboratory Press, Cold Spring Harbor, NY, pp 42–52
4. Stoneley M, Willis AE (2004) Cellular internal ribosome entry segments: structures, trans-acting factors and regulation of gene expression. Oncogene 23:3200–3207
5. Lal A, Mazan-Mamczarz K, Kawai T et al (2004) Concurrent versus individual binding of HuR and AUF1 to common labile target mRNAs. EMBO J 23:3092–3102
6. Windbichler N, Schroeder R (2006) Isolation of specific RNA-binding proteins using the streptomycin-binding RNA aptamer. Nat Protoc 1:638–641
7. Iioka H, Loiselle D, Haystead TA et al (2011) Efficient detection of RNA–protein interactions using tethered RNAs. Nucleic Acids Res 39:1–7
8. Baltz AG, Munschauer M, Schwanhausser B et al (2012) The mRNA-bound proteome and its global occupancy profile on protein-coding transcripts. Mol Cell 46:674–690

Chapter 18

De Novo Approach to Classify Protein-Coding and Noncoding Transcripts Based on Sequence Composition

Haitao Luo, Dechao Bu, Liang Sun, Runsheng Chen, and Yi Zhao

Abstract

Each day, more and more transcripts are being discovered along the genome (especially in poorly annotated species) thanks to the rapid progress of high-throughput technology such as RNA sequencing. However, this situation unravels the challenge of how to classify the newly identified transcripts into protein coding or noncoding. Here, we describe a de novo approach named coding–noncoding index (CNCI), a powerful signature tool by profiling adjoining nucleotide triplets (ANT) to effectively distinguish between protein-coding and noncoding sequences independently of known annotations. The main advantage of CNCI is its ability to accurately classify transcripts assembled from whole-transcriptome sequencing data in a cross-species manner, which allowed it to be used for all vertebrates and invertebrates based on the training data of well-annotated species (such as human and *Arabidopsis*). In this chapter, we illustrate the CNCI method in detail through an example of RNA-sequencing data generated from six biological replicates of six mouse tissues. CNCI software is available at http://www.bioinfo.org/software/cnci.

Key words Long noncoding RNA, Coding–noncoding index, Classification, Identification, RNA sequencing

1 Introduction

Long noncoding RNAs (lncRNAs), a kind of noncoding RNAs with a length longer than 200 nucleotides, have been recently paid great attention for their role in the regulation of many biological processes [1, 2]. Last year, the ENCODE project published 30 papers (including some about extensive characterization of lncRNAs), suggesting that more than 70 % of human genome is transcribed into RNA molecules but not translated into proteins [3]. The rapid progress of high-throughput technology such as RNA sequencing (RNA-Seq) allowed scientists to analyze transcriptome comprehensively and with an unprecedented resolution, providing a way to study lncRNAs across the whole genomes [4]. Although almost all expressed transcripts could be identified based

M. Lucrecia Alvarez and Mahtab Nourbakhsh (eds.), *RNA Mapping: Methods and Protocols*, Methods in Molecular Biology, vol. 1182, DOI 10.1007/978-1-4939-1062-5_18, © Springer Science+Business Media New York 2014

on RNA-Seq, it is still a challenge to classify these transcripts into protein-coding and noncoding transcripts.

Most of the current computational methods for protein-coding and noncoding transcript classification rely on the information of open reading frame (ORF), known protein database, evolutionary signatures, or other sequence search approaches [5, 6]. Given the fact that some reconstructed transcripts from RNA-Seq experiments may be incomplete generally by losing their 3′ or 5′ exons due to sequencing bias or computational methods, it is not surprising that many transcripts have incomplete ORF. Additionally, the use of conservation status to evaluate coding potential is also problematic because most functional noncoding RNAs have higher evolutionary conservation relative to introns. Thus, certain conserved element exists in these noncoding sequences and, in contrast, many newly evolved proteins do not contain a conserved ORF [7, 8]. Currently, only some species are well annotated and have whole-genome sequence information, but it is hard to define the transcript status using existing tools for the large portion of poorly annotated species.

Recently, we developed the coding–noncoding index (CNCI) method, a de novo approach that can effectively distinguish between protein-coding and noncoding sequences in a given transcript set constructed from RNA-Seq data, independently of known annotations [9]. In this chapter, we describe the theory and use of CNCI software tool.

2 Methods

2.1 Training Set Preparation (See Note 1)

To obtain reliable lncRNA and protein-coding genes as a training set (a knowledge dataset used to fit our prediction model), we used an lncRNA catalog from GENCODE database [10], which annotated all evidence-based gene features in the entire human genome at a high accuracy and human protein-coding genes from RefSeq database [11]. Our previous study has showed that CNCI can classify protein-coding and noncoding genes in a cross-species manner due to the ubiquitous sequence feature that we used across all vertebrates. That means the human training set could be used to classify protein-coding and noncoding sequences of all species in vertebrates [9]. Accordingly, the sequence information of human lncRNA and protein-coding transcripts was prepared to train the support vector machine (SVM) model, which is a supervised learning model that analyzes data as well as recognizes patterns used for classification or regression analysis.

2.2 Training Model Evaluation

CNCI evaluates the sequence coding potential of the training genes based on five features, length, sequence score, length percentage, score distance, and codon bias, and assigns a CNCI for each gene. Specifically, to extract the five features that are incorporated into the SVM to train the classification model, CNCI uses first the coding domain sequence (CDS) of human protein-coding genes as a positive training set and the whole sequence of human lncRNA genes as a negative training set. Next, for a given sequence, the most likely coding domain sequence (MLCDS) of each of the six reading frames is identified according to the sequence score, and the five features of MLCDS are extracted and incorporated into SVM to perform its classification. The results of tenfold cross-validation of a training set demonstrate the high accuracy of CNCI (97.3 %). To run CNCI, users should first download the CNCI software from the website (http://www.bioinfo.org/software/cnci/) and use the following command in linux operating system: "./CNCI -f ./example.fa -l ./libsvm-3.0 -o result". The parameter "-f" is a training set with Fasta format, "-l" is the path of SVM you have installed, and "-o" is the output file.

2.3 CNCI Classification of a Testing Transcript Set Constructed from RNA-Seq Data

The RNA-Seq data used as test dataset was downloaded from the Mouse Genomes Project at the Wellcome Trust Sanger Institute (http://www.sanger.ac.uk/resources/mouse/genomes/) under the accession number ERP000591. The RNA-Seq data contains six biological replicates of six mouse tissues: heart, hippocampus, liver, lung, spleen, and thymus. For each RNA-Seq dataset, all sequenced reads were mapped to the mouse genome (genome assembly version mm9) and the matched reads were assembled into transcripts by utilizing Scripture [12] and Cufflinks [13] tools. Transcripts reconstructed by the two assemblers were separately merged into combined sets of transcripts using the Cuffcompare utility provided by Cufflinks. After filtering the transcript based on the exon number and transcript length, a total of 95,519 reliably expressed multi-exon transcripts longer than 200 nucleotides are obtained as a testing transcript set. After comparison of these 95,519 transcripts with a known gene annotation database, they are categorized into four types: annotated protein-coding genes (14,683), known noncoding genes (379), ambiguous genes (overlap but not perfectly match with known genes; 70,439), and unknown genes (no overlap with any known annotated genes; 10,018). Next, CNCI assigned an index (coding or noncoding) for each of the 95,519 transcripts based on the training model. Among 14,683 transcripts that were previously annotated as protein-coding genes, 13,775 (94 %) were classified as protein-coding transcripts by CNCI, while among 379 transcripts that were previously annotated as lncRNA genes, 358 (94 %) were classified as noncoding transcripts by CNCI (mean of accuracy is 94 %). Moreover, we compared the performance of CNCI with the most commonly used tools, such as Coding Potential Calculator (CPC) [5] and phyloCSF [6], using the same dataset. CPC is a web server (http://cpc.cbi.pku.edu.cn), which uses SVM-based classifier to assess the protein-coding potential of a transcript. PhyloCSF is a comparative genomics method that analyzes a multispecies

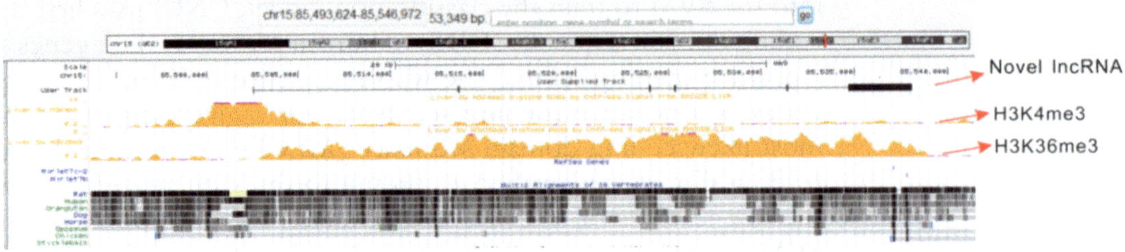

Fig. 1 Genomic contexts of a novel lncRNA gene. The genomic structure of lncRNA is showed at the *top* of the browser; exons are showed as *black blocks*. Other information includes H3K4me3 marker, H3K36me3 marker, and conservation status

nucleotide sequence alignment to distinguish protein-coding and noncoding regions. The accuracy of CPC for protein-coding and lncRNA gene classification was 87 % and 99 % (mean 93 %), respectively. The accuracy of phyloCSF for protein-coding and lncRNA gene classification was 86 % and 72 % (mean 79 %), respectively. These results strongly demonstrated the high accuracy of transcript classification by CNCI.

2.4 Identification of Novel lncRNAs Using CNCI (See Note 2)

The reconstructed transcripts that do not overlap with known genes are the pool of novel lncRNAs. Because novel lncRNAs such as antisense transcripts, promoter-associated transcripts, 3′ UTR-associated transcripts, and intronic transcripts are likely the extension or the transcriptional noise of known genes, we only focused on the novel transcripts located in the intergenic regions. For these novel intergenic transcripts, CNCI was used to evaluate their CNCI. As a result, 9,165 out of 10,018 intergenic transcripts were identified as novel lncRNAs. The mean transcript length of these novel lncRNAs was 1.6 kb. Previous study has demonstrated that trimethylation of lysine4 of histone H3 (H3K4me3) in the promoter and trimethylation oflysine36 ofhistone H3 (H3K36me3) in the transcribed region are markers of genes actively transcribed by RNA polymerase II [8]. Chromatin signature analysis revealed that a large portion of novel lncRNAs had both H3K4me3 peaks in the promoter region and H3K36me3 peaks in the gene body (histone modification data were generated by ENCODE project) [3], which is consistent with the fact that lncRNAs had similar transcriptional signature with protein-coding genes [8]. For example, a novel lncRNA located in chromosome 15 included six exons, a clear H3K4me3 peak in the promoter, and a series of H3K36me3 peaks in the body region (Fig. 1).

3 Notes

1. Although CNCI can be used in cross-species manner, it is unsuitable to use a human training set for lncRNA classification in invertebrates. For example, to classify the transcripts of

plants, the gene annotation of *Arabidopsis thaliana* may be more suitable as a training set.

2. As demonstrated by our previous study, CNCI not only is useful for classifying incomplete transcripts from RNA-Seq data but also has good performance at classifying sense–antisense transcript pairs, which is also one of the main advantages of CNCI. However, some limitations still exist. For example, in mammalian genomes, the exons in 3′ end of protein-coding transcripts may extend for several kilobases away from the coding regions (CDS) of transcripts. This is a common phenomenon in most of the RNA-Seq libraries, and these exons may be deemed as independent transcript units by most assembly tools. Thus, CNCI may misclassify these partial sequences as noncoding RNAs in the following analysis. To avoid the misclassification, accurate query sets containing high-quality assembled transcripts are required.

References

1. Liao Q, Liu C, Yuan X et al (2011) Large-scale prediction of long non-coding RNA functions in a coding-non-coding gene co-expression network. Nucleic Acids Res 39:3864–3878
2. Bu D, Yu K, Sun S et al (2012) NONCODE v3.0: integrative annotation of long noncoding RNAs. Nucleic Acids Res 40:D210–D215
3. Bernstein BE, Birney E, Dunham I et al (2012) An integrated encyclopedia of DNA elements in the human genome. Nature 489:57–74
4. Ozsolak F, Milos PM (2011) RNA sequencing: advances, challenges and opportunities. Nat Rev Genet 12:87–98
5. Kong L, Zhang Y, Ye ZQ et al (2007) CPC: assess the protein-coding potential of transcripts using sequence features and support vector machine. Nucleic Acids Res 35: W345–W349
6. Lin MF, Jungreis I, Kellis M (2011) PhyloCSF: a comparative genomics method to distinguish protein coding and non-coding regions. Bioinformatics 27:i275–i282
7. Guttman M, Donaghey J, Carey BW et al (2011) lincRNAs act in the circuitry controlling pluripotency and differentiation. Nature 477:295–300
8. Guttman M, Amit I, Garber M et al (2009) Chromatin signature reveals over a thousand highly conserved large non-coding RNAs in mammals. Nature 458:223–227
9. Sun L, Luo H, Bu D et al (2013) Utilizing sequence intrinsic composition to classify protein-coding and long non-coding transcripts. Nucleic Acids Res 41:e166
10. Derrien T, Johnson R, Bussotti G et al (2012) The GENCODE v7 catalog of human long noncoding RNAs: analysis of their gene structure, evolution, and expression. Genome Res 22:1775–1789
11. Pruitt KD, Tatusova T, Brown GR et al (2012) NCBI Reference Sequences (RefSeq): current status, new features and genome annotation policy. Nucleic Acids Res 40:D130–D135
12. Guttman M, Garber M, Levin JZ et al (2010) Ab initio reconstruction of cell type-specific transcriptomes in mouse reveals the conserved multi-exonic structure of lincRNAs. Nat Biotechnol 28:503–510
13. Trapnell C, Williams BA, Pertea G et al (2010) Transcript assembly and quantification by RNA-Seq reveals unannotated transcripts and isoform switching during cell differentiation. Nat Biotechnol 28:511–515

Chapter 19

Computational Methods to Predict Long Noncoding RNA Functions Based on Co-expression Network

Yi Zhao, Haitao Luo, Xiaowei Chen, Yi Xiao, and Runsheng Chen

Abstract

Although long noncoding RNAs (lncRNAs) have been recognized in recent years to constitute a significant portion of mammalian transcriptome, and the functional impact of several lncRNAs has been characterized by a few studies, yet it is still difficult to large-scale ascertain their functions from lab experiment or structure prediction. To address this deficit, we have developed a computational pipeline to large-scale annotate the functions of lncRNA based on coding–noncoding gene co-expression network. In this network, a node (circle) represents the protein-coding gene or lncRNA, and an edge (connecting line between nodes) indicates that the two genes are co-expressed (the correlation coefficients of connected genes reached the defined cutoff). In this chapter, we show how to use an lncRNA functional annotation pipeline to construct a co-expression network based on gene expression profiles in prostate cancer and how to further predict lncRNA functions using model-based and hub-based sub-networks.

Key words Long noncoding RNA, Co-expression network, Functional annotation, Microarray

1 Introduction

Long noncoding RNA (lncRNA) is a kind of RNAs with length longer than 200 nucleotides and sequence characteristics similar to mRNAs, but lacking protein encoding capacity [1, 2]. lncRNAs such as antisense, intronic, and intergenic transcripts have once been thought as the "dark matter" of the genome. However, increasing evidences show that lncRNAs are abundant in mammalian and play key roles in a wide variety of biological processes such as imprinting control, cell differentiation, immune responses, human diseases, and tumorigenesis by regulating the expression, activity, and localization of protein-coding genes [1, 3]. There are many mechanisms of lncRNA regulation. For example, antisense noncoding RNAs, which are transcribed from the opposite DNA strand of mRNAs, may negatively regulate the expression of the corresponding sense mRNAs [4]. In addition, the transcript process of lncRNAs can be regulated by microRNAs and transcription factors [5, 6].

M. Lucrecia Alvarez and Mahtab Nourbakhsh (eds.), *RNA Mapping: Methods and Protocols*, Methods in Molecular Biology, vol. 1182, DOI 10.1007/978-1-4939-1062-5_19,

Although the functions of several lncRNAs have been characterized, the roles of most lncRNAs are still unknown. Thus, it is necessary to annotate the functions of lncRNAs at large scale.

The prediction of lncRNA functions by computational methods is still a challenge because the primary sequence and secondary structure of lncRNAs are not conserved, and the high-throughput information about lncRNAs is limited. To address this deficit, we developed a computational pipeline to large-scale annotate lncRNA functions based on two-color network, which included both coding and noncoding genes [1, 7, 8]. Here, we present a case study of microarray data from human prostate cancer to describe in detail how to use our computational pipeline to predict lncRNA functions.

2 Methods

2.1 Re-annotation of Array Probes to Calculate Expression of Both lncRNA and Protein-Coding Genes (See Note 1)

Human protein-coding genes were collected from ENSEMBL database [9] and human lncRNA genes from Human Body Map lncRNA set [10] and GENCODE database [11]. Genes with one or more common transcripts between ENSEMBL and HAVANA [12] are deposited in GENCODE database and deemed as golden genes. The lncRNA transcripts of the two sets that belong to the same genes were assigned a new ID beginning with "TCL" using the Cuffcompare utility provided by Cufflinks [13].

To simultaneously calculate the expression of both human lncRNA and protein-coding genes, we first used our previously well-established computational workflow to re-annotate the probes of Human Genome U133 Plus 2.0 Array [1]. Detail explanation about the process of probe re-annotation has been described before [1]. In total, 323,512 and 34,858 probes were mapped to 19,980 protein-coding and 15,050 lncRNA genes (corresponding to 77,682 and 27,799 transcripts), respectively. Then, a new chip description file (CDF) containing information about the relationship between probes and genes was obtained. Accordingly, expression level of both lncRNA and protein-coding genes was calculated based on the information of re-annotated probe deposited in new CDF (*see* **Note 2**).

2.2 Calculation of Gene Expression and Construction of Cancer-Specific Co-expression Network

As an example of lncRNA function predictions under specific experimental conditions, we used microarray data from a case study of prostate cancer. Ten human prostate cancer associate U133 Plus 2.0 array data sets were obtained from the Gene Expression Omnibus (GEO) database (GSE159472, GSE174822, GSE330542, GSE114282, GSE179512, GSE279142, GSE284032, GSE324482, GSE33252, GSE87022). Preprocessing of microarray data consisted of Robust Multichip Average (RMA) background correction, constant normalization, and expression summarization.

These processes were implemented using the *affy* package of the R Bioconductor software. Then expression intensity was log2 transformed.

Ten data sets were used to construct a prostate cancer-related coding–noncoding (CNC) gene co-expression network. For each data set, the data processing is briefly described as follows:

1. Genes with expressional variance ranked in the top 75 percentile of each data set were retained.
2. A set of Pearson correlation coefficient (PCC) *p*-values for each gene pair was estimated through Fisher's asymptotic test and adjusted by the Bonferroni multiple test correction. Only gene pairs with an adjusted *p*-value of 0.01 or less and with a PCC value ranked in the top or the bottom 0.5 percentile for each gene were regarded as co-expressed in the given data set.
3. Each gene pair was assigned a score according to the number of data sets in which the gene pair was co-expressed in the same "direction" (i.e., positively or negatively).
4. Only gene pairs co-expressed in three or more data sets were included in the co-expression network.

The resulting network included 1,324 lncRNA genes and 13,446 coding genes, which composed 55,305 coding–coding edges, 4,262 CNC edges, and 483 noncoding–noncoding edges (*see* **Note 3**).

2.3 Prediction of lncRNA Functions Based on lncRNA-Centered Sub-networks

The method based on lncRNA-centered sub-networks (hub-based method) assigns functions to central unannotated lncRNAs according to the functional enrichment of its neighboring protein-coding genes. First, we parsed the co-expression network into sub-networks, each of which consisted of a central lncRNA and its directly connected genes. Then, for each sub-network, the functions of central lncRNA were predicted according to the enriched Gene Ontology (GO) terms [14] of neighboring protein-coding genes. To obtain more accuracy results, we acquired that the *p*-value of the enriched function is no more than 0.05 and the number of coding neighboring genes annotated with the enriched GO terms is no less than 10. As a result, the functions of 134 central lncRNAs in sub-networks were predicted based on the above processing (*see* **Note 3**). For example, one of lncRNAs in our collection named TCL0003844 as a central gene in a sub-network is connected with 27 protein-coding genes (Fig. 1a). The enriched functions of these 27 genes could be briefly separated into three kinds of biological processes: cell cycle, chromosome organization, and cellular response (Fig. 1b, c). Then, the lncRNA TCL0003844 should be annotated as these biological functions. Interestingly, the two transcripts (ENST00000546421 and ENST00000548760) of TCL0003844 gene are the antisense of TMPO gene, which may

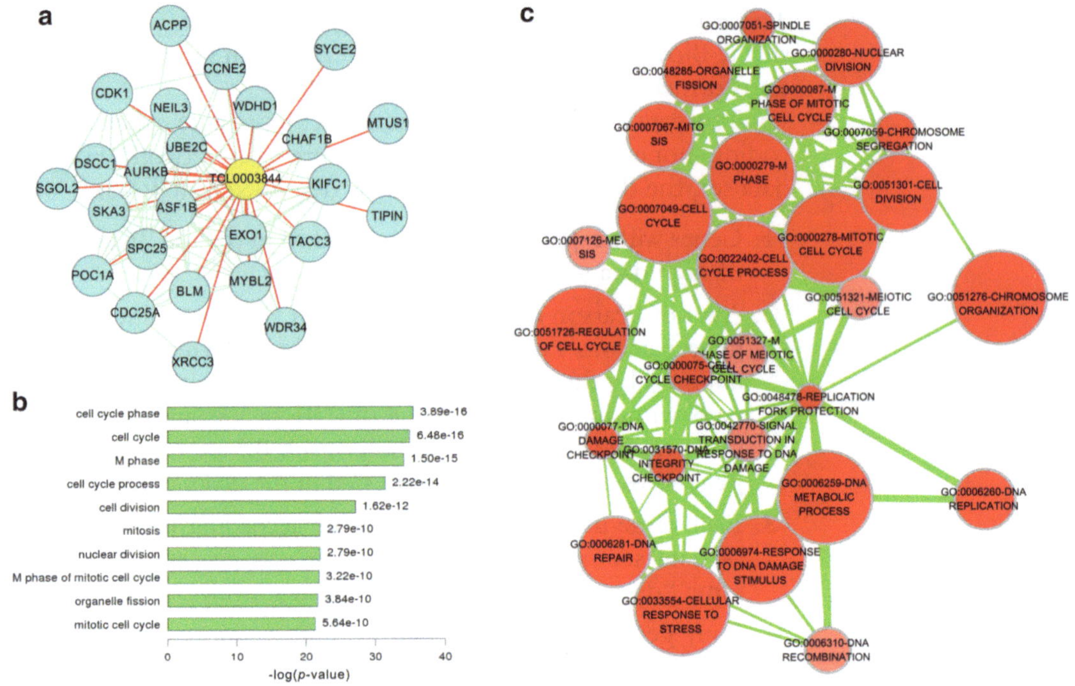

Fig. 1 An example of lncRNA-centered sub-network. (**a**) The lncRNA TCL0003844 sub-network consists of TCL0003844 (*yellow, center node*) and its 27 direct neighbors (*light blue nodes*). Edges colored in *light green* are coding–coding correlations; edges colored in *red* are lncRNA-coding correlations. (**b**) GO enrichment analysis result of 27 coding genes. (**c**) Functional enrichment map of 27 coding genes [17]. The *notes* denote GO terms; the *node size* denotes the number of genes in the GO term, and the *width of lines* denotes the number of genes shared between two GO terms

play a role in the assembly of the nuclear lamina and in cell cycle pathway [15]. The similar functional annotation and sense–antisense location between TCL0003844 and TMPO may suggest that TMPO is the target gene of TCL0003844.

2.4 Prediction of lncRNA Functions Based on Network Modules

We defined modules in co-expressed network as a group of genes more densely connected to each other than to genes outside the group. Given the fact that genes within a co-expressed module commonly have similar functions, lncRNA roles could be deduced from the protein-coding genes within the same modules. Accordingly, we first used the Markov cluster algorithm (MCL) [16] to parse the network into different modules. Next, the modules that were significantly enriched for at least one GO term and were composed of both coding and noncoding genes were retained. Finally, the enriched functions of protein-coding genes in the module were assigned to the lncRNAs in the same module. As a result, the functions of 171 lncRNAs could be annotated based on 100 selected modules (*see* **Note 3**). One of the modules, named "translational elongation and RNA processing" (according to the most significantly enrichment function), includes 54 genes (52 protein-coding genes and 2 lncRNA genes; Fig. 2a). Using the annotation method described above, we found that the predicted functions of the two lncRNA genes (named TCL0010193 and TCL0004671) selected were RNA processing and translation-related functions (Fig. 2b, c). Noteworthy, both lncRNAs are long intergenic noncoding RNAs (lincRNAs). One of them named TCL0010193 includes three transcripts (TCONS_00008410, TCONS_00008409, and TCONS_00009008) and is located near the NELFA gene, which participates in the regulation of RNA polymerase II transcription elongation (Fig. 2d).

2.5 lncRNA Functional Annotation Based on Global Prediction Method

The methods described above for function prediction of lncRNAs were exclusively based on a local strategy (hub-based and module-based sub-network) that annotated only a small part of lncRNAs. We recently developed a long noncoding RNA global function predictor ("lnc-GFP") to predict functions of lncRNAs at a large scale (http://bioinfo.org/ncfans/) [8]. Lnc-GFP utilizes the same co-expression network to infer functions for all the lncRNAs involved in network. Given a function category (for example, a GO term), the association score for each gene is calculated. Then, top-ranked genes are assigned to the given function. Here, we assign the corresponding functions to the lncRNAs ranked within top 1,000. Using this method, the function of 471 lncRNAs was successfully characterized, about three times more than the number of lncRNAs analyzed using module-based and hub-based methods (*see* **Note 3**).

2.6 A Simple and Quick Way to Predict Mouse lncRNA Functions Using ncFNAs Web Server

The process to predict human lncRNA functions based on microarray data described above include three main parts: microarray processing, network construction, and function prediction. Below we describe step by step how to use ncFANs (noncoding RNA Function ANnotation Server, available at http://bioinfo.org/ncfans/) to predict mouse lncRNA functions [7].

Step 1: First, upload your own microarray raw data to ncFANs in the tab "Microarray Raw Data" of Part I. Here, we used example data set to do the following analysis by selecting the "use example data" in the page (Fig. 3a). Next, "RMA" was selected as microarray processing method, which includes four steps: (1) background correction, which separates out true probe signal from background signal; (2) probe normalization, which removes specific probe signal bias; (3) probe correction, which corrects probes within a probe set for nonspecific binding; and (4) summarization, which summarizes the probe signals into one probe set expression level (or gene expression level). Then, the gene expression level is log2 transformed.

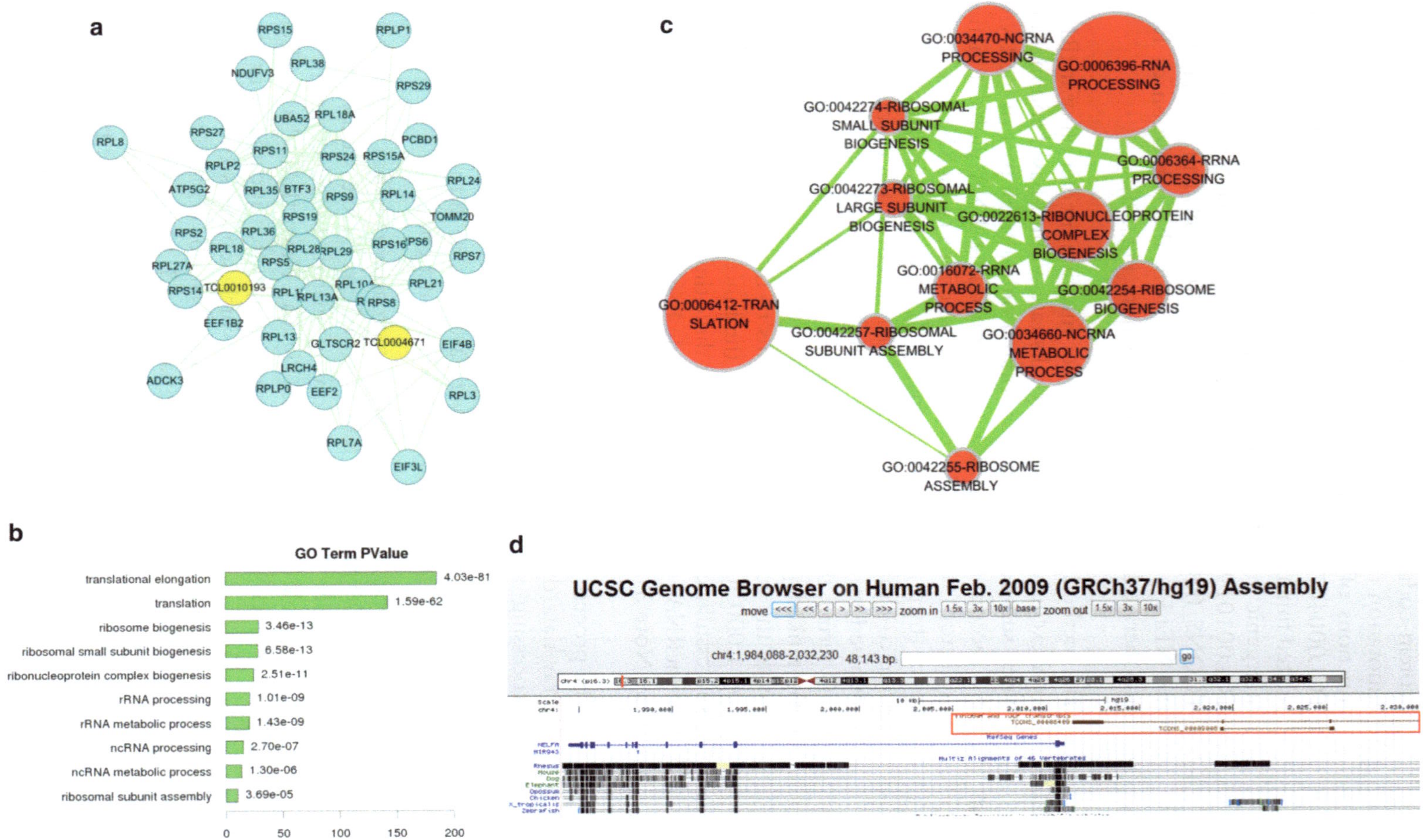

Fig. 2 An example of lncRNA functional prediction based on functional module. (**a**) The module consists of 52 protein-coding genes (*light blue*) and 2 lncRNA genes (*yellow*). (**b**) GO enrichment analysis result of 52 coding genes. (**c**) Functional enrichment map of 52 coding genes [17]. The *notes* denote GO terms; the *node size* denotes the number of genes in the GO term, and the *width of lines* denotes the number of genes shared between two GO terms. (**d**) Genomic context of the two lncRNAs (*highlighted by red box*)

Step 2: Using gene expression profile obtained from step 1 to construct "two-color" co-expression network: The potential functions of lncRNAs are annotated mainly based on the topology of network. Four parameters are needed for user to selection (Fig. 3b) including the following: (1) filter genes with expression variance (default is 75 %), which allow filtering out the genes with low variance to avoid the false-positive correlations due to the unreliable expression measures; (2) correlation calculated method: select using PCC or Spearman rank correlation coefficient method to calculate the correlations between genes; (3) the *p*-value cutoff of the correlation coefficients (default is 0.01); and (4) for each gene, all genes with a PCC value ranked in the selected upper percentile were regarded as co-expressed with that gene (default is 1 %).

Step 3: lncRNA function annotation algorithm setting (Fig. 3c): Based on topology of co-expression network, two methods including module-based and hub-based method could be used for lncRNA function prediction as described above. In default, the minimum number of genes in each module or hub is set to be 3 or 10, respectively. In the module-based and hub-based methods, the functions significantly enriched by the protein-coding genes within the same module or its co-expressed coding gene neighbors will be assigned to the lncRNA. Four types of functional categories can be selected: the three sub-ontology terms (BP, CC, MF) of GO and the KEGG pathways. Furthermore, the significantly enriched functions can be filtered by the user-defined cutoffs of the minimal number of the genes annotated with the function terms (3 for default), and the *p*-value is evaluated by hypergeometric test and adjusted by Bonferroni for multiple statistic correction (0.01 for default).

Step 4: Submit and data processing: After submitting the object, ncFANs will go to a page which shows the state of the job and the address of the coming results.

Step 5: Results output: The results of lncRNA function annotation mainly include five parts as follows. The first part is the overview of parameters selected in steps 1–3 (Fig. 3d). The second is "Expression Profiles Download" that allows to download the expression profile of coding and noncoding genes. The third part is the basic information of co-expression network (Fig. 3e). The fourth part is the result of module-based method, which as table format shows the information of each module (Fig. 3f). Detail information (including all the enriched functions, the noncoding and coding genes within the module) is shown in another page by clicking the name of the module (Fig. 3g). The fifth part is the result of hub-based method which includes the overview information of hub lncRNA. Similar with module-based results, detail information of each hub lncRNA is shown in another page by clicking the noncoding ID of the lncRNA.

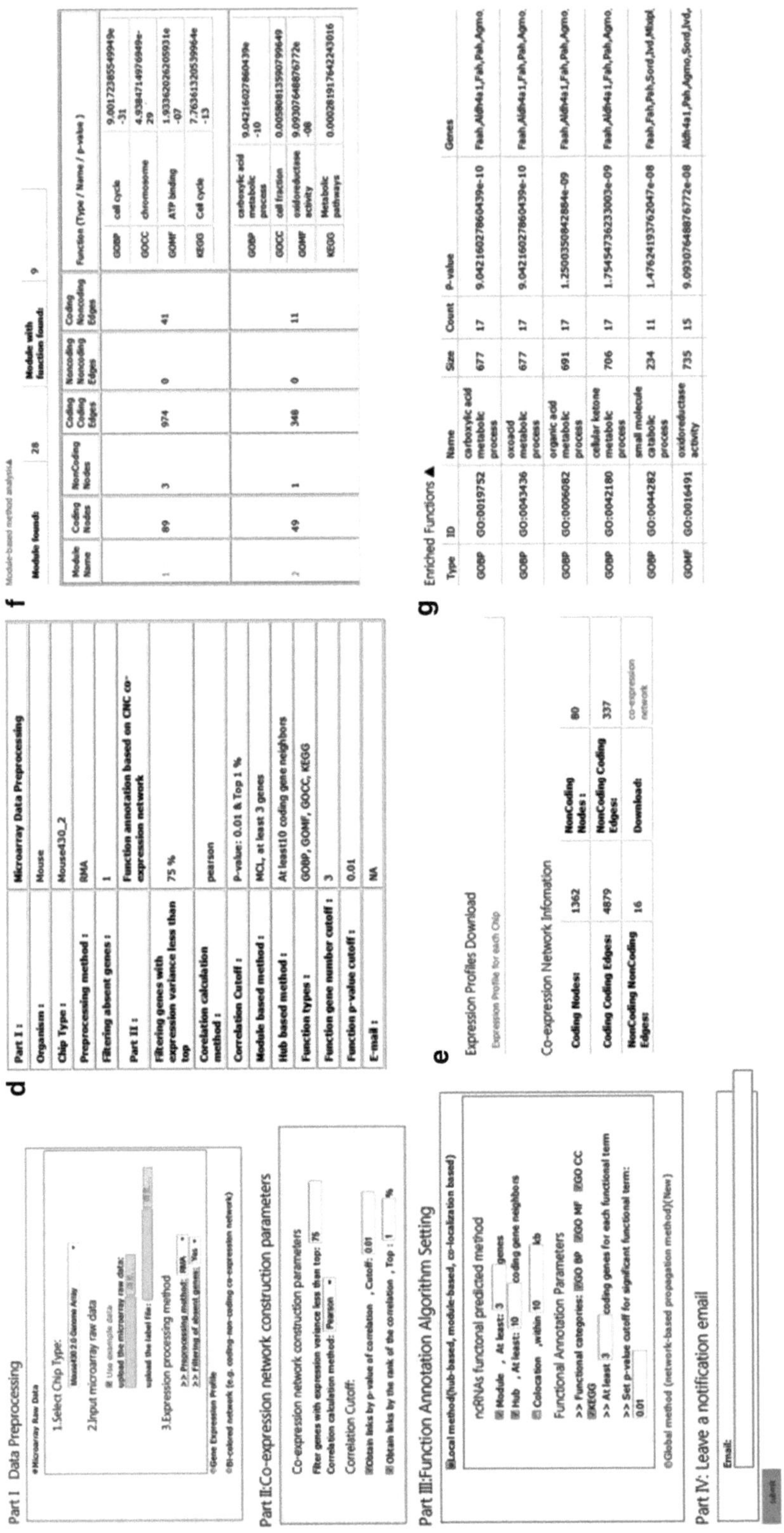

Fig. 3 Overview of ncFNA web server workflow

3 Notes

1. Currently, RNA sequencing (RNA-seq) technology provides a possibility for the comprehensive analysis of gene expression with an unprecedented resolution. Therefore, in addition to the re-annotation of microarray, RNA-seq is also an effective means to calculate the expression level of genes. Combining the databases that we collected above, the expression values of tens of thousands of lncRNA and protein-coding genes could be calculated based on RNA-seq reads. The resulting co-expression network would be more powerful to predict lncRNA functions.
2. Probe re-annotation process is the most important part in the whole workflow when using microarray data. In our website (http://bioinfo.org/ncfans/), currently a total of ten types of arrays have re-annotated CDF and can be automatically used for expression calculation. For those species without the re-annotated CDF, the expression value of lncRNA and protein-coding genes can be obtained from other means such as RNA-seq. As in the case, gene RPKM/FPKM values should be directly as the input file to upload to perform the following functional prediction process.
3. The co-expression network and functional annotation results of lncRNAs can be downloaded from our website (http://www.bioinfo.org/MCB/).

References

1. Liao Q, Liu C, Yuan X et al (2011) Large-scale prediction of long non-coding RNA functions in a coding-non-coding gene co-expression network. Nucleic Acids Res 39:3864–3878
2. Bu D, Yu K, Sun S et al (2012) NONCODE v3.0: integrative annotation of long noncoding RNAs. Nucleic Acids Res 40:D210–D215
3. Mercer TR, Dinger ME, Mattick JS (2009) Long non-coding RNAs: insights into functions. Nat Rev Genet 10:155–159
4. Pandey RR, Mondal T, Mohammad F et al (2008) Kcnq1ot1 antisense noncoding RNA mediates lineage-specific transcriptional silencing through chromatin-level regulation. Mol Cell 32:232–246
5. Zhao Y, He S, Liu C et al (2008) MicroRNA regulation of messenger-like noncoding RNAs: a network of mutual microRNA control. Trends Genet 24:323–327
6. Huarte M, Guttman M, Feldser D et al (2010) A large intergenic noncoding RNA induced by p53 mediates global gene repression in the p53 response. Cell 142:409–419
7. Liao Q, Xiao H, Bu D et al (2011) ncFANs: a web server for functional annotation of long non-coding RNAs. Nucleic Acids Res 39:W118–W124
8. Guo X, Gao L, Liao Q et al (2013) Long non-coding RNAs function annotation: a global prediction method based on bi-colored networks. Nucleic Acids Res 41:e35
9. Flicek P, Amode MR, Barrell D et al (2012) Ensembl 2012. Nucleic Acids Res 40:D84–D90
10. Cabili MN, Trapnell C, Goff L et al (2011) Integrative annotation of human large intergenic noncoding RNAs reveals global properties and specific subclasses. Genes Dev 25:1915–1927
11. Derrien T, Johnson R, Bussotti G et al (2012) The GENCODE v7 catalog of human long noncoding RNAs: analysis of their gene structure, evolution, and expression. Genome Res 22:1775–1789
12. Ashurst JL, Chen CK, Gilbert JG et al (2005) The Vertebrate Genome Annotation (Vega) database. Nucleic Acids Res 33:D459–D465

13. Trapnell C, Williams BA, Pertea G et al (2010) Transcript assembly and quantification by RNA-Seq reveals unannotated transcripts and isoform switching during cell differentiation. Nat Biotechnol 28:511–515
14. Ashburner M, Ball CA, Blake JA et al (2000) Gene ontology: tool for the unification of biology. The Gene Ontology Consortium. Nat Genet 25:25–29
15. Dechat T, Vlcek S, Foisner R (2000) Review: lamina-associated polypeptide 2 isoforms and related proteins in cell cycle-dependent nuclear structure dynamics. J Struct Biol 129:335–345
16. Borodovsky M, Peresetsky A (1994) Deriving non-homogeneous DNA Markov chain models by cluster analysis algorithm minimizing multiple alignment entropy. Comput Chem 18: 259–267
17. Merico D, Isserlin R, Stueker O et al (2010) Enrichment map: a network-based method for gene-set enrichment visualization and interpretation. PLoS One 5:e13984

Chapter 20

MicroRNA Biogenesis: Dicing Assay

Carlos A. Melo and Sonia A. Melo

Abstract

RNA-induced silencing complex is the cytoplasmic effector machine of the microRNA (miRNA) pathway and contains a single-stranded miRNA guiding it to its target mRNAs. The biogenesis of mature miRNAs is a regulatory process achieved by complex machinery composed of few protein components, among which the ribonuclease III Dicer plays a central role. Dicer is essential for miRNA maturation and catalyzes one of the rate-limiting steps of miRNA production. In this chapter, we describe a protocol to study the catalytic activity of Dicer in cell extracts by measuring their ability to process precursor RNA (pre-miRNAs) into functional mature miRNAs. Impairment of the miRNA biogenesis machinery due to loss-of-function mutations in effectors of the pathway such as Dicer has been demonstrated before. Dicing assay can be used in cancer to assess Dicer enzymatic activity compared with healthy controls. Therefore, this experimental approach is likely to be useful to researchers investigating the main steps of miRNA biogenesis and function in human health and diseases.

Key words Dicer, Precursor microRNA, MicroRNA, Enzymatic activity

1 Introduction

MicroRNAs (miRNAs) are small (17–25 nucleotides, nt), single-stranded noncoding RNAs that control gene expression in animals, plants, and unicellular eukaryotes that are key posttranscriptional regulators of gene expression [1]. These small RNAs have emerged as key posttranscriptional regulators of gene expression controlling ~60 % of all protein-coding genes in mammals. Thus, miRNAs participate in the regulation of almost every cellular event such as development, differentiation, proliferation, and apoptosis. The biosynthesis of miRNAs is a tightly regulated multistep process that starts in the nucleus of the cell, following transcription, and continues through the cytoplasm where finally the mature miRNA molecule exerts its main function (Fig. 1). Each one of the multiple steps that compose miRNA biosynthesis seems to be remarkably well coordinated. Dicer is an RNase III enzyme that together with its catalytic partner trans-activator RNA (tar)-binding protein (TRBP) is a key enzyme that dices the pre-miRNA into a miRNA

M. Lucrecia Alvarez and Mahtab Nourbakhsh (eds.), *RNA Mapping: Methods and Protocols*, Methods in Molecular Biology, vol. 1182, DOI 10.1007/978-1-4939-1062-5_20, © Springer Science+Business Media New York 2014

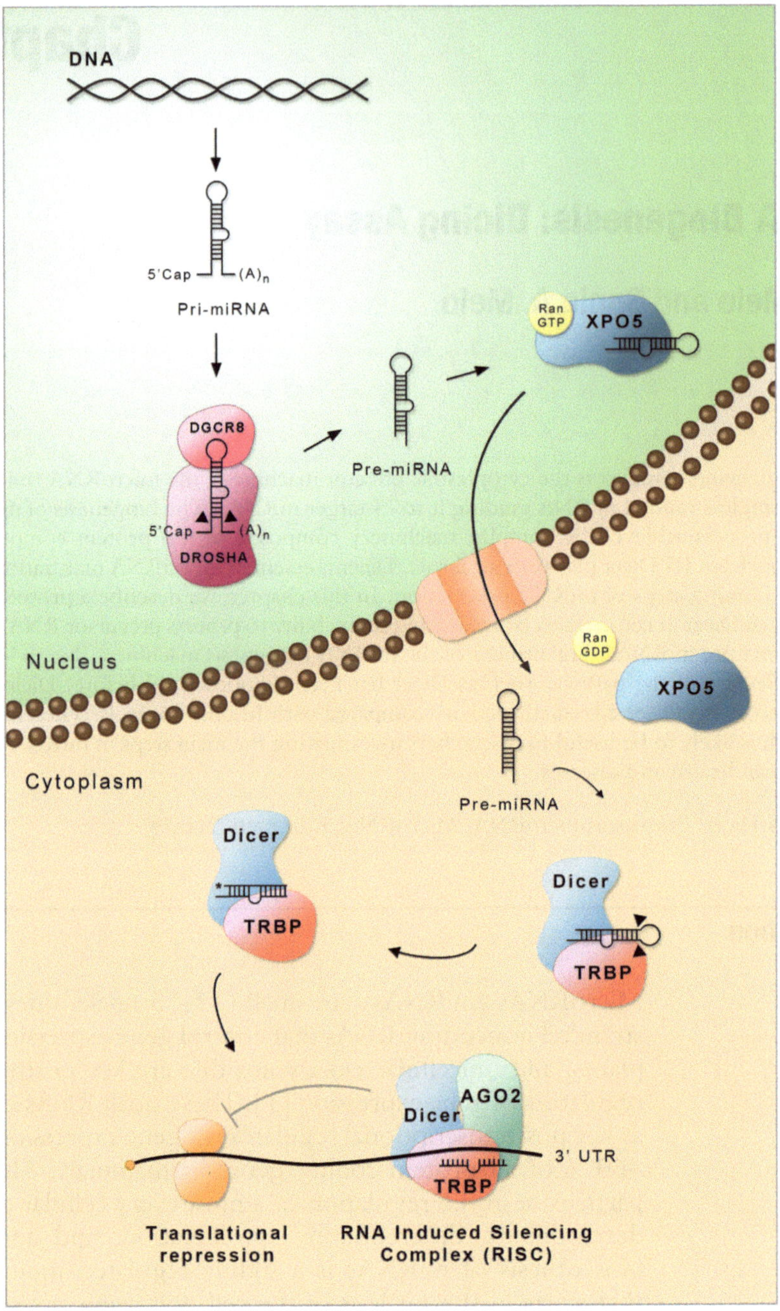

Fig. 1 MicroRNA biogenesis pathway. Long primary transcripts (pri-miRNAs) containing one or several miRNAs are transcribed by RNA polymerase II and cleaved by the microprocessor complex containing at least Drosha (RNAase III endonuclease) and DiGeorge syndrome critical region gene 8 (DGCR8) in humans. This complex recognizes the double-stranded RNA structure of the pri-miRNA and specifically cleaves at the base of the stem–loop, hence releasing a 60- to 70-nucleotide precursor (pre)-miRNA. This pre-miRNA is then exported through the Exportin-5 pathway into the cytoplasm where it is further processed into a mature miR/miR* duplex by DICER1, a second RNase III endonuclease together with its catalytic partner the human immunodeficiency virus transactivating response RNA-binding protein (TRBP). The miR/miR* duplex is then loaded into a multicomponent complex, the RNA-induced silencing complex (RISC), constituted of at least TRBP, DICER1, and one Argonaute inhibiting mRNA translation

duplex giving rise to the mature functional miRNA molecule [2–9]. Dicer has a single processing center with intramolecular dimerization of the two RNase III domains. Each RNase domain cuts independently one RNA strand of the duplex and generates products with 2 nt 3′ overhangs [7]. Mammalian cells have a single Dicer protein, with a molecular weight of ~200 kDa. Dicer contains an ATPase/RNA helicase domain, a DUF domain (double-stranded RNA-binding domain), a PAZ domain (a nucleic acid-binding domain, with a strong preference for single-stranded nucleic acids or RNA duplexes with single-stranded 3′ overhangs), two RNase III domains, and a dsRNA-binding domain (dsRBD) [2]. The PAZ domain of Dicer, also found in Argonaute proteins, is thought to bind to the 2 nt 3′ overhang present at the base of the pre-miRNA hairpin, and the dsRBD of Dicer binds the stem and defines the distance of the cleavage from the base [10–12]. The two strands of the duplex are separated by Dicer, and the strand whose 5′ end forms the more unstable duplex with its partner seems to preferentially survive as the mature miRNA [8, 9, 11, 12].

The dicing activity of Dicer in the cytoplasm is essential to the biogenesis of mature functional miRNAs that can further control gene expression posttranscriptionally. Dicer recognizes its pre-miRNA substrates via its PAZ domain through the extremity harboring of pre-miRNA formed by a 5′ phosphate and a 3′ hydroxylated end with 2 nt overhang [13]. The miRNA:miRNA* duplex is then excised by Dicer through the intramolecular homodimerization of the two RNase III domains [10, 14].

The dicing assay or Dicer activity assay described in this chapter is a protocol aimed to study and assess the specific catalytic activity of Dicer under various situations and/or conditions. The protocol can be transposed to different experimental contexts; therefore, it is possible to compare different forms, mutant and wild-type Dicer, as well as cells and tissues related to human diseases. We first describe the preparation of Dicer extracts from cells followed by an incubation period, the "dicing step," during which pre-miRNAs are processed to miRNAs. After RNA extraction, we describe the method for analysis of pre-miRNAs and miRNAs by denaturing urea polyacrylamide gel electrophoresis, which is required to visualize the results.

2 Materials

Use RNase/DNase-free material throughout the protocol. It is also important to use diethyl pyrocarbonate (DEPC)-treated water (*see* **Note 1**).

2.1 Cell Culture

1. Dulbecco's modified Eagle's medium (DMEM) or Roswell Park Memorial Institute (RPMI) medium supplemented with 10 % (v/v) fetal bovine serum (FBS), 1 mM sodium pyruvate,

100 units/ml penicillin, 100 μg/ml streptomycin, and 2 mM L-glutamine. Store at 4 °C.

2. 10× Phosphate-buffered saline (PBS): Dissolve 80 g NaCl, 11.5 g $Na_2HPO_4 \cdot 7H_2O$, 2 g KCl, and 2 g KH_2PO_4 in 800 ml of H_2O, adjust pH to 7.4, and add H_2O up to 1 l. Autoclave and store at room temperature.
3. 1× Trypsin–ethylenediaminetetraacetic acid (EDTA) solution.

2.2 Preparation of Dicer Extracts

1. 80 % Glycerol.
2. 50× Stock solution of protease inhibitor: Dissolve one complete protease inhibitor cocktail tablet (Roche) in 1 ml H_2O to make a 50× stock solution. Store aliquots at −20 °C.
3. Hypotonic buffer: 2 mM $MgCl_2$, 0.1 % β-mercaptoethanol, 10 mM Hepes–KOH, pH 7.0, 1× protease inhibitor. To make 10 ml of the buffer, mix 0.2 ml of 100 mM $MgCl_2$, 0.1 ml of 1.0 M Hepes–KOH, pH 7.0, 10 μl of β-mercaptoethanol, 0.2 ml of 50× stock solution of protease inhibitor, and 9.5 ml of H_2O. Should be prepared fresh every time.

2.3 Dicer Activity Assay: "Dicing" Assay

1. 10× Buffer K: 200 mM Hepes–KOH, pH 7.0, 20 mM dithiothreitol (DTT), 20 mM $MgCl_2$. Make 10 ml of the 10× buffer by mixing 2 ml of 1 M Hepes, pH 7.0, 0.2 ml of 1 M DTT, 0.2 ml of 1 M $MgCl_2$, and 7.6 ml of H_2O together. Prepare fresh every time.
2. 0.5 M Creatine Phosphate (disodium salt, Sigma): Dissolve 80 mg creatine phosphate in 500 μl H_2O, and store aliquots at −20 °C.
3. Storage buffer: 50 mM Tris–HCl, pH 7.0, 1 mM DTT, 1 μM ATP, 0.1 mM EDTA, and 50 % glycerol.
4. 1 mg/ml Creatine Phosphate Kinase (Sigma): Dissolve creatine phosphate kinase in storage buffer, and save at −20 °C.
5. RNasin Ribonuclease Inhibitor, 40 units/μl (Ambion).
6. 2× Protease K (PK) buffer: 1.0 % SDS, 10 mM Tris–HCl, pH 7.5, 1 mM EDTA, and 1 mg/ml protease K. Prepare fresh every time from individual stock solutions (20 % SDS, 1 M Tris–HCl, pH 7.5, 500 mM EDTA, pH 8.0, 20 mg/ml protease K).
7. Glycogen (20 μg/μl, Invitrogen): Store at −20 °C.
8. TE buffer: 10 mM Tris–HCl, pH 8.0, 1 mM EDTA.
9. Sample buffer: Deionized formamide/TE (1:1); prepare fresh every time.

2.4 RNA Extraction

1. Acid Phenol:$CHCl_3$ 5:1 solution (Ambion).
2. 70 and 100 % ethanol (ETOH).
3. Glycogen (20 mg/ml) (Sigma).

4. 0.5 M EDTA pH 8.0.
5. Proteinase K (20 mg/ml) (Ambion).
6. Proteinase K buffer: 200 mM NaCl, 20 mM Tris–HCl pH 8.0, 2 mM EDTA, 1 % SDS.

2.5 Denaturing Urea-Polyacrylamide Gel Electrophoresis

1. Acrylamide solution 40 % (Acrylamide:Bis-Acrylamide 19:1): Store at 4 °C.
2. 5× TBE (electrophoresis running buffer): Dissolve 54.0 g of Tris base and 27.5 g of boric acid in 800 ml of H_2O, add 20 ml of 0.5 M EDTA (pH 8.0), and adjust the volume to 1 l. Autoclave.
3. 10 % Ammonium persulfate (APS): Store at 4 °C.
4. Tetramethylethylenediamine (TEMED): Store at 4 °C.
5. 6× Gel loading buffer: 0.25 % Xylene cyanol, 0.25 % bromophenol blue, and 30 % glycerol. Prepare with DEPC-H_2O, autoclave, and store at –20 °C.
6. Vertical electrophoresis system: Glass plates (16 cm width × 20 cm length), spacer (1 mm thick), and comb (20-well).
7. N^+-BrightStar Membrane (Ambion).
8. BrightStar BioDetect Kit (Ambion).
9. X-ray film (BioMax MS Film, Kodak) or a screen of PhosphorImager (Molecular Dynamics).

3 Methods

3.1 Dicing Assay Using Mammalian Cell Extracts

3.1.1 Preparation of Dicer Extracts from Mammalian Cells

1. When cells are about 80 % confluent, collect cell culture into two 15 ml centrifuge tubes.
2. Centrifuge at 1,000 × *g* for 5 min at room temperature to pellet the cells, and discard the supernatant.
3. Gently resuspend the cells in 10 ml 1× PBS. Centrifuge at 1,000 × *g* for 5 min to pellet the cells.
4. Repeat **step 3** twice.
5. Resuspend the cell pellets in 4 ml of ice-cold hypotonic buffer containing 60 mM KCl, and combine both centrifuge tubes into one ice-cold 15 ml centrifuge tube. Centrifuge at 1,000 × *g* for 5 min at 4 °C, and discard the supernatant (*see* **Note 2**).
6. Add 50–80 % volume of ice-cold hypotonic buffer (without KCl). Gently resuspend the cells.
7. Sonicate samples in an ice-cold water bath for 30 s.
8. Transfer the lysate to a 1.5 ml microcentrifuge tube.
9. Centrifuge at 20,000 × *g* at 4 °C for 30 min (*see* **Note 3**).

10. Transfer the supernatant into a new 1.5 ml microcentrifuge tube, and discard the pellets.
11. At this point, you can add 80 % glycerol stock to the supernatant at a final concentration of 10 % and store the extracts at −80 °C.

3.2 Dicer Activity Assay

Hannon et al. first described processing of the dsRNA into siRNAs by Dicer extract [2]. The miRNA biogenesis machinery is ubiquitously expressed. Therefore, Dicer activity should be present in all cellular extracts. Hence, extracts of cells are used to incubate with any pre-miRNA under appropriate chemical conditions that should allow efficient conversion of a pre-miRNA into its corresponding mature functional miRNA.

1. Assemble the Dicer assay reaction in a volume of 10 μl: 1 μl 10× Buffer K, 1 μl 10 mM ATP, 0.5 μl 0.5 M creatine phosphate, 0.3 μl 1 mg/ml creatine phosphate kinase, 0.2 μl ribonuclease inhibitor (40 units/μl), 1 μl biotin-labelled dsRNA (50 nM, Ambion), and 6 μl of cell extract (*see* Subheading 3.1).
2. Incubate at 30 °C for 2 h.
3. Add 90 μl of 2× PK buffer to the 10 μl Dicer assay reaction, and incubate at 65 °C for 15 min.
4. Add 100 μl of TE buffer to a volume of 200 μl, and mix with an equal volume of phenol/chloroform (1:1). Vortex for 10 s.
5. Centrifuge at 16,000 × *g* for 15 min at room temperature, and transfer the aqueous phase to a new centrifuge tube.
6. Add 1 μl of 20 μg/μl glycogen to help in RNA precipitation.
7. Add 3× of ice-cold ethanol, and keep at −20 °C for 4 h (*see* **Note 4**).
8. Centrifuge at 20,000 × *g* for 15 min at 4 °C, and discard the supernatant (*see* **Note 5**).
9. Air-dry the RNA pellet at room temperature for 10 min (*see* **Note 6**).
10. Dissolve RNAs in 20 μl sample buffer. Heat at 95 °C for 5 min to denature the RNA, and leave it on ice for 1 min.
11. Add 6× gel loading buffer, and the sample is ready for loading on urea-polyacrylamide gel electrophoresis (PAGE).

3.3 Denaturing Urea-Polyacrylamide Gel Electrophoresis

1. Prepare a 15 % denaturing urea-polyacrylamide gel: Mix 9 ml of H_2O, 4.5 ml of 5× TBE, and 18.9 g of urea in a 100 ml beaker. Heat in a microwave oven to completely dissolve urea. Add 16.9 ml of 40 % acrylamide solution while continuously stirring with a magnetic bar. Adjust the volume to 45 ml with H_2O, and add 240 μl of 10 % APS and 10 μl of TEMED.

2. Immediately pour the gel between the two glass plates, place the comb, and keep at room temperature for about 30 min until polymerized.
3. Pre-run the gel in 0.5× TBE buffer for 15 min at 200 V.
4. Load the heat-denatured RNA sample prepared from Dicer activity assay (*see* Subheading 3.2) and a size marker (*see* **Note 7**).
5. Run the gel at 200 V until bromophenol blue reaches about 2/3 of the gel. Small RNAs might migrate between the bromophenol blue and xylene cyanol.
6. Carefully remove the gel from the glass plate, and transfer it during 2 h at 300 mA in 1× TBE buffer using an N$^+$-BrightStar Membrane (Ambion) (*see* **Note 8**).
7. Cross-link the membrane under UV light for 15 min (*see* **Note 9**).
8. Develop the membrane using the BrightStar BioDetect Kit (Ambion) following exactly the detection procedure.
9. Expose the membrane to a sheet of X-ray film. Exposure time will depend on the strength of the signal (*see* **Note 10**).

4 Notes

1. DEPC water can be prepared by adding 1 ml of DEPC to 1 l of milliQ water, mixing overnight, and autoclaving. DEPC is carcinogenic and should be handled with care under a chemical hood.
2. The addition of KCl in the hypotonic buffer is to prevent the cells from swelling and burst when they are rinsed.
3. If the centrifugation is carried out at 100,000 × *g* for 30 min, the supernatant could give rise to a better result in the processing assay. However, the quality of the lysate may be compromised if 4 °C is not kept at all times.
4. Four hours is the minimum period of time to get a good yield of RNA. If the yield is lower than expected, overnight precipitation may be a better choice.
5. To minimize the loss of RNA, washing the RNA pellet is not recommended.
6. The RNA pellet can be kept at −80 °C if the urea-PAGE is not immediately performed.
7. Biotinylated size marker is commercially available. Because overexposed image of the marker can happen, it is recommended to leave one or two lanes blank between the sample and size marker.
8. The use of this specific membrane for biotinylated RNA is essential to reduce background.

9. If commercial cross-linker is used, the manufacturer's recommendations should be followed (i.e., UV exposure of approximately 1.2 mJ).
10. To achieve a good balance between signal intensity and background, it is recommended to use three films on top of each other for the first exposure. If signal of the probe is good but there is high background (something very common using biotin), the second and/or third films should be clean of background and still have a good signal intensity from the desired band.

References

1. Bartel DP (2004) MicroRNAs: genomics, biogenesis, mechanism, and function. Cell 116:281–297
2. Bernstein E, Caudy AA, Hammond SM et al (2001) Role for a bidentate ribonuclease in the initiation step of RNA interference. Nature 409:363–366
3. Grishok A, Pasquinelli AE, Conte D et al (2001) Genes and mechanisms related to RNA interference regulate expression of the small temporal RNAs that control C. elegans developmental timing. Cell 106:23–34
4. Hutvagner G, McLachlan J, Pasquinelli AE et al (2001) A cellular function for the RNA-interference enzyme Dicer in the maturation of the let-7 small temporal RNA. Science 293:834–838
5. Ketting RF, Fischer SE, Bernstein E et al (2001) Dicer functions in RNA interference and in synthesis of small RNA involved in developmental timing in C. elegans. Genes Dev 15:2654–2659
6. Knight SW, Bass BL (2001) A role for the RNase III enzyme DCR-1 in RNA interference and germ line development in Caenorhabditis elegans. Science 293:2269–2271
7. Filipowicz W (2005) RNAi: the nuts and bolts of the RISC machine. Cell 122:17–20
8. Han J, Lee Y, Yeom KH et al (2006) Molecular basis for the recognition of primary microRNAs by the Drosha-DGCR8 complex. Cell 125: 887–901
9. Khvorova A, Reynolds A, Jayasena SD (2003) Functional siRNAs and miRNAs exhibit strand bias. Cell 115:209–216
10. Zhang H, Kolb FA, Jaskiewicz L et al (2004) Single processing center models for human Dicer and bacterial RNase III. Cell 118:57–68
11. Salzman DW, Shubert-Coleman J, Furneaux H (2007) P68 RNA helicase unwinds the human let-7 microRNA precursor duplex and is required for let-7-directed silencing of gene expression. J Biol Chem 282:32773–32779
12. Schwarz DS, Hutvagner G, Du T et al (2003) Asymmetry in the assembly of the RNAi enzyme complex. Cell 115:199–208
13. Elbashir SM, Lendeckel W, Tuschl T (2001) RNA interference is mediated by 21- and 22-nucleotide RNAs. Genes Dev 15:188–200
14. Macrae IJ, Zhou K, Li F et al (2006) Structural basis for double-stranded RNA processing by Dicer. Science 311:195–198

Chapter 21

Faster Experimental Validation of microRNA Targets Using Cold Fusion Cloning and a Dual Firefly-*Renilla* Luciferase Reporter Assay

M. Lucrecia Alvarez

Abstract

Different target prediction algorithms have been developed to provide a list of candidate target genes for a given animal microRNAs (miRNAs). However, these computational approaches provide both false-positive and false-negative predictions. Therefore, the target genes of a specific miRNA identified in silico should be experimentally validated. In this chapter, we describe a step-by-step protocol for the experimental validation of a direct miRNA target using a faster Dual Firefly-*Renilla* Luciferase Reporter Assay. We describe how to construct reporter plasmids using the simple, fast, and highly efficient cold fusion cloning technology, which does not require ligase, phosphatase, or restriction enzymes. In addition, we provide a protocol for co-transfection of reporter plasmids with either miRNA mimics or miRNA inhibitors in human embryonic kidney 293 (HEK293) cells, as well as a description on how to measure Firefly and *Renilla* luciferase activity using the Dual-Glo Luciferase Assay kit. As an example of the use of this technology, we will validate glucose-6-phosphate dehydrogenase (G6PD) as a direct target of miR-1207-5p.

Key words Luciferase reporter assay, Cold fusion cloning, microRNA, miRNA target, miR-1207-5p, G6PD

1 Introduction

MicroRNAs (miRNAs)—small RNA molecules that decrease the levels of a specific protein by destabilizing mRNA or by repressing their translation into proteins—play an important role in complex physiological process such as embryogenesis, proliferation, differentiation, and apoptosis [1]. MicroRNAs also have a role in the development of different diseases including cancer, diabetes, cardiovascular diseases, obesity, and viral infections [1–3].

Functional characterization of a specific miRNA requires the identification of its target gene/s. In plants, this is quite straightforward primarily due to the perfect base-pairing between miRNAs and their target genes. Thus, a genomic-wide search for complementary

M. Lucrecia Alvarez and Mahtab Nourbakhsh (eds.), *RNA Mapping: Methods and Protocols*, Methods in Molecular Biology, vol. 1182, DOI 10.1007/978-1-4939-1062-5_21, © Springer Science+Business Media New York 2014

regions to the miRNAs is generally a very effective procedure to identify miRNA target genes in plants [4–6]. In contrast, animal miRNAs rarely comprise a perfect sequence match to their targets. Mature metazoan miRNAs recognize their target mRNAs by base-pairing between nucleotides 2–7 or 2–8 at the 5′ of the miRNA (the seed region) and a complementary sequence in the 3′ untranslated region (3′UTR) of mRNAs [7].

Different target prediction algorithms have been developed to provide a list of candidate target genes for a given animal miRNA including TargetScan [8–11], miRanda [12, 13], PicTar [14], Diana-microT [15], etc. Unfortunately, these computational approaches will often predict different target genes for the same miRNA, suggesting the presence of both false-positive and false-negative target genes [16]. The fact that all target prediction algorithms use distinct parameters and have different performance values (i.e., sensitivity and specificity in target prediction) [17] may explain the discrepancies between the results obtained. Therefore, it is recommended that at least two different algorithms predict the same target/s for a specific miRNA before validation experiments are conducted [17].

Different approaches are commonly used for experimental validation of miRNA targets including miRNA overexpression using miRNA mimics, miRNA downregulation with miRNA inhibitors, western blot or ELISA quantification of the protein generated from the target mRNA, etc. However, the main drawback of all these strategies is that they cannot distinguish whether the effect of the miRNA on the target gene is *direct* (i.e., miRNA binds to the target mRNA) or *indirect* (i.e., miRNA binds to an unknown gene that is the one that downregulates the identified target). The only method for experimental validation of direct miRNA targets is based on reporter assays. In this chapter, we describe a step-by-step protocol for experimental validation of a direct miRNA target using a 3′UTR Dual Firefly-*Renilla* Luciferase Reporter Assay (Fig. 1). In the reporter plasmid used in this assay, the 3′UTR of the putative target gene is placed downstream the gene encoding Firefly luciferase (*luc2*) to determine if a specific miRNA is able to bind the 3′UTR causing a decrease in luciferase activity (Fig. 2). The reporter plasmid also contains the gene encoding *Renilla* luciferase to normalize luciferase activity and make it independent of the number of cells—otherwise it is hard to distinguish between cell death and cellular downregulation. In addition, the *Renilla* luciferase activity can be used as a control for transfection efficiency. Here we describe how to construct the reporter plasmids using the simple, fast, and highly efficient cold fusion cloning technology, which does not require ligase, phosphatase, or restriction enzymes. We also provide a protocol for cells co-transfection with reporter plasmids and either miRNA mimics or miRNA inhibitors, as well as a description of how to measure Firefly and *Renilla* luciferase

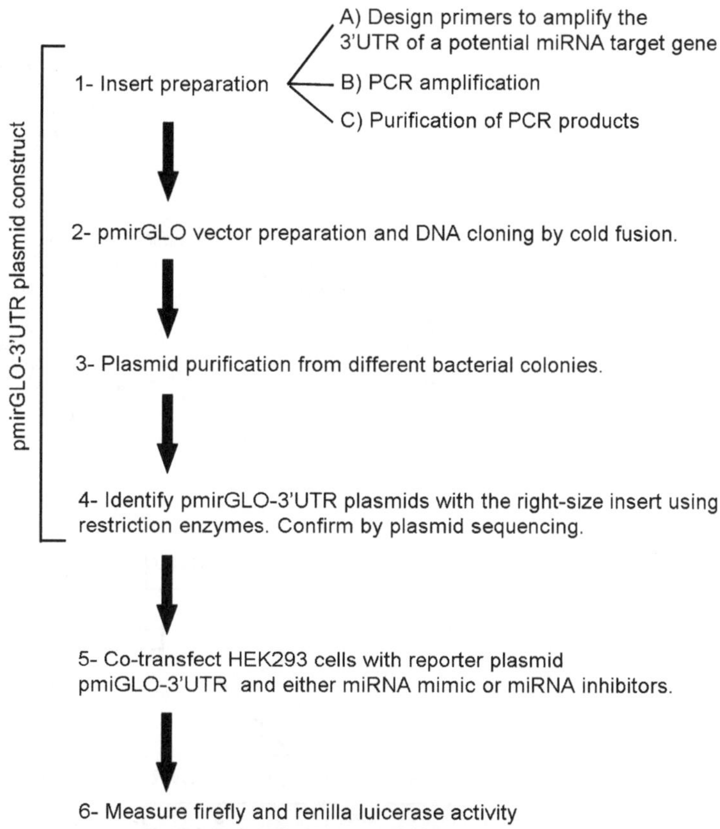

Fig. 1 Outline of the main steps for the experimental validation of a direct miRNA target using a Dual Firefly-*Renilla* Luciferase Reporter Assay. pmirGLO is a commercial dual-luciferase miRNA target expression vector (Promega). pmirGLO-3′UTR is the reporter vector containing the 3′UTR (untranslated region) of the putative miRNA target gene downstream the Firefly luciferase gene. Cold fusion is a cloning technology that does not require ligase, phosphatase, and restriction enzymes, and is faster than the classical cloning. HEK293, human embryonic kidney 293 cells

activity using the fast Dual-Glo Luciferase Assay kit. As an example of the use of this technology, we will validate glucose-6-phosphate dehydrogenase (G6PD) as a direct target of miR-1207-5p.

2 Materials

2.1 Plasmid Constructs

2.1.1 Insert Preparation: DNA Amplification by PCR

1. CloneAmp HiFi PCR Premix (Clonetech) (*see* **Notes 1** and **2**).
2. Specific primers to amplify the whole 3′UTR (untranslated region) of the potential microRNA target gene (*see* **Note 3** and Subheading 3).
3. Optional: Quick Change II Site-Directed Mutagenesis kit (Agilent) (*see* **step 6** in Subheading 3.1.1).

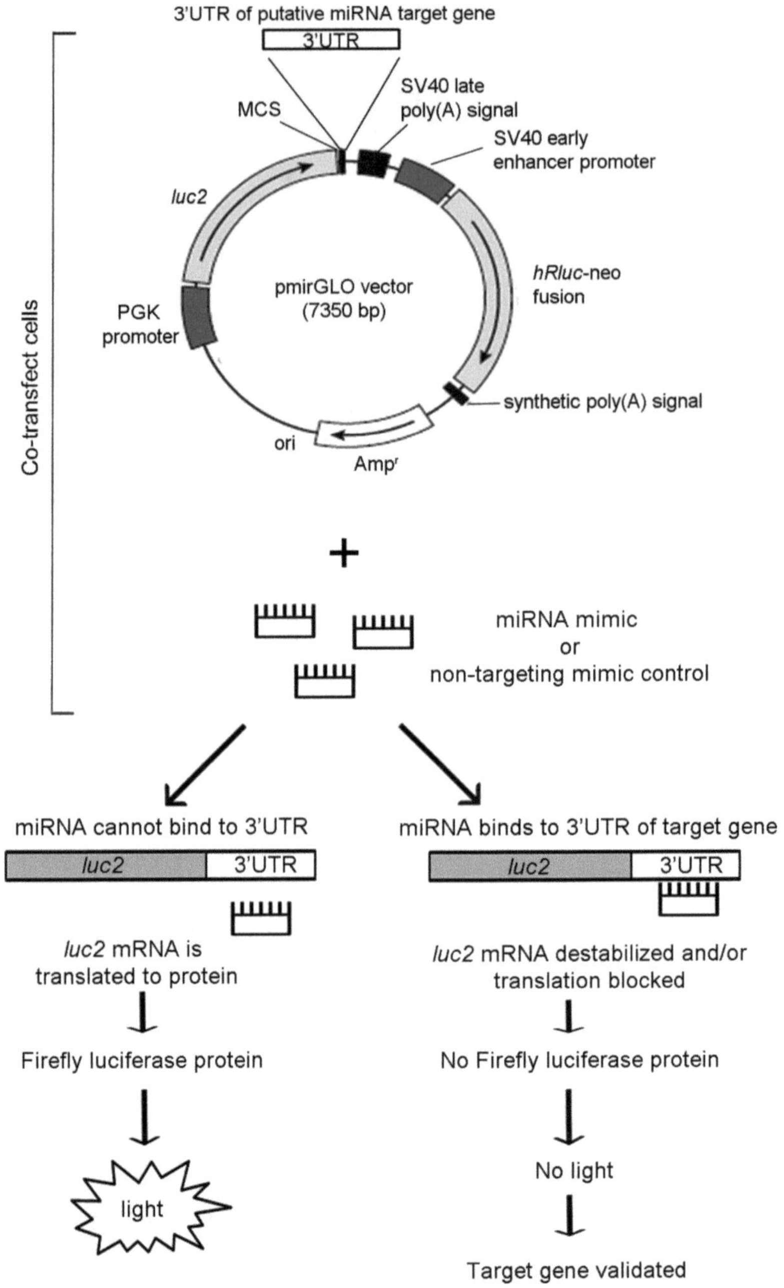

Fig. 2 How to use a 3′UTR Dual Firefly-*Renilla* Luciferase Reporter Assay to validate miRNA targets. The assay described in this chapter is based on Promega dual-luciferase technology, with Firefly luciferase (*luc2*) used as the primary reporter to monitor mRNA regulation and *Renilla* luciferase (*hRluc-neo*) acting as a control reporter for normalization and selection. *hRluc-neo* fusion, codon-optimized synthetic *Renilla* luciferase for expression in mammalian cells fused with the neomycin-resistance gene. *luc2*, Firefly luciferase gene engineered to remove most cryptic transcription factor binding sites and improve mammalian expression through codon optimization. *Amp^r* ampicillin resistance, β-lactamase coding region; *PGK* human phosphoglycerate kinase; *3′UTR* 3′ untranslated region of the putative miRNA target

4. Thermo cycler.
5. Gel electrophoresis apparatus to run agarose gels.
6. Agarose (molecular biology grade).
7. 1× TAE (Tris–acetate–EDTA) buffer (*see* **Note 4**).
8. QIAquick PCR Purification Kit (Qiagen).

2.1.2 DNA Cloning

1. pmirGLO Dual Luciferase miRNA Target Expression Vector (Promega).
2. Restriction enzymes: PmeI and XbaI (New England Biolabs).
3. Cold Fusion Cloning kit (System Biosciences).
4. S.O.C medium: 2 % tryptone, 0.5 % yeast extract, 10 mM NaCl, 2.5 mM KCl, 10 mM $MgCl_2$, 10 mM $MgSO_4$, and 20 mM glucose (*see* **Notes 5** and **6**).
5. Ampicillin for bacterial selection.
6. 100 ml LB liquid medium: 10 g/l tryptone, 5 g/l yeast extract, 10 g NaCl.
7. Sterile 15 ml tubes for bacterial culture.
8. 10-cm bacterial culture plates.
9. LB Ampicillin plates: LB liquid media plus 8 g/l agar. Autoclave, cool down to about 50 °C in a water bath, and add 1 ml ampicillin [50 mg/ml]. Mix well and distribute 15–20 ml of medium per 10-cm plate.
10. Miniprep kit such as QIAprep Spin Miniprep Kit (Qiagen) for isolation of high purity plasmid from bacteria.

2.2 Cell Transfection and 3′ UTR Luciferase Reporter Assay

1. HEK293 (human embryonic kidney cells).
2. EMEM (Eagle's Minimum Essential Medium) (ATCC).
3. Fetal Bovine Serum (FBS).
4. 96-well plate, flat bottom, *white*, with lid (plate for luminescence).
5. Clear flat bottom plates with lid for tissue culture and microscopy (plate to check cell growth).
6. Multichannel pipette.
7. Nuclease-free 1.5 ml tubes and pipette tips.
8. Sterile and nuclease-free water.
9. Transfection reagent Lipofectamine 2000 (Invitrogen).
10. 5× siRNA Buffer (Thermo Scientific).
11. miRIDIAN microRNA Mimic Negative Control 1, 5 nmol (Dharmacon).
12. hsa-miR-1205-5p miRIDIAN Mimic, 5 nmol (Dharmacon).
13. miRCURY LNA Power Inhibitor Negative Control (Exiqon).

14. miRCURY LNA Power Inhibitor miR-1207-5p (Exiqon).
15. Dual-Glo Luciferase Assay Kit (Promega).
16. Plate reader able to detect luminescence.

3 Methods

Figure 1 shows an outline of the sequential procedures for the experimental validation of a potential miRNA target gene using a 3′UTR dual Firefly-*Renilla* Luciferase Reporter Assay. First we describe the step-by-step method to construct the pmirGLO-3′UTR plasmid by cloning the 3′UTR of a potential miRNA target gene downstream the Firefly luciferase gene using a fast cold fusion cloning system. The pmirGLO-3′UTR plasmid is then co-transfected with either miRNA mimics or miRNA inhibitors in HEK293 cells. Finally, luciferase activity is measured to determine whether the specific miRNA is able to bind the 3′UTR of the potential target gene. A decreased luciferase signal compared with the controls confirms that the specific miRNA directly targets and downregulates the gene under study (Fig. 2).

3.1 Plasmid Constructs

3.1.1 Primers Design

1. Use TargetScan (www.targetscan.org) to identify potential miRNA target genes for experimental validation [8–11] (*see* **Note** 7).
2. In the Website of the National Center for Biotechnology Information (NCBI), obtain the 3′UTR sequence of the potential target gene as follows:
   (a) Go to http://www.ncbi.nlm.nih.gov/nucleotide.
   (b) Type the specie and name of the potential target gene. For example, "Homo sapiens G6PD" (glucose-6-phosphate dehydrogenase).
   (c) Select the option for mRNA sequence. For example: "Homo sapiens glucose-6-phosphate dehydrogenase (G6PD), transcript variant 1, mRNA" (NM_000402).
   (d) After the references and under "Features", select "CDS", which is the sequence that is translated to protein. The required 3′UTR sequence is immediately downstream of the CDS. For example, for the human *G6PD* gene (mRNA, transcript variant 1), the CDS is located between the 149 and 1,786 bp. Thus, the 3′UTR is located between 1,787 and 2,377 bp (the rest of the mRNA sequence without the polyA tail). Select and copy the 3′UTR sequence and paste it in a Word file (Fig. 3a).
3. Use TargetScan (www.targetscan.org) to identify potential miRNA-binding site/s in the 3′UTR of the target gene, and mark it in the sequence that you have in Word (in yellow in Fig. 3a).

a

```
                                                            gccc tgggcaccca
1801 cctccacccc cgccacggcc accctccttc ccgccgcccg accccgagtc gggaggactc
1861 cgggaccatt gacctcagct gcacattcct ggccccgggc tctggccacc ctggcccgcc
1921 cctcgctgct gctactaccc gagcccagct acattcctca gctgccaagc actcgagacc
1981 atcctggccc ctccagaccc tgcctgagcc caggagctga gtcacctcct ccactcactc
2041 cagcccaaca gaaggaagga ggagggcgcc cattcgtctg tcccagagct tattggccac
2101 tgggtctcac tcctgagtgg ggccagggtg ggagggaggg acaaggggga ggaaaggggc
2161 gagcacccac gtgagagaat ctgcctgtgg ccttgcccgc cagcctcagt gccacttgac
2221 attccttgtc accagcaaca tctcgagccc cctggatgtc ccctgtccca ccaactctgc
2281 actccatggc caccccgtgc cacccgtagg cagcctctct gctataagaa aagcagacgc
2341 agcagctggg accccctccca acctcaatgc cctgccatta aatccgcaaa cagcc
```

b

Forward primer: start 1922 (136), length 19 bp, Tm 58.8C, GC: 63.2C

5'CTCGCTGCTGCTACTACCC3'

Reverse primer: start 2378 (592); length 18, Tm: 61.5C, GC: 55.6%

5'ATGGCAGGGCATTGAGGT3'

Amplification product: 457 bp

c

Final G6PD forward primer

5' GTGTAATTCTAGTTGGTTTAAACTCGCTGCTGCTACTACCC 3'

15 bp vector pmirGLO + PmeI + specific sequence (18-20 bp)

Final G6PD reverse primer

5' TGCCTGCAGGTCGACTCTAGAATGGCAGGGCATTGAGGT 3'

15 bp vector pmirGLO (comp.) + XbaI + specific sequence (18-20 bp)

Fig. 3 Primer design for cloning the 3′UTR of the human glucose-6-phosphate dehydrogenase (G6PD), a potential target gene for miR-1207-5p. (**a**) 3′UTR (untranslated region) of G6PD (NM_000402). In *yellow*, miR-1207-5p putative binding site; in *grey* and *green*, forward and reverse primers to amplify the 3′UTR region by PCR, respectively. (**b**) Sequence of forward and reverse primers obtained using the Primer 3 Web tool (http://primer3.ut.ee/). (**c**) Final sequence of forward and reverse primers after adding a 15 bp sequence from the pmirGLO vector (or complementary) and restriction sites for XbaI or PmeI. These are the primers used to generate the 3′UTR insert by PCR. The short sequence from pmirGLO is required for the homologous recombination between insert and vector during the cold fusion cloning

4. Using the Primer3 web tool (http://primer3.ut.ee/), design 18–20 bp primers to amplify the whole 3′UTR of your target gene or at least most of it. Make sure that the amplification product includes the potential miRNA-binding site/s. Figure 3b shows the forward and reverse primers selected to amplify the 3′UTR of G6PD.
5. To each primer, add a restriction site for PmeI (only forward primer) and XbaI (only reverse primer) as well as a sequence of 15 bp that is homologous to the vector sequence and required

Table 1
Master mix for PCR amplification

Reagent	Volume [μl]	Final concentration
CloneAmp Hi Fi PCR premix 2×	12.5	1×
Forward primer [20 μM]	0.37	0.3 μM[a]
Reverse primer [20 μM]	0.37	0.3 μM[a]
Genomic DNA [200 ng/μl][b]	1	8 ng/μl
Water	up to 25 μl	

[a]The recommended final primer concentration is 0.2–0.3 μM
[b]50–200 ng of DNA template

for homologous recombination using the cold fusion cloning technology. Figure 3c shows the primers used to amplify the 3′UTR of G6PD. Use the same primers for the target gene, but replace the sequences marked in grey and pink background corresponding to the forward and reverse primers, respectively, with the 18–20 bp sequence of the specific primers obtained by using Primer 3.

6. Negative control for luciferase assays: clone the insert in antisense direction, or mutate the seed region on the 3′UTR using a site-directed mutagenesis kit such as Quick Change II Site-Directed Mutagenesis kit (Agilent) (*see* **Note 8**).

3.1.2 DNA Amplification by PCR

We strongly recommend the *CloneAmp HiFi PCR Premix* to amplify the 3′UTR of the potential miRNA target gene (*see* **Note 9**). This kit provides exceptionally accurate and efficient DNA amplifications, especially for high GC content and/or repetitive sequences. The kit is a convenient 2× master mix containing the DNA polymerase, dNTPs, and $MgCl_2$.

1. Prepare a Master Mix as described in Table 1.
2. Mix the Master Mix by pipetting up and down and centrifuge briefly.
3. Program your thermal cycler: 5 min at 94 °C followed by 35 cycles of 98 °C for 10 s, and 72 °C for 30–60 s/kb (*see* **Notes 10** and **11**). Add a final step of 7 min at 72 °C.
4. Analyze the PCR amplification product using agarose gel electrophoresis (*see* **Notes 12** and **13**).
5. Purify the amplification product using the QIAquick PCR purification kit according to the manufacturer's instructions.

3.1.3 Vector Linearization and Cold Fusion Cloning

Cold fusion technology is a faster and simpler alternative to classical cloning. The plasmid insert is generated by PCR with primers that are designed to have ~15 bases of homology with the vector plus a specific sequence to amplify the region of interest. The insert is rapidly and accurately fused with the linearized vector in the right orientation after only 5 min incubation at room temperature followed by 10 min on ice. The cold fusion cloning is highly effective—with a more than 95 % positive cloning rate—and does not require restriction enzyme, phosphatase, or ligase.

1. Digest pmiRGlo vector with PmeI and XbaI restriction enzymes as follows:
   - 2 µg pmiRGlo.
   - 2 µl PmeI [10 U/µl].
   - 1 µl XbaI [20 U/µl].
   - 5 µl 10× Cut Smart Buffer.
   - ddH_2O up to 50 µl.
2. Mix by pipetting up and down and incubate at 37 °C from 3 h to overnight to decrease linearization of the vector and decrease background during cloning (*see* **Note 14**).
3. Purify the linearized vector using the QIAquick PCR purification kit according to the manufacturer's instructions.
4. Set up the *cold fusion reaction* as follows:
   - 10–100 ng linearized pmiRGlo vector (*see* **Notes 15** and **16**).
   - 20–200 ng PCR insert (*see* **Notes 15** and **16**).
   - 2 µl 5× Master Mix (Cold Fusion Cloning kit).
   - ddH_2O up to 10 µl.
5. Prepare the *positive control* reaction as follows:
   - 1 µl linearized vector (positive control, Cold Fusion Cloning kit).
   - 1 µl 500 bp PCR insert (positive control, Cold Fusion Cloning kit).
   - 6 µl ddH_2O.
   - 2 µl 5× Master Mix (Cold Fusion Cloning kit).
6. Prepare the *negative control* reaction as follows:
   - 10–100 ng linearized pmiRGlo vector (same amount as in **step 4**).
   - 2 µl 5× Master Mix (Cold Fusion Cloning kit).
   - ddH_2O up to 10 µl.
7. Incubate for 5 min at room temperature followed by 10 min on ice all the reactions prepared in **steps 4–6**.
8. Add 50 µl of competent cells (Cold Fusion kit) to the cloning mixture.

9. Incubate on ice for 20 min.
10. Heat shock at 42 °C for 50 s and immediately place on ice for 2 min.
11. Add 250 μl S.O.C medium.
12. Incubate at 37 °C for an hour.
13. Spread 100 μl of the bacterial culture on a pre-warmed (37 °C) culture plate containing ampicillin.
14. Incubate the plate overnight at 37 °C.
15. *Culture bacterial colonies*: touch one individual colony with the tip of a sterile pipette tip and release the tip in a sterile bacterial culture tube containing 5 ml of liquid LB medium + ampicillin. Do the same with at least ten different colonies.
16. Place the tubes with the bacterial culture in a 37 °C incubator with agitation.
17. Isolate bacterial plasmid using the QIAprep Spin Miniprep Kit.
18. Digest the plasmids obtained from the ten colonies with restriction enzymes PmeI and XbaI.
19. Analyze the digested plasmids using agarose gel electrophoresis.
20. Select the plasmids with the expected size of the insert and send them for sequencing.

3.2 Cells Transfection

1. Culture HEK293 cells in EMEM medium + 10 % FBS in a 37 °C/5 % CO_2 incubator until cells are ~80 % confluent.
2. In an Excel spreadsheet, draw a 96-well plate and design your experiment (cell treatment per well). Figure 4a shows an example of an experimental design to validate G6PD as target gene of miR-1207-5p using both miRNA mimics and LNA (lock nucleic acids) anti-miR-1207-5p (*see* **Note 17**).
3. Plate 100 μl per well of a cell suspension of 2×10^5 HEK293 cells/ml (20,000 cells per well) in a *white* 96-wells plate in the

Fig. 4 Experimental design for cell co-transfection of a reporter plasmid and either miRNA mimics or miRNA inhibitors. (**a**) Example of an experimental design to validate G6PD as target gene of miR-1207-5p. *Numbers* from 1 to 12 and *letters* from A to E correspond to the location of wells in a 96-well plate where the cell co-transfection takes place. In each well, plate 100 μl of a 2×10^5 HEK293 cell suspension per ml (20,000 cells per well). The four replicates per treatment are indicated in the same color. Cells are co-transfected with 100 ng/well of plasmid and either 30 nM miRNA mimics or 50 nM miRNA inhibitors (LNA). *Empty vector* 100 ng/well plasmid pmirGLO; *pG6PD* pmirGLO plasmid containing the 3′UTR of the G6PD gene (putative target gene of miR-1207-5p), *pG6PDmut* pG6PD plasmid with a mutated miR-1207-5p binding site. *NC mimic* negative control mimic (*Caenorhabditis elegans* miR-67-3p or cel-miR-67). *LNA miR-1207-5p* power inhibitor lock nucleic acid anti-miR-1207-5p (Exiqon), *LNA NC* negative control, *NT* no-treated cells, *HEK293* human embryonic kidney 293 cells. (**b**) Preparation of mixture 1 (M1 = mix of plasmids with either miRNA mimics or miRNA inhibitors in serum-free antibiotic-free medium). T1–T9, treatments 1–9. (**c**) Loading plan of different treatments: 100 μl of transfection mix (T1–T9 treatments) per well

a

	1	2	3	4	5	6	7	8	9	10	11	12
A	EMPTY VECTOR NT	EMPTY VECTOR NT	EMPTY VECTOR NT	EMPTY VECTOR NT	EMPTY VECTOR NC mimic	EMPTY VECTOR NC mimic	EMPTY VECTOR NC mimic	EMPTY VECTOR NC mimic	EMPTY VECTOR miR-1207 mimic	EMPTY VECTOR miR-1207 mimic	EMPTY VECTOR miR-1207 mimic	EMPTY VECTOR miR-1207 mimic
B	pG6PD + NC mimic	pG6PD + NC mimic	pG6PD + NC mimic	pG6PD + NC mimic	pG6PD + miR-1207 mimic	pG6PD + miR-1207 mimic	pG6PD + miR-1207 mimic	pG6PD + miR-1207 mimic	pG6PDmut + miR-1207 mimic	pG6PDmut + miR-207 mimic	pG6PDmut + miR-1207 mimic	pG6PDmut + miR-1207 mimic
C	pG6PD + LNA NC	pG6PD + LNA NC	pG6PD + LNA NC	pG6PD + LNA NC	pG6PD + LNA miR-1207	pG6PD + LNA miR-1207-5p	pG6PD + LNA miR-1207-5p	pG6PD LNA miR-1207-5p	pG6PD	pG6PD	pG6PD	pG6PD
D	NT	NT	NT	NT								
E												

b

Reagents	M1 T1 [µl]	M1 T2 [µl]	M1 T3 [µl]	M1 T4 [µl]	M1 T5 [µl]	M1 T6 [µl]	M1 T7 [µl]	M1 T8 [µl]	M1 T9 [µl]
Empty vector pmiR-Glo [307.4 ng/µl]	1.63	1.63	1.63	0	0	0	0	0	0
pmiRGlo-G6PDmut [240 ng/ml]						2.08	0	0	0
pmiRGlo-G6PD [268.4 ng/ml]	0	0	0	1.86	1.86		1.86	1.86	1.86
miR-1207-5p mimic [2 µM]	0	0	10.5	0	10.5	11.25	0	0	0
NC mimic [2 µM]	0	10.5	0	10.5	0	0	0	0	0
LNA miR-1207-5p [2 µM]	0	0	0	0	0	0	0	17.5	0
LNA NC [2 µM]	0	0	0	0	0	0	17.5	0	0
Serum-free media	148.4	137.9	137.9	137.6	137.6	136.7	130.6	130.6	148.1
Total	150.0	150.0	150.0	150.0	150.0	150.0	150.0	150.0	150.0

c

	1	2	3	4	5	6	7	8	9	10	11	12
A	T1	T1	T1	T1	T2	T2	T2	T2	T3	T3	T3	T3
B	T4	T4	T4	T4	T5	T5	T5	T5	T6	T6	T6	T6
C	T7	T7	T7	T7	T8	T8	T8	T8	T9	T9	T9	T9
D	NT	NT	NT	NT								

wells in which you are planning to treat the cells (for example, in wells Al to D4, Fig. 4a). Make sure to use only white plates for detection of luciferase activity.

4. Plate 100 μl per well of the same HEK293 cell suspension used in **step 3** but in a CLEAR 96-well plate to assess cell confluence.
5. Incubate cells overnight in an incubator at 37 °C with 5 % CO_2.
6. Thaw the miRNAs mimics and LNA inhibitors (specific and nontargeting controls) at room temperature and briefly centrifuge.
7. In two RNAse-free 1.5 ml tubes, prepare 1× RNA Buffer by mixing 800 μl of RNAse-free water with 200 μl of 5× siRNA Buffer.
8. Using the 1× siRNA Buffer, make a 1/10 dilution of the miRNA mimic, LNA anti-miRs, and negative control stock solutions [2 μM].
9. Prepare *mixture 1* (*M1*) by combining 500 ng of reporter plasmid (i.e., pG6PD) with either 30 nM miRNA mimic, 30 nM negative control (NC) mimic, 50 nM LNA anti-miR, or 50 nM LNA NC in sterile RNase-free 1.5 ml tubes (*see* example of M1 preparation in Fig. 4b).
10. Prepare the following negative controls:
    - *Negative control A*: mix 500 ng of empty vector (pmiRGlo) with either 30 nM miRNA mimic or negative control (NC).
    - *Negative control B*: 500 ng of empty vector (pmiRGlo).
    - *Negative control C*: 500 ng of reporter plasmid (i.e., pG6PD) (*see* **Note 18**).
    - *Negative control D*: no-treated cells.
    - *Negative control E*: reporter plasmid mutated in the seed region or with insert in the antisense direction (i.e., pG6PD mut.).
11. Add 150 μl of serum-free EMEM medium to each tube containing mixture 1.
12. Prepare *mixture 2* (M2) by adding 1 μl of Lipofectamine 2000 to 49 μl of serum-free EMEM medium per treatment. In the example shown in Fig. 4, we performed 9 different treatments. Therefore, we prepared a M2 cocktail for all the treatments by adding 10 μl of Lipofectamine 2000 to 490 μl of serum-free EMEM medium NC in a sterile RNase-free 1.5 ml tube. Mix well by pipetting up and down many times.
13. Incubate for 5 min at room temperature.
14. Add 50 μl of M2 to each of the tubes containing M1, and mix by pipetting up and down may times.

15. Incubate for 20 min at room temperature.
16. Add 550 μl of pre-warmed (37 °C) antibiotic-free complete media (EMEM + 10 % FBS) to each tube containing M1 and M2 (transfection mixture).
17. Remove 96-well plate with HEK293 cells from the incubator, and discard the cell culture medium from the plate using a multichannel pipette.
18. Add 100 μl of the transfection mixture to each well according to the experimental design. In the example presented here, each treatment was performed in quadruplicate (Fig. 4a). Therefore, the content of each of the nine transfection mixture tubes (T1–T9) was loaded on the plate according to Fig. 4c.
19. Place the plate back in the 37 °C/5 % CO_2 incubator for 24 h.

3.3 3′ UTR Luciferase Reporter Assay

1. Completely thaw the Dual Glow Luciferase Buffer and the Dual Glow stop buffer (Dual-Glo Luciferase Assay Kit) in a water bath at room temperature (*see* **Note 19**).
2. Transfer the contents of one bottle of Dual-Glo Luciferase Buffer to one bottle of Dual-Glo Luciferase Substrate to create the *Dual-Glo Luciferase Reagent*. Mix thoroughly by inversion and make sure the substrate is completely dissolved.
3. Calculate the amount of Dual-Glo Stop & Glo Reagent needed to perform the desired experiments (75 μl per well plus 10 %). Dilute the Dual-Glo Stop & Glo Substrate 1:100 into an appropriate volume of Dual-Glo Stop & Glo Buffer in a new container (*see* **Note 20**). For a full plate, combine 10 ml of Dual-Glo Stop & Glo Buffer and 100 μl of Dual-GloStop & Glo Substrate.
4. *Measuring Firefly luciferase activity*: Add a volume of Dual-Glo Luciferase Reagent equal to the culture medium volume to each well and mix. For 96-well plates, typically 75 μl of reagent is added to cells grown in 75 μl of medium (*see* **Notes 21–24**). This reagent induces cell lysis and acts as a substrate for Firefly luciferase, which has a half-life of approximately 2 h.
5. Incubate for at least 10 min at room temperature, and then measure the Firefly luminescence in a plate reader set up for 96-well plates (*see* **Note 25**).
6. *Measuring Renilla luciferase activity*: Add 75 μl of Dual-Glo Stop & Glo Reagent to each well and mix (*see* **Notes 26** and **27**).
7. Incubate for at least 10 min at room temperature, and then measure the *Renilla* luminescence in a plate reader.
8. Calculate the ratio of luminescence from the experimental reporter (Firefly) to the control reporter (*Renilla*) for each well, and then determine the average of the four replicates. Normalize the results for the specific miRNA mimic and LNA

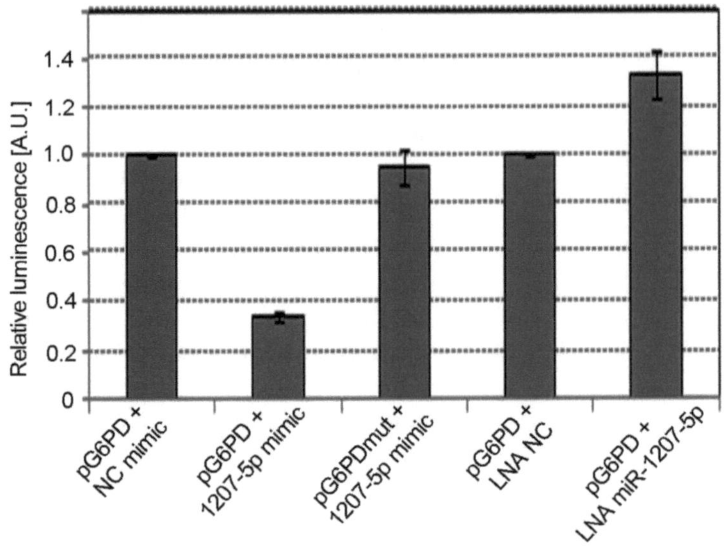

Fig. 5 Experimental Validation of G6PD as a direct target of miR-1207-5p using the Dual Firefly/*Renilla* Luciferase Reporter Assay described in this chapter. pG6PD, pmirGLO plasmid containing the 3′UTR of the G6PD gene (putative target gene of miR-1207-5p); pG6PDmut pG6PD plasmid with a mutated miR-1207-5p binding site; *NC mimic* negative control mimic (*Caenorhabditis elegans* miR-67-3p or cel-miR-67); *LNA miR-1207-5p* power inhibitor lock nucleic acid anti-miR-1207-5p (Exiqon); *LNA NC* negative control; *A.U.* arbitrary units

anti-miR with the control treatment (i.e., HEK293 cells treated with reporter plasmid plus either NC mimic or LNA NC). As an example, Fig. 5 shows the results that confirm G6PD as a direct target gene of miR-1207-5p [18] using the 3′UTR Dual Firefly/*Renilla* Luciferase Reporter Assay described in this chapter.

4 Notes

1. Although many different companies offer high fidelity polymerases for PCR amplifications, we strongly recommend CloneAmp HiFi PCR Premix kit for the PCR amplification of the 3′UTR of the putative miRNA targets. In our hands, this kit showed the highest accuracy and efficiency for DNA amplification of difficult sequences (high GC content, repetitive sequences, etc.).
2. The CloneAmp HiFi PCR Premix is a convenient 2× Master Mix that already contains the CloneAmp HiFi (high fidelity) Polymerase, nucleotides, and $MgCl_2$.
3. Restriction sites for PmeI and XbaI as well as a 15 bp sequence homologous to the vector should be added to the PCR primers (*see* Fig. 3c and Subheading 3).

4. Prepare 1 l of 50× TAE Buffer stock by mixing 242 g/l of Tris base, 57.1 ml of glacial acetic acid, and 100 ml of 0.5 M EDTA pH 8.
5. Alternatively, LB medium (Luria Bertani medium: 10 g/l peptone, 5 g/l yeast extract, 10 g NaCl) can be used but S.O.C. is recommended to obtain a maximal transformation efficiency of *E. coli* [19].
6. Some companies such as Life Technologies offer a convenient ready-to-use S.O.C. medium.
7. Other target prediction algorithms can also be used to identify potential miRNA targets. For example, miRanda (http://www.microrna.org/), PicTar (http://pictar.mdc-berlin.de/), and Diana-microT (http://diana.pcbi.upenn.edu/cgi-bin/micro_t.cgi/). It is recommended that at least two different algorithms predict the same target for a specific miRNA before performing the experimental validation of the target.
8. It is important to use negative controls in the luciferase reporter assays to discard any unspecific or off-target effect of miRNAs.
9. Alternatively, other PCR kits and DNA polymerases can be used as long as the polymerase is a high fidelity enzyme.
10. The recommended range of annealing temperature is 55–72 °C. The annealing temperature (Ta) chosen depends directly on the length and composition of the primers used. Generally, the Ta is about 5 °C below the lowest Tm (melting temperature) of the pair of primers used for PCR amplification [20]. The optimal Ta for the PCR amplification using any given pair of primers on a particular target can be calculated as follows [21]:

    Ta opt. = 0.3 × (Tm of the less-stable primer-template pair) + 0.7 × (Tm of the product).

    Alternatively, some websites offer Tm calculators (Applied Biosystems: http://www6.appliedbiosystems.com/support/techtools/calc/; Thermo Scientific: http://www.thermo-scientificbio.com/webtools/tmc/, Promega: http://www.promega.com/techserv/tools/biomath/calc11.htm).
11. The melting temperature (Tm) should be calculated based on the gene-specific ends of the primer, NOT the entire primer with the 15 pb sequence homologous to the vector and the restriction sites.
12. Make sure that only one band of the expected molecular weight is present in the agarose gel. If there is more than one band, repeat the PCR but increasing the annealing temperature 2–3 °C (or run a gradient PCR at different annealing temperatures between 55 and 72 °C). Alternatively, repeat the PCR but using a lower amount of DNA template. If after PCR

optimization you still continue having extra undesirable bands, design new primers and repeat the PCR, or cut the specific band from the agarose gel and purify it using QIAquick Gel Extraction kit (Qiagen).

13. If no amplification bands are visible in the agarose gel, repeat the PCR but decreasing the annealing temperature 2–3 °C (or run a gradient PCR at different annealing temperatures between 55 and 72 °C). Alternatively, increase the amount of DNA template and repeat PCR.
14. The background of the vector can be assessed by transforming 1 μg linearized and purified vector into competent cells. If many colonies appear in the plate (high background), continue digestion the remaining vector for longer time or add more restriction enzymes.
15. A molar ratio of 2:1 or 1:1 insert:vector is recommended for fusion reaction.
16. For reactions with larger volumes of vector and insert (>8 μl of vector + insert), double the amount of reaction buffer and enzyme, and add ddH_2O up to 20 μl.
17. Alternatively, reporter plasmids can be co-transfected with different amounts of miRNA mimics and LNA anti-miRs (i.e., 10, 25, and 50 nM).
18. If the miRNA of interest is naturally present in HEK293 cells, a decrease in luciferase activity is expected in the negative control C.
19. To avoid the need to temperature equilibrate reagents before use, store the Dual-Glo Luciferase Buffer and the Dual-Glo Stop & Glo Buffer at room temperature.
20. The Dual-Glo Stop & Glo Reagent should be prepared immediately before use.
21. The Luciferase Assay is affected by temperature. Therefore, to achieve maximum reproducibility equilibrate cells in media to room temperature before performing luciferase measurements.
22. After overnight incubation at 37 °C, the initial 100 μl of media per well is generally reduced to 75 μl by evaporation.
23. Add the Dual-Glo Luciferase Reagent directly to cells in growth medium without removing the medium or washing the cells. The Dual-Glo Luciferase Assay System is designed to work in growth media commonly used for mammalian cells with or without added serum.
24. Save the rest of the Dual-Glow Luciferase *reagent* at –70 °C. The reagent can be exposed to five freeze-thaw cycles with approximately a 15 % loss in Firefly RLU (relative luciferase activity).

Holding or storing the reconstituted reagent may cause *Renilla* RLU to rise.

25. For optimal results, measure luminescence within 2 h of the addition of the Dual-Glo Luciferase Reagent.
26. The Dual-Glo Stop & Glo Reagent quenches the luminescence from the Firefly reaction by at least 10,000-fold and provides the substrate for the *Renilla* luciferase.
27. Optimal results will be generated if the luminescence is measured within 2 h of the addition of Dual-Glo Stop & Glo Reagent.

References

1. Chang TC, Mendell JT (2007) microRNAs in vertebrate physiology and human disease. Annu Rev Genomics Hum Genet 9:215–239
2. Tang X, Tang G, Ozcan S (2008) Role of microRNAs in diabetes. Biochim Biophys Acta 1179:697–701
3. Lanford RE, Hildebrandt-Eriksen ES, Petri A et al (2010) Therapeutic silencing of microRNA-122 in primates with chronic hepatitis C virus infection. Science 327:198–201
4. Jones-Rhoades MW, Bartel DP (2004) Computational identification of plant microRNAs and their targets, including a stress-induced miRNA. Mol Cell 14:787–799
5. Zhang Y (2005) miRU: an automated plant miRNA target prediction server. Nucleic Acids Res 33:W701–W704
6. Rhoades MW, Reinhart BJ, Lim LP et al (2002) Prediction of plant microRNA targets. Cell 110:513–520
7. Bartel DP (2009) MicroRNAs: target recognition and regulatory functions. Cell 136: 215–233
8. Lewis BP, Burge CB, Bartel DP (2005) Conserved seed pairing, often flanked by adenosines, indicates that thousands of human genes are microRNA targets. Cell 120:15–20
9. Friedman RC, Farh KK, Burge CB et al (2009) Most mammalian mRNAs are conserved targets of microRNAs. Genome Res 19:92–105
10. Grimson A, Farh KK, Johnston WK et al (2007) MicroRNA targeting specificity in mammals: determinants beyond seed pairing. Mol Cell 27:91–105
11. Garcia DM, Baek D, Shin C et al (2011) Weak seed-pairing stability and high target-site abundance decrease the proficiency of lsy-6 and other microRNAs. Nat Struct Mol Biol 18: 1139–1146
12. Betel D, Wilson M, Gabow A et al (2008) The microRNA.org resource: targets and expression. Nucleic Acids Res 36:D149–D153
13. Griffiths-Jones S, Grocock RJ, van Dongen S et al (2006) miRBase: microRNA sequences, targets and gene nomenclature. Nucleic Acids Res 34:D140–D144
14. Krek A, Grun D, Poy MN et al (2005) Combinatorial microRNA target predictions. Nat Genet 37:495–500
15. Maragkakis M, Vergoulis T, Alexiou P et al (2011) DIANA-microT Web server upgrade supports fly and worm miRNA target prediction and bibliographic miRNA to disease association. Nucleic Acids Res 39:W145–W148
16. Sthupathy P, Megraw M, Hatzigeorgiou AG (2006) A guide through present computational approaches for the identification of mammalian microRNA targets. Nat Methods 3:881–886
17. Kuhn DE, Martin MM, Feldman DS (2008) Experimental validation of miRNA targets. Methods 44:47–54
18. Alvarez ML, Khosroheidari M, Eddy E et al (2013) Role of MicroRNA 1207-5P and its host gene, the long non-coding RNA Pvt1, as mediators of extracellular matrix accumulation in the kidney: implications for diabetic nephropathy. PLoS One 8:e77468
19. Hanahan D (1983) Studies on transformation of Escherichia coli with plasmids. J Mol Biol 166:557–580
20. Innis MA, Gelfand DH, Sninsky JJ et al editors (1990) PCR Protocols: A guide to methods and applications. Academic Press Inc, San Diego, CA
21. Rychlik W, Spencer WJ, Rhoads RE (1990) Optimization of the annealing temperature for DNA amplification *in vitro*. Nucleic Acids Res 18:6409–6412

Chapter 22

Experimental Validation of Predicted Mammalian MicroRNAs of Mirtron Origin

Anita Schamberger and Tamás I. Orbán

Abstract

MicroRNAs (miRNAs) are ~22 nucleotide-long noncoding RNAs influencing many cellular processes by their regulatory functions on gene expression. MiRNAs of mirtron origin represent the most prominent group of the alternatively processed miRNAs. They reside in short introns, which are essentially equivalent to the precursor form of the given miRNA. Consequently, their maturation is independent of the Drosha/DGCR8 complex, while depends on the mechanism of mRNA splicing. The number of predicted human mirtron sequences increases as a consequence of the growing deep sequencing data and refined bioinformatics tools. However, experimental validations of particular sequences are also essential. In this chapter, we intend to provide detailed protocols for the investigation of predicted mirtron sequences. First, we use the *Sleeping Beauty* transposon-based gene-delivery system for the development of cell lines stably overexpressing mirtrons. The processing of functional mature miRNAs is then detected by a luciferase assay using a very strict "triple control" system. In addition, *bona fide* mirtron features are confirmed by demonstrating splicing dependency through splice site mutations, while Drosha/DGCR8 independency is assessed in DGCR8 deficient cell line. Finally, the presence of mirtron-derived mature miRNAs is detected by quantitative real-time PCR.

Key words miRNA, Mirtron, *Sleeping Beauty* transposon, Stable miRNA expression, Luciferase assay, qRT-PCR, Real-time PCR, Splicing, DGCR8, Drosha

1 Introduction

MicroRNAs (miRNAs) are short, single-stranded regulatory RNAs about 20–24 nucleotides in length. These small noncoding RNA molecules form a gene regulatory network comparable to that of transcription factors by targeting mRNA molecules via sequence complementarity [1–4]. MiRNAs often control several mRNAs simultaneously, and have an impact on most cellular processes, e.g., on the regulation of ontogenesis, cell proliferation, cell differentiation, and apoptosis, and could play a role in the pathomechanisms of cancer and metabolic diseases [2, 5, 6]. Most of these small miRNA molecules are generated via a canonical pathway,

M. Lucrecia Alvarez and Mahtab Nourbakhsh (eds.), *RNA Mapping: Methods and Protocols*, Methods in Molecular Biology, vol. 1182, DOI 10.1007/978-1-4939-1062-5_22, © Springer Science+Business Media New York 2014

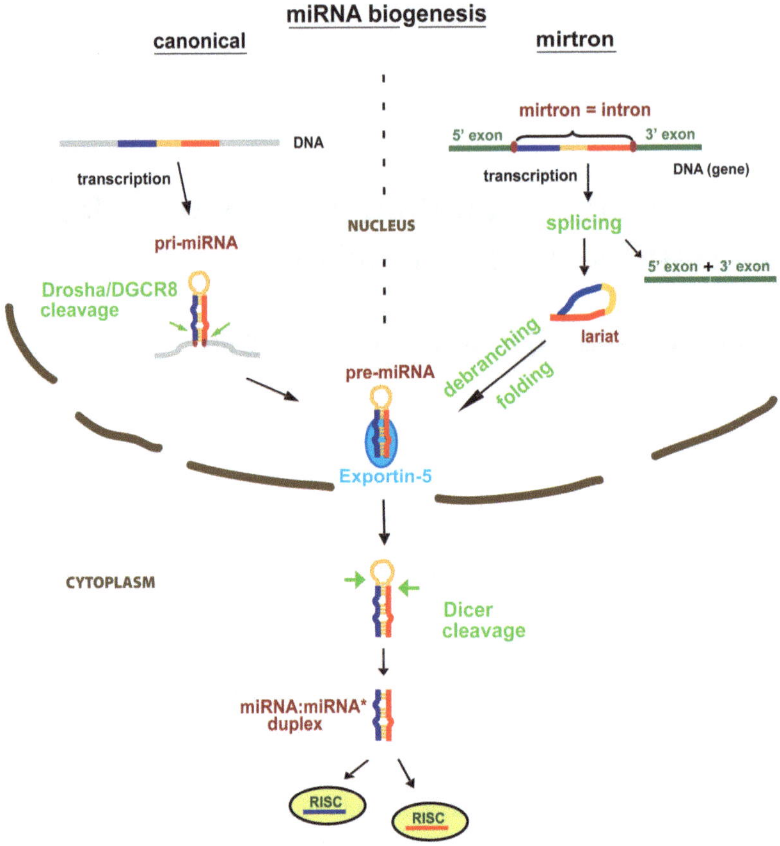

Fig. 1 Comparison of the canonical and mirtron biogenesis pathways. These two miRNA maturation pathways differ in their first cleavage step in which a pre-miRNA is liberated from the primary transcript. In the canonical pathway, the Drosha/DGCR8 microprocessor complex carries out the first cleavage reaction, whereas the splicing machinery is used in the mirtron-biogenesis pathway

but there are also emerging data on various alternative miRNA maturation routes.

In the canonical miRNA biogenesis pathway (Fig. 1), after the transcription of the primary miRNA transcript (pri-miRNA), an RNaseIII-type enzyme called Drosha cuts out the hairpin structured precursor miRNA form (pre-miRNA). This cleavage step is mediated by DGCR8, a partner protein of Drosha. The pre-miRNA is then transported from the nucleus to the cytoplasm, where Dicer, another RNaseIII-type enzyme cuts the precursor. This second cleavage step liberates a small RNA (miRNA:miRNA*) duplex, from which one strand is incorporated into an Argonaute (Ago) protein containing RNA-induced silencing complex (RISC), while the other strand is degraded. The miRNA-mediated gene silencing effect can be manifested by translation inhibition and/or mRNA decay [7, 8]. The alternative biogenesis pathways could bypass certain steps of the canonical process, typically one of the two cleavage steps. Therefore, there are Drosha-independent and Dicer-independent alternative maturation pathways [9, 10].

The most prominent Drosha-independent miRNA biogenesis is the mirtron pathway (Fig. 1). It was first described in *Drosophila melanogaster* and *Caenorhabditis elegans* [11, 12], and experimentally demonstrated also in mammals [13–15]. MiRNAs of mirtron origin are localized in short introns where the whole intron is essentially equivalent to the pre-miRNA form. Thus, the first step of the mirtron processing is different from the canonical one: the pre-miRNA is cleaved out from the primary transcript by the splicing machinery instead of the Drosha/DGCR8 complex [16]. Apart from the originally described type of mirtrons, there are other closely related groups: the 3′- and 5′-tailed mirtrons. Their ends are also defined by the intronic splicing donor and acceptor sites, but they have extended ends in the pre-miRNA form either in 3′ or in 5′ direction. These "overhangs" are further processed by exonucleases [17–19].

There are emerging data on mirtrons indicating the importance of this alternative miRNA maturation pathway in various organisms, including plant species [20, 21]. However, beside small RNA data analyses and bioinformatics predictions, experimental validation of predicted mirtron sequences is also an important issue. In this chapter, we provide detailed protocols for the experimental validation of mammalian mirtrons. First, we describe an efficient way for the development of mirtron-expressing stable cell lines by using the *Sleeping Beauty* transposon-based gene-delivery system. Next, to test the presence of mirtron-derived functional miRNA species, we describe a protocol for luciferase assay using a very strict "triple control" analysis. To confirm mature miRNA processing by the mirtron pathway, details of testing splicing-dependent and Drosha/DGCR8-independent maturation are provided. Finally, we describe a refined protocol of the real-time quantitative stem-loop PCR to accurately detect and quantify mature miRNAs.

2 Materials

2.1 Generation of Mirtron-Encoding Vectors

2.1.1 Annealing of Oligonucleotides and Ligation

1. DNA oligonucleotides (sense and antisense strands of mirtron coding sequence) can be purchased from commercial source (Sigma; *see* **Note 1**).
2. NEBuffer2 (New England BioLabs): 50 mM NaCl, 10 mM Tris–HCl, 10 mM $MgCl_2$, 1 mM dithiothreitol (pH = 7.9).
3. Heating block.
4. GFP-encoding plasmid suitable for cloning (*see* **Note 2**).
5. T4 DNA Ligase (New England BioLabs).

2.1.2 Transformation

1. Competent *Escherichia coli* cells.
2. Sterile Luria-Bertani (LB) medium: 10 g tryptone, 5 g yeast extract, 10 g NaCl in 1 l distilled H_2O.
3. LB-agar (1 l LB + 15 g agar) plated in Petri dishes.
4. Antibiotics matching with the transformed plasmid (ampicillin, 100 mg/ml).
5. Water bath.
6. 37 °C incubator/shaker.

2.1.3 Colony PCR, Gel Electrophoresis, and Plasmid Isolation

1. PCR Master Mix: 50 U/ml Taq DNA polymerase in reaction buffer of pH = 8.5, 400 μM dNTP Mix, 3 mM $MgCl_2$ (Promega).
2. Primers flanking the intronic mirtron sequences: forward (5′– TTCTTCAAGTCCGCCATGCC) and reverse (5–ACT TGTACAGCTCGTCCATGCCG) (*see* **Note 3**).
3. PCR machine.
4. Agarose.
5. 10× TBE buffer: 890 mM Tris–HCl, 890 mM boric acid, 20 mM EDTA (pH = 8).
6. Ethidium bromide.
7. DNA ladder and loading dye.
8. Gel electrophoresis equipment.
9. Plasmid isolation kit (QIAprep®Spin Miniprep Kit, Qiagen).

2.2 Development of Stable Mirtron-Expressing HeLa Cell Lines

2.2.1 Cell Line Maintenance and Transfection

1. HeLa cell line (ATCC).
2. Fetal bovine serum.
3. Dulbecco's Modified Eagle Medium (DMEM).
4. Phosphate-buffered saline (PBS).
5. Trypsin.
6. *Sleeping Beauty* transposon and transposase vectors.
7. FuGENE® HD reagent (Promega).

2.2.2 Cell Sorting

1. Tubes suitable for flow cytometry.
2. Flow cytometry cell sorter (FACS Aria High Speed Cell Sorter, Beckton-Dickinson).

2.3 Detection of Mirtron Function by Luciferase Assay

1. FuGENE® HD reagent (Promega).
2. psiCHECK2 vector (Promega; *see* **Note 4**).
3. Dual Luciferase® Reporter Assay System Kit (Promega), containing special reagents such as the "Passive Lysis Buffer" and "LARII" and "Stop & Glo" reagents.

4. Rocking platform or orbital shaker.
5. Opaque plate (OptiPlate-96, PerkinElmer).
6. Glass tube.
7. Luminometer (2030 Multilabel Reader luminometer, PerkinElmer).

2.4 Investigation of Mirtron Features

2.4.1 Splicing Dependency

1. Splice site mutant mirtron oligonucleotides (*see* **Note 5**).
2. Materials and equipment for the generation of vectors (*see* Subheading 2.1).
3. Transfection reagent (*see* Subheading 2.2.1).
4. Luciferase assay (*see* Subheading 2.3).

2.4.2 DGCR8 Independency

1. Wild type and DGCR8-deficient mouse embryonic fibroblast (MEF) cell lines.
2. Fetal bovine serum.
3. Dulbecco's Modified Eagle Medium (DMEM).
4. Phosphate-buffered saline (PBS).
5. Trypsin.
6. Lipofectamine® 2000 (Invitrogen).
7. Opti-MEM (Gibco®, Invitrogen).
8. Luciferase assay (*see* Subheading 2.3).

2.5 Detection of Mature miRNAs by qRT-PCR

Use RNase-free reagents and equipment.

2.5.1 Total RNA Isolation

1. Trizol reagent (Invitrogen; *see* **Note 6**).
2. Chloroform.
3. Isopropyl alcohol.
4. 75 % ethanol.
5. Formamide.
6. Ethidium bromide.
7. Spectrophotometer (NanoDrop 2000 Spectrophotometer, Thermo Scientific).
8. Gel electrophoresis equipment.

2.5.2 miRNA-Specific Reverse Transcription

1. DNase (2 U/μl) and 10× DNase buffer (New England BioLabs; *see* **Note 7**).
2. RNase inhibitor (20 U/μl, Life Technologies).
3. Spectrophotometer (NanoDrop 2000 Spectrophotometer, Thermo Scientific).

4. miRNA Reverse Transcription Kit (TaqMan® MicroRNA, Life Technologies): 100 mM dNTP Mix, Reverse Transcriptase (50 U/μl), Reverse Transcriptase buffer, RNase inhibitor (20 U/μl).
5. miRNA-specific primer (provided in TaqMan® MicroRNA Assays, Life Technologies or can be synthesized).
6. PCR machine.

2.5.3 Quantitative Real-Time PCR

1. Master Mix (TaqMan® Universal Master Mix II with UNG, Life Technologies).
2. miRNA-specific probe (20× probe provided in TaqMan® MicroRNA Assays, Life Technologies or can be synthesized).
3. cDNA.
4. Real-time PCR machine (StepOne™ or StepOnePlus™ platforms, Life Technologies).
5. Equipments for real-time PCR (MicroAmp Fast Optical 48 or 96-well Reaction plate, MicroAmp Optical Adhesive Film or MicroAmp Fast Reaction Tubes and Cap Strip, Life Technologies).
6. Software for data analysis (StepOne version 2.1, Life Technologies).

3 Methods

3.1 Generation of Mirtron-Encoding Vectors

Because of its ability to make stable cell lines, we use the *Sleeping Beauty* transposon-based vector system for the expression of mirtrons (Fig. 2a; [22, 23]). We clone mirtron sequences in the previously modified EGFP coding sequence (a PvuII site was introduced) as artificial introns [15, 24].

3.1.1 Mirtron Sequence

Mirtron sequence can be designed based on miRNA databases, e.g., miRBase (http://www.mirbase.org). Subsequently, mirtron-encoding (intronic) sequences can be obtained by annealing DNA oligonucleotides (sense and its respective antisense counterpart).

3.1.2 Annealing of Sense and Antisense Oligonucleotides and Ligation

1. Mix 5 μl of mirtron-encoding sense oligonucleotide [100 μM] and 5 μl of antisense oligonucleotide [100 μM]. Add 5 μl of NEB2 buffer and 85 μl of dH_2O.
2. Mix and incubate for 3 min at 96 °C, then turn off heating block, and let the tube cool down slowly in block to room temperature (typically overnight). The annealed double-stranded oligos can be used next day for ligation or stored at –20 °C.

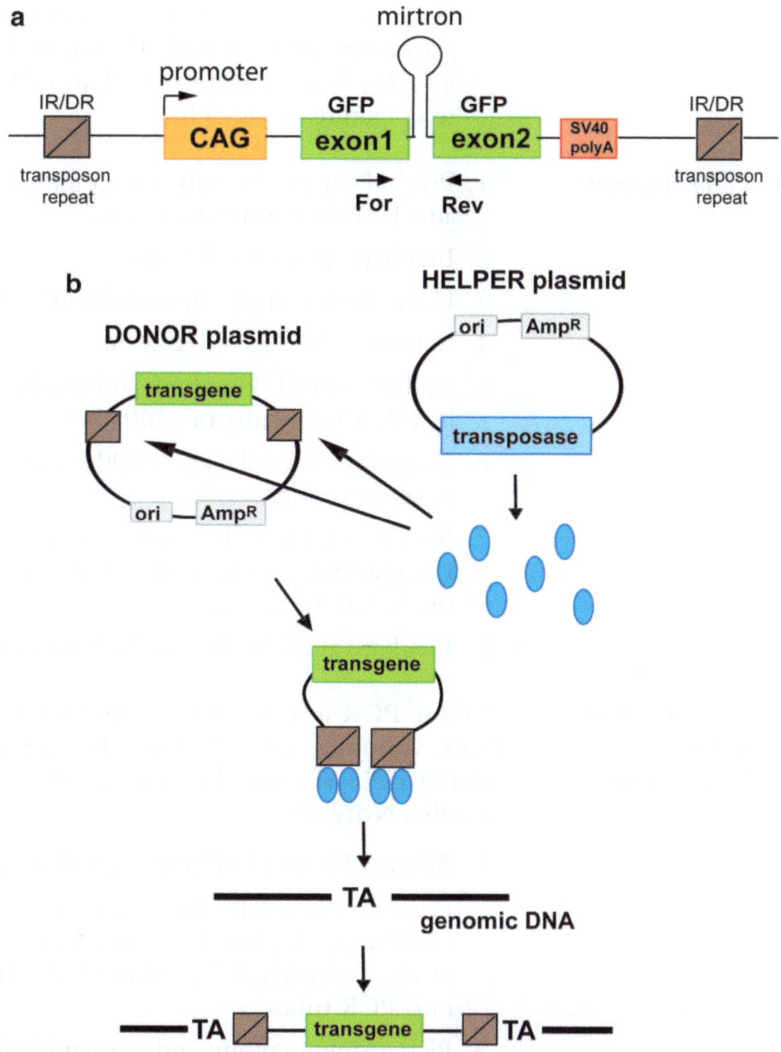

Fig. 2 Overview of the *Sleeping Beauty* transposon-based gene-delivery system. (**a**) Schematic illustration of a transposon vector, suitable for stable mirtron expression. *IR/DR* stands for inverted repeat-direct repeat sequences, *For* and *Rev* indicate binding sites for the PCR primers used for subsequent analysis (colony PCR, detection of splicing). (**b**) The "cut-and-paste" transposition mechanism of the *Sleeping Beauty* system. Following cotransfection, the transposase protein (*blue ellipses*) is expressed from the helper plasmid. Subsequently, the transposase recognizes the IR/DR repeats on the donor plasmid, cleaves the DNA, and liberates the transgene carrying transposon. Then, the transposase/transgene complex scans the genomic DNA for a target sequence, and integrates the transposon at a "TA" dinucleotide. During the integration process, the "TA" motif is duplicated, leaving a typical "footprint" of the *Sleeping Beauty* system. *ori* bacterial replication origin sequence, *Amp*^{R} ampicillin resistance gene

3. For ligation, mix 1 μl of 10× ligase buffer, 0.5 μl (~10 ng) of plasmid backbone, 0.5 μl of hybridized oligos (previously diluted in 1:6), 7 μl of H_2O, and 1 μl of ligase. Incubate at room temperature for ~2–5 h.

3.1.3 Transformation

1. Prechill on ice the tube containing the ligation reaction, and add 100 μl of competent cells.
2. Incubate on ice for 30 min.
3. Heat shock competent cells for 40 s at 42 °C.
4. Return cells to ice for 1–2 min.
5. Add 890 μl of LB medium and incubate at 37 °C for 20–60 min in a shaking incubator (200–300 rpm; *see* **Note 8**).
6. Centrifuge the tube at ~8,000 × *g* for 1 min and take off 900 μl of supernatant.
7. Resuspend the pellet and spread onto antibiotic-containing LB-agar plates (e.g.: 100 mg/ml ampicillin, corresponding to the backbone resistance).
8. Incubate at 37 °C for 16–20 h for colony formation.

3.1.4 Colony PCR, Gel Electrophoresis, and Plasmid Isolation

Colony PCR is a suitable method for screening successful ligations. Colonies are picked one by one, their plasmids are liberated by cell lysis, and then they can be used as template for PCR (*see* also **Note 3**).

1. Aliquot 10–10 μl of H_2O into PCR tubes.
2. Prepare the PCR mix: 12.5 μl of PCR 2× Master Mix (Promega), 1 μl of forward primer [10 μM], 1 μl of reverse primer [10 μM], 9.5 μl of H_2O. Aliquot 24–24 μl of PCR mix into PCR tubes.
3. Pick a single colony and suspend it in 10 μl of H_2O. Pick as many colonies as you wish (approx. 5–10 per ligation).
4. Lyse bacteria at 95 °C for 3 min, then cool down to 4 °C or put on ice.
5. Add 1 μl of lysate (from supernatant) to an aliquoted PCR mix (24 μl).
6. Perform PCR reaction: 95 °C for 3 min; 35 cycles of: 95 °C for 30 s (denaturation), 59 °C for 30 s (annealing), 72 °C for 45 s (extension); 72 °C for 5 min (final extension).
7. Run 3 μl of PCR samples on 3 % agarose gel (*see* **Note 9**).
8. PCR amplicons in the appropriate size can be purified and sequenced.
9. Selected colonies can be grown overnight in 3–5 ml of LB (containing the appropriate antibiotics) in a shaking incubator.
10. Plasmid DNA can be prepared by any commercially available method (e.g.: QIAprep®Spin Miniprep Kit).

3.2 Development of Stable Mirtron-Expressing HeLa Cell Lines

The *Sleeping Beauty* transposon-based gene-delivery system is a simple and effective method for the generation of stable transgene-expressing cell lines. It requires two plasmids: a donor plasmid (encoding the gene of interest flanked by the transposon repeats) and a helper plasmid (encoding the transposase; Fig. 2b). The transposase inserts the transposon from the plasmid to a genomic location by a "cut-and-paste" mechanism [25, 26].

3.2.1 Transfection

1. Plate 3×10^6 HeLa cells/well in a 6-well plate in 2 ml of medium.
2. Next day mix 1 μg of transposon and 100 ng of transposase-encoding plasmid with 100 μl of serum- and antibiotics-free DMEM.
3. Add 3.3 μl of FuGENE® HD reagent (avoid the contact between the plastic wall of the tube with undiluted transfection reagent), mix gently by tapping the tube, and incubate for 45–60 min at room temperature (*see* **Note 10**).
4. Add the transfection complex to the cells in a drop-wise manner, swirl the plate gently to mix.
5. After 48 h, check transfection efficiency by fluorescence microscope (EGFP fluorescence). Ideally, about 70–100 % of cells should be positive for EGFP.

3.2.2 Cell Sorting

1. 2 days after transfection, transfer the cells into a 25 cm^2 cell culture flask.
2. When cells are confluent, passage them to a 75 cm^2 flask.
3. When cells are confluent (normally at day 8 after transfection; *see* **Note 11**), collect 3/4 of the cell population in 0.5–1 ml of medium for sorting, and save the rest as a backup cell culture.
4. Sort the cells based on marker gene expression (GFP fluorescence), gating the cells rigorously (*see* **Note 12**).
5. Put the sorted cells back into a 6-well plate or into a 25 cm^2 cell culture flask (*see* **Note 13**).
6. To achieve 100 % positive cell population, sort the cells again for marker gene (GFP) expression at day 17 after transfection.

3.3 Detection of Mirtron Function by Luciferase Assay

The assay can be measured either in stably or transiently transfected mirtron-expressing HeLa cell line. We test every mirtron for both 5′- and 3′- arms with sensor and mutant sensor constructs. Sensor constructs contain two fully complementary antisense copies of either the 5′- or the 3′-arm of a particular mirtron, cloned into the 3′-untranslated region (3′-UTR) of Renilla luciferase in the psiCHECK2 vector. Mutant sensors differ in three mismatched nucleotides in the predicted miRNA seed region (*see* also **Note 4**). For luciferase assay, prepare three parallels for each setup (*see* **Note 14**). To exclude nonspecific effects, we compare mirtron-derived data with a negative control (non-cognate miRNA) using the particular mirtron-specific sensors (*see* also Subheading 3.3.3). We describe here a single-sample measurement in a plate-reading luminometer; however, luciferase assays can be measured also in a multi-sample manner (*see* also **Note 15**).

3.3.1 Transfection

1. Plate 0.5×10^6 stable mirtron-expressing HeLa cells or 0.5×10^6 "parental" HeLa cells per well in a 24-well plate (0.5 ml of medium per well).
2. Next day, transfect cells using the FuGENE® HD reagent, (*see* Subheading 3.2.1). For stable cell lines, use 15 ng of luciferase sensor or mutant sensor construct (*see* **Notes 4** and **14**) and 0.045 μl of transfection reagent per well (typically, prepare a mastermix for several reactions because of the small volumes). For transient expression, use 300 ng of mirtron-encoding plasmid and 15 ng of luciferase sensor (or mutant sensor) construct with 0.945 μl of transfection reagent per well.
3. For transient transfection, check transfection efficiency after 24 h by fluorescence microscope (EGFP fluorescence).

3.3.2 Luciferase Assay

1. 48 h after transfection, remove the culture medium, and rinse the cells carefully with PBS. Remove the PBS completely.
2. Lyse the cells by passive lysis: add 100 μl of 1× Passive Lysis Buffer (1:10 dilution of the 10× Passive Lysis Buffer provided in the Dual Luciferase® Reporter Assay System).
3. Place the culture plate on an orbital shaker or rocking platform for 15–30 min to ensure complete lysis.
4. Prepare LARII and Stop & Glo reagents (provided in the Dual Luciferase® Reporter Assay System) according to the manufacturer's instructions (*see* **Note 16**).
5. Pipette 100 μl of LARII (provided in the Dual Luciferase® Reporter Assay System) into a well of a 96-well OptiPlate (*see* also **Note 15**).
6. Add 20 μl of lysate, mix gently by pipetting.
7. Measure luminescence immediately for 10 s, and record the data (firefly activity).
8. Add 100 μl of Stop & Glo reagent (Dual Luciferase® Reporter Assay System), mix thoroughly to quench luminescence from firefly luciferase reaction.
9. Immediately measure luminescence (Renilla luciferase activity) for 10 s, record the data.

3.3.3 Evaluation of the Data

The psiCHECK2 vector encodes two luciferases. The sensor element for the mirtron is cloned in the 3′ UTR of the Renilla luciferase (the level of this protein can be altered by silencing), while the firefly luciferase serves as transfection control. Consequently, the raw data of Renilla activity is always normalized with the data of the firefly activity.

To ensure that the silencing effect is specific for a given mirtron, we use two additional controls: a non-cognate control and a sensor/seed-mutant sensor pair (*see* also **Note 14**). After normalization of the data with the transfection control (firefly luciferase), we calculate the sensor/mutant sensor ratio for the mirtron, and compare it to the sensor/mutant sensor ratio of a non-cognate miRNA control. Using this "triple control" system, we maximize the exclusion of nonspecific effects, which can result either from the different plasmid constructs and/or from endogenous small RNAs. In addition, we use three replicates for each experimental setup.

3.4 Investigation of Mirtron Features: Splicing Dependency and DGCR8 Independency

3.4.1 Generation of 5′ Splice Site-Mutated Mirtron-Encoding Vectors

Splicing dependency of mature mirtron generation can be investigated by abolishing the splicing of mirtrons. The generation of splice site-mutated mirtron-encoding vectors is similar to that of the wild type construct (*see* Subheading 3.1), except that a 5′ dinucleotide mutated oligonucleotide (corresponding to the mutant splicing donor site) is ligated into the marker gene as an artificial intron (*see* **Note 17**).

3.4.2 Functional Testing of Mutant Mirtrons by Luciferase Assay

A luciferase assay is used to examine if splicing influences the biogenesis of functional miRNAs of mirtron origin. In this assay, we compare the capability of a wild type and a 5′ splice (donor site) mutated mirtron to produce a functional mature miRNA. For mutated mirtrons, we use the same setup described earlier, including a non-cognate control and sensor/mutant-sensor pairs (*see* Subheading 3.3).

3.4.3 Functional Testing of the Generation of a Mature miRNA of Mirtron Origin in a DGCR8-Deficient Cell Line

To prove that a mature miRNA is processed by the mirtron biogenesis pathway, a Drosha/DGCR8-independent processing has to be demonstrated. We investigate Drosha/DGCR8-independent mature miRNA generation using a luciferase assay in a DGCR8-deficient mouse embryonic fibroblast (MEF) cell line (*see* **Note 18**). For this measurement, we recommend to use a canonical miRNA as a control [15]. In wild type MEFs, functional mature miRNAs are processed normally from both the mirtron and the canonical pathway. However, in DGCR8-deficient MEFs, the detection of a functional mature miRNA is expected only in the case of *bona fide* mirtrons, since functional mature miRNA generation by the canonical pathway is abolished.

1. Plate 0.3×10^5 DGCR8-deficient MEF cells or 0.6×10^5 wild type MEF cells in 0.4 ml of antibiotic-free DMEM per well in a 24-well plate.

2. Next day prepare the transfection mix by adding 0.33 μl of Lipofectamine® 2000 to 50 μl of serum- and antibiotics-free Opti-MEM. Mix gently. Incubate for 15 min.
3. Mix 50 μl of serum- and antibiotics-free Opti-MEM with 300 ng of mirtron-encoding plasmid and 15 ng of sensor (or mutant-sensor) construct.
4. After incubation, combine transfection mix generated in **step 2** with the DNA mix prepared in **step 3**. Mix gently and incubate for 20 min.
5. Add the transfection complex to the cells in a drop-wise manner, swirl the plate gently to mix.
6. After 48 h, check transfection efficiency by fluorescence microscope (EGFP fluorescence).

3.5 Detection of Mature miRNAs by qRT-PCR

Stem-loop quantitative PCR is a good method of choice for directly detecting low levels of mature miRNAs (*see* **Note 19**; Fig. 3).

3.5.1 Total RNA Isolation

1. Add 1 ml of Trizol per well to confluent cells (100 % confluency) in a 6-well plate, mix thoroughly by pipetting. A minimum of 1–2 × 10^6 cells/ml Trizol is strongly recommended (*see* **Note 6**). Samples can be processed immediately or stored at −80 °C.
2. Incubate for 2–3 min at room temperature before adding 0.2 ml of chloroform.
3. Shake for 15 s by hand and incubate for additional 2–3 min at room temperature.
4. Centrifuge at 12,000 × *g* for 15 min at 4 °C.
5. Transfer about 3 × 150 μl of the aqueous phase to a fresh tube (*see* **Note 20**).
6. Add 0.5 ml of isopropanol, mix by inverting the tubes, and incubate at room temperature for 10 min.
7. Centrifuge at 12,000 × *g* for 10 min at 4 °C.
8. Remove the supernatant, and wash the pellet by adding 1 ml of 75 % ethanol. Vortex the samples.
9. Centrifuge at 7,500 × *g* for 10 min at 4 °C.
10. Remove the supernatant, and let the pellet dry (do not let it dry completely because it would decrease its solubility).
11. Dissolve the RNA in 50 μl of preheated (60–70 °C) RNase-free H_2O.
12. Measure purity and concentration by spectrophotometry (e.g.: NanoDrop). Ideally, RNA concentration should be 1–1.5 μg/μl.

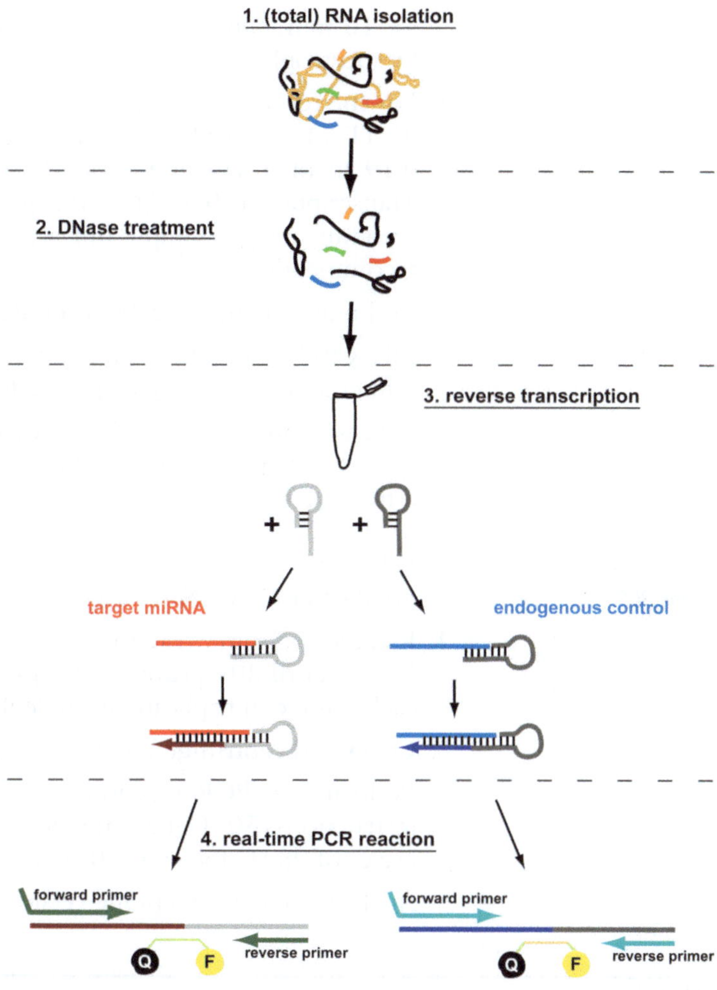

Fig. 3 Overview of methods used for the detection of mature miRNAs by qRT-PCR. *F* fluorophore, *Q* quencher

13. Check integrity of the samples by gel electrophoresis: Mix 1 μg (~1–2 μl) of RNA and 10 μl of formamide, incubate at 70 °C for 10 min, and chill on ice. Add 1 μl of ethidium bromide [10 mg/ml] and 2.5 μl of 6× loading dye. Analyze samples by electrophoresis in a 1 % agarose gel. The ratio of the intensity of the bands corresponding to the 28S and 18S ribosomal RNA reflects the integrity of the RNA sample (ideally it should be about 2:1).

3.5.2 miRNA-Specific Reverse Transcription

1. DNase treatment of the total RNA is strongly recommended before reverse transcription (*see* **Note 7**): Mix ~5 μg of total RNA, 2 μl (4 U) of DNase, 2 μl of 10× DNase buffer, 1 μl (20 U) of RNase inhibitor in the total volume of 20 μl. Incubate at 37 °C for 1 h, inactivate the enzyme at 75 °C

for 10 min, and place it on ice. Quantify RNA concentration by spectrophotometry.

2. For the reverse transcription reaction, mix 0.15 μl of 100 mM dNTP Mix, 1.5 μl of 10× buffer, and 1.16 μl of H_2O. Add 0.19 μl of RNase inhibitor [20 U/μl] and 1 μl of Reverse Transcriptase [50 U/μl]. Mix gently, do not vortex. These amounts are per sample but a preparation of a mastermix is recommended.
3. Add 5 μl of DNase treated total RNA [2 ng/μl].
4. Mix gently and add 3 μl of endogenous control-specific RT primer and 3 μl of target-specific RT primer.
5. Reverse transcribe the RNA according to the manufacturer's instructions (16 °C for 30 min, 42 °C for 30 min, 85 °C for 5 min; *see* **Note 21**).

3.5.3 Quantitative Real-Time PCR

1. Dilute five times the total volume of cDNA (15 μl) by adding 60 μl of H_2O (*see* **Note 22**).
2. Prepare real-time reactions by mixing 10 μl of 2× Reaction Mix, 1 μl of 20× probe, and 9 μl of diluted cDNA. (Perform each sample in triplicate in singleplex reactions.)
3. Cover and centrifuge the plate or strip at 500 ×*g* for 1 min.
4. Perform the PCR reaction according to the manufacturer's instructions (50 °C for 2 min, 95 °C for 10 min followed by 40 cycles of 95 °C for 15 s, 60 °C for 1 min).
5. Analyze the data (StepOne version 2.1, *see* **Note 23**).

4 Notes

1. Mirtron-sized long oligonucleotides must be purified by at least HPLC although the most suitable method is PAGE separation. To get sufficient yield after purification, the starting amount for synthesis should be a minimum of 100 nmol. Depending on the cloning plan, oligonucleotides can be designed to be complementary to each other for blunt end ligation, or with 5′ or 3′ overhangs for the formation of sticky ends. In the case of blunt end ligation, 5′ phosphorylated oligos should be purchased if the removal of phosphate groups of the backbone's ends is planned.
2. Annealed mirtron sequences can be inserted into a desired marker gene (e.g., EGFP) as artificial introns [12, 13, 15]. Alternatively, mirtrons can be amplified from genomic DNA with their flanking exons and then this minigene can be cloned behind a marker gene [11, 12, 14, 15].
3. To screen for successful ligations by colony PCR, usually one primer is designed to anneal in the region of the insert and the

other one on the backbone sequence. In the case of numerous different mirtron ligations, employment of a backbone-specific primer pair is more practical. However, the primers should anneal as close to the insert as possible for easy distinction between the plasmids containing the insert and the empty vector during gel electrophoresis.

4. To test the mirtron function using a luciferase assay, make sensor and mutant-sensor plasmids. Sensor constructs contain two fully complementary antisense copies of the particular predicted mature mirtron species, which is cloned downstream of Renilla luciferase in the psiCHECK2 vector. The mutant-sensor plasmid differs from the sensor plasmid in the three mismatched nucleotides in the predicted miRNA seed region (every second nucleotide). Annealed double-stranded oligonucleotides are ligated between the XhoI/NotI restriction sites of Renilla luciferase 3′-UTR according to the manufacturer's instructions.

5. To investigate splicing dependency, generation of a splicing-mutant mirtron construct is suitable. In such constructs, intronic splicing-signal sequences are mutated to block the removal of the intron by splicing, thereby also inhibiting the first step of mirtron biogenesis. Mutations can be introduced at both 5′ (donor) and 3′ (acceptor) splice sites of the intron; however, the mutation of only the 5′ splice site is usually sufficient (e.g., change GT to TG but see also refs. [11–15]). Mutant oligonucleotides can be purchased from a commercial source (*see* also **Note 1**).

6. There are several total and miRNA isolation methods and reagents; however, total RNA extraction by Trizol reagent is still the most commonly used method for miRNA studies. On the other hand, it is important to keep in mind that the efficient isolation of miRNAs with low GC content or stable secondary structure is sensitive to the initial number of cells [27]. Therefore, for reproducible measurements, we recommend to use a relatively high cell number (a minimum of $1\text{–}2 \times 10^6$ cells/ml) for this type of isolation.

7. Before reverse transcription, we strongly recommend DNase treatment of RNA samples. First, contamination of RNA samples with genomic and transiently transfected DNA plasmid might interfere with the accurate determination of RNA concentration, since both DNA and RNA are measured by absorbance at 260 nm. Second, DNA contamination might influences mature miRNA detection by stem-loop qRT-PCR by giving false positive signals.

8. Alternatively, depending on the nature of antibiotics, competent cells can be spread onto LB-agar plates immediately after heat shock. Typically, this is possible for ampicillin, which is an inhibitor of cell wall synthesis. However, for antibiotics inhibiting

translation such as kanamycin, a short time of growth in a nonselective medium is required to obtain resistant cells.

9. Amplicon length differs depending on the localization of the primer pair. To distinguish an empty vector from the longer (70–100 nucleotides) mirtron-containing amplicon, gel electrophoresis using different agarose concentrations can be applied. As an example, we use 3 % agarose gels to distinguish between 470 bp and 550 bp amplicons.
10. Although 0–15 min is recommended to obtain DNA molecules efficiently packaged in lipid complexes, longer incubation of DNA with FuGENE® HD reagent (45–60 min) usually improves transfection efficiency.
11. Sorting the cells about a week after transfection is preferable because marker gene expression detected at that time originates mostly from the inserted genomic copies; transiently residing plasmids are usually lost by then. Another advantage is that the higher number of sorted cells obtained about a week after transfection improves cell viability.
12. It is possible to separate cells with different fluorescent intensity corresponding to different levels of transgene expression, but it depends on the type of the cell sorter. After the sorting procedure, there is a heterogeneous cell population with different marker gene copy numbers. If it is desired, it can be cloned further to get a homogenous one-cell-derived population.
13. Generally, a significant proportion of the cells (approximately 30–50 % of the total) die during the sorting process. Keep this in mind to select the appropriate cell culture dish size for plating the sorted cells, since cells generally do not grow well in low cell density.
14. A typical experimental setup for testing the function of a potential miRNA derived from a mirtron (either of its 5′ or 3′ arm):
    - mirtron + sensor (three parallel transfections)
    - mirtron + seed-mutant sensor (three parallel transfections)
    - non-cognate miRNA (negative control, having unrelated sequence) + mirtron sensor (three parallel transfections)
    - non-cognate miRNA + mirtron seed-mutant sensor (three parallel transfections).

    You have to perform 12 transfections for testing one arm of the mirtron, and 12 transfections for testing the other arm of potential miRNA species. In summary, there are 24 transfections in one experiment for a particular mirtron.
15. Samples can be measured one by one in a single sample or in a plate-reading luminometer. Alternatively, more samples can be measured simultaneously in a plate-reading luminometer by

pipetting no more than 2–4 samples in the plate at the same time. Another option is to culture, transfect, and measure cells directly in a specific opaque plate. In this case, a plate-reading luminometer has to be equipped with two injectors. When measuring single samples in multiwell opaque plate, we advise using only every second well to prevent cross talk between them. It is important to keep the samples away from bright lights and sunlight.

16. LARII is provided in lyophilized form and, according to the manufacturer's instructions, should be resuspended in 10 ml of Luciferase Assay Buffer II. However, the supplied volume of buffer is usually about 11 ml, and all can be added directly to the LARII. It is also important to note that the Stop & Glo reagent has to be prepared in a glass vial or in a siliconized polypropylene tube.
17. Splicing deficiency of the splice site-mutated construct can be verified by RT-PCR with the same pair of primers used for colony detection by PCR (localized outside of the mirtron insert, in the region of the marker gene). Moreover, in case of a fluorescent reporter (EGFP), mutant constructs can also be tested for splicing deficiency based on fluorescence, since no fluorescent signals should be detected after transfection.
18. The independency of mirtron processing from the Drosha/DGCR8 complex can be examined in Drosha- or DGCR8-deficient cells (knocking down Drosha or DGCR8 with siRNA or shRNA, or in genetically knocked out cell lines). Although in HeLa cells the knockdown of Drosha with siRNAs was demonstrated at both mRNA and protein levels, this was not sufficient to abolish the canonical miRNA biogenesis pathway. Alternatively, a transdominant-negative Drosha (TN-Drosha) [14], or as we did, DGCR8-deficient mouse embryonic fibroblasts can be used [15, 28]. Moreover, a DGCR8-knockout mouse embryonic stem cell line can also be purchased from commercial source.
19. There are several methods for direct detection of mature miRNAs such as Northern blot and qRT-PCR. The former method is more traditional and can detect the different processed miRNA forms. Although its sensitivity can be strongly increased by locked nucleic acids (LNAs; [29, 30]), the detection of ~20–22-nucleotide-long mature miRNAs by qRT-PCR is a more sensitive technique [15]. For this measurement, we use stable mirtron-expressing cell lines and two negative controls to check the specificity of the reactions. One negative control is a sample from the original, genetically non-modified (parental) cell line and the other one is from another, unrelated miRNA-overexpressing cell line-derived sample.

20. When removing the aqueous phase during the phase separation step of RNA isolation, approximately 500–600 μl can be recovered from 1 ml Trizol. However, to avoid touching the interphase and thereby contaminating with DNA, we recommend removing a smaller volume.
21. Reverse transcription efficiency may vary between samples when using different type of PCR tubes.
22. The final dilution of cDNA in the qPCR reaction is 11× (it has to be a minimum of 10×). Always use non-template controls in all the assays.
23. Always check the baseline and threshold values since big differences in Ct values between samples as well as a little contamination in the non-template control might cause a wrong autofit by the program.

Acknowledgements

We would like to thank Károly Fátyol for helpful discussions. Anita Schamberger is a recipient of the Jedlik Ányos predoc fellowship. This research was supported by the European Union and the State of Hungary, co-financed by the European Social Fund in the framework of TÁMOP 4.2.4. A/1-11-1-2012-0001 "National Excellence Program" and also supported by the TransRat grant KMR_12-2012-0112 given to Tamás I. Orbán.

References

1. Filipowicz W, Bhattacharyya SN, Sonenberg N (2008) Mechanisms of post-transcriptional regulation by microRNAs: are the answers in sight? Nat Rev Genet 9:102–114
2. Carthew RW, Sontheimer EJ (2009) Origins and mechanisms of miRNAs and siRNAs. Cell 136:642–655
3. Ghildiyal M, Zamore PD (2009) Small silencing RNAs: an expanding universe. Nat Rev Genet 10:94–108
4. Slezak-Prochazka I, Durmus S, Kroesen BJ et al (2010) MicroRNAs, macrocontrol: regulation of miRNA processing. RNA 16:1087–1095
5. Bartel DP (2009) MicroRNAs: target recognition and regulatory functions. Cell 136:215–233
6. Krol J, Loedige I, Filipowicz W (2010) The widespread regulation of microRNA biogenesis, function and decay. Nat Rev Genet 11:597–610
7. Pasquinelli AE (2012) MicroRNAs and their targets: recognition, regulation and an emerging reciprocal relationship. Nat Rev Genet 13: 271–282
8. Yates LA, Norbury CJ, Gilbert RJ (2013) The long and short of microRNA. Cell 153:516–519
9. Miyoshi K, Miyoshi T, Siomi H (2010) Many ways to generate microRNA-like small RNAs: non-canonical pathways for microRNA production. Mol Genet Genomics 284:95–103
10. Yang JS, Lai EC (2011) Alternative miRNA biogenesis pathways and the interpretation of core miRNA pathway mutants. Mol Cell 43:892–903
11. Ruby JG, Jan CH, Bartel DP (2007) Intronic microRNA precursors that bypass Drosha processing. Nature 448:83–86
12. Okamura K, Hagen JW, Duan H et al (2007) The mirtron pathway generates microRNA-class regulatory RNAs in Drosophila. Cell 130: 89–100
13. Sibley CR, Seow Y, Saayman S et al (2012) The biogenesis and characterization of mammalian

microRNAs of mirtron origin. Nucleic Acids Res 40:438–448
14. Havens MA, Reich AA, Duelli DM et al (2012) Biogenesis of mammalian microRNAs by a non-canonical processing pathway. Nucleic Acids Res 40:4626–4640
15. Schamberger A, Sarkadi B, Orban TI (2012) Human mirtrons can express functional microRNAs simultaneously from both arms in a flanking exon-independent manner. RNA Biol 9:1177–1185
16. Westholm JO, Lai EC (2011) Mirtrons: microRNA biogenesis via splicing. Biochimie 93:1897–1904
17. Flynt AS, Greimann JC, Chung WJ et al (2010) MicroRNA biogenesis via splicing and exosome-mediated trimming in Drosophila. Mol Cell 38:900–907
18. Ladewig E, Okamura K, Flynt AS et al (2012) Discovery of hundreds of mirtrons in mouse and human small RNA data. Genome Res 22:1634–1645
19. Curtis HJ, Sibley CR, Wood MJ (2012) Mirtrons, an emerging class of atypical miRNA. Wiley Interdiscip Rev RNA 3: 617–632
20. Meng Y, Shao C (2012) Large-scale identification of mirtrons in Arabidopsis and rice. PLoS One 7:e31163
21. Joshi PK, Gupta D, Nandal UK et al (2012) Identification of mirtrons in rice using MirtronPred: a tool for predicting plant mirtrons. Genomics 99:370–375
22. Ivics Z, Hackett PB, Plasterk RH et al (1997) Molecular reconstruction of Sleeping Beauty, a Tc1-like transposon from fish, and its transposition in human cells. Cell 91:501–510
23. Ammar I, Izsvak Z, Ivics Z (2012) The Sleeping Beauty transposon toolbox. Methods Mol Biol 859:229–240
24. Lacy-Hulbert A, Thomas R, Li XP et al (2001) Interruption of coding sequences by heterologous introns can enhance the functional expression of recombinant genes. Gene Ther 8: 649–653
25. Feschotte C, Pritham EJ (2007) DNA transposons and the evolution of eukaryotic genomes. Annu Rev Genet 41:331–368
26. Izsvak Z, Ivics Z (2004) Sleeping beauty transposition: biology and applications for molecular therapy. Mol Ther 9:147–156
27. Kim YK, Yeo J, Kim B et al (2012) Short structured RNAs with low GC content are selectively lost during extraction from a small number of cells. Mol Cell 46:893–895
28. Wang Y, Medvid R, Melton C et al (2007) DGCR8 is essential for microRNA biogenesis and silencing of embryonic stem cell self-renewal. Nat Genet 39:380–385
29. Varallyay E, Burgyan J, Havelda Z (2008) MicroRNA detection by northern blotting using locked nucleic acid probes. Nat Protoc 3:190–196
30. Varallyay E, Burgyan J, Havelda Z (2007) Detection of microRNAs by Northern blot analyses using LNA probes. Methods 43:140–145

Chapter 23

A Guide for miRNA Target Prediction and Analysis Using Web-Based Applications

Ana Lúcia Leitão, Marina C. Costa, and Francisco J. Enguita

Abstract

MiRNAs are small noncoding RNAs which act by binding to the 3′UTR of mRNA transcripts to exert a negative regulatory effect. The miRNA binding to its target follows rules based on the base complementarity of the seed sequence (2–9 first nucleotides of the miRNA sequence). Several algorithms have been developed to predict miRNA binding to genomic targets and its physiological consequences. This chapter will describe several practical aspects for the use of miRNA target prediction algorithms taking advantage of their web interfaces as well as how to produce integrative results in a graphical manner.

Key words Target predictor, miRNA, Seed sequence, Web server

1 Introduction

Micro-RNAs (miRNAs) are an abundant class of small noncoding RNAs generated from specific transcription units in complex eukaryotic genomes [1]. These tiny RNAs (19–23 nucleotides) are negative posttranscriptional regulators that exert their effect by binding to the 3′UTR of a mRNA transcript mediated by a family of RNA-binding proteins designed as argonautes [2]. The protein complex responsible for the negative regulatory activity over mRNA transcripts is known as RISC (RNA Induced Silencing Complex) and contains Ago2, an argonaute protein, as a main effector. The binding of a miRNA to its cognate target is driven by Watson–Crick complementarity of a small stretch of its sequence called the "seed," which comprises the nucleotides 2–9 of the 5′ end. The size of the seed sequence is directly dependent on the molecular dimensions of the substrate binding pocket of Ago2 protein [3]. In addition to Ago2, other argonaute proteins appear to be involved in miRNA function [4].

M. Lucrecia Alvarez and Mahtab Nourbakhsh (eds.), *RNA Mapping: Methods and Protocols*, Methods in Molecular Biology, vol. 1182, DOI 10.1007/978-1-4939-1062-5_23, © Springer Science+Business Media New York 2014

2 Target Prediction Algorithms

Target prediction algorithms are essential tools for discovery and characterization of miRNA function because experimental data is still limited and always derived from a particular biological process. A particular miRNA target can be recognized by a single miRNA in a set of different ways (Fig. 1). Canonical binding sites are characterized by a direct base pairing between the miRNA seed sequence and the mRNA target region comprising seven or eight nucleotides (7-mer and 8-mer target recognition) [5]. Noncanonical mechanisms for target recognition are based on the imperfect base pairing at the seed sequence, which is at least partially compensated by a base pairing at the 3′-end of the miRNA sequence [6].

Target prediction remains an important challenge in computational biology because the rules of this process are still not well understood and the size of miRNAs can lead to potential off-target predictions in big genomes. Typically, the majority of target prediction algorithms are restricted to the 3′UTR segment of mRNAs; however some of them are also designed to search for targets in the 5′UTR region and coding sequence [7]. Indeed, the biological relevance of miRNA binding to regions distinct from the 3′UTR is still a matter of discussion [8].

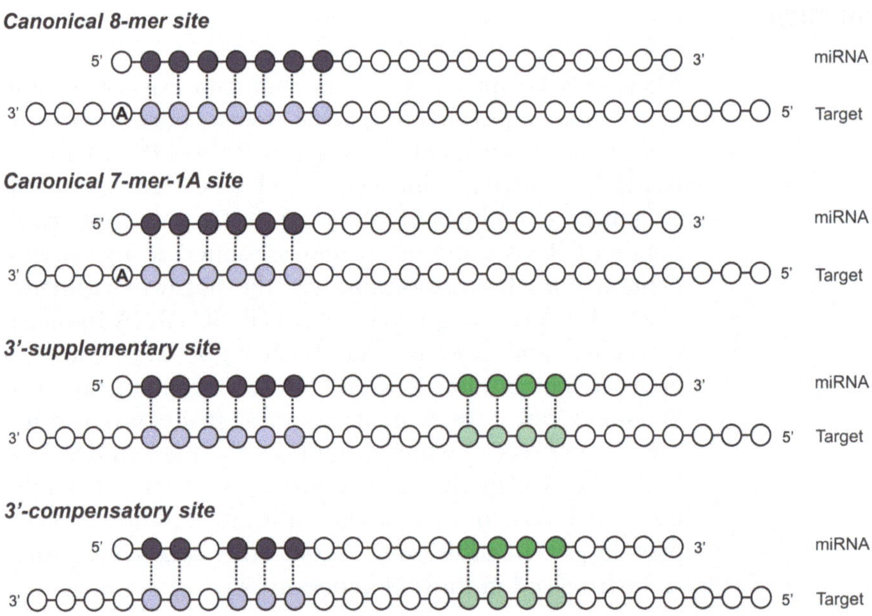

Fig. 1 Types of miRNA target recognition. Canonical sites with 7 or 8 mer are based on the base pairing of the miRNA seed sequence with the mRNA target in a region flanked by a 3′ adenosine nucleotide. Noncanonical sites present either an imperfect base pairing in the seed sequence, which is compensated by binding of the miRNA 3′-end to the target, or just supplementary binding of the 3′-end of the miRNA. Noncanonical binding sites also have the characteristic of not requiring the presence of an adenosine nucleotide in the targeted mRNA at the beginning of the seed sequence binding region

Target prediction algorithms are mainly based in the complementarity of the seed sequence of a miRNA with its cognate target. However, because of the limited size of miRNA sequences, other factors are added to filter the results such as interspecies conservation of the binding site, free energy of the RNA–RNA hybrid, secondary structure of the target that may prevent the binding of a miRNA, and possible existence of compensatory binding in the 3′ end of the miRNA [9]. Characteristics, advantages, and drawbacks for the main algorithms are summarized in Table 1. Initially developed algorithms such as PicTar (http://pictar.mdc-berlin.de/) and TargetScan (http://www.targetscan.org/) are very stringent only allowing a perfect match between the seed sequence and the target, taking into account not only the free energy of the hybrid but also evolutionary conservation of the seed region [10, 11]. Other algorithms such as miRanda (http://www.microrna.org/), PITA (http://genie.weizmann.ac.il/pubs/mir07), and RNA22 (http://cbcsrv.watson.ibm.com/rna22.html) are more inclusive and relaxed, taking into account the possibility of compensatory binding of the 3′ end of the miRNA, but also showing a higher rate of false positive predictions. MiRanda algorithm is a part of the microRNA.org web resource, and has the advantage of working also as a local computer application that can be used to tackle miRNA targets in genomes not available in general purpose databases [12]. MiRanda and PITA algorithms take into account the free energy of the miRNA–mRNA duplex in a different way. While miRanda only uses ΔG Gibbs function to score the more relevant targets, PITA relies on the presence of potential secondary structures in the target mRNA that may prevent the miRNA binding [13]. Composite predictors such as Diana micro-T use a combination of assumptions derived from the classic algorithms, conservation of interspecies and free energy of the RNA–RNA hybrid. This web server is a component of a whole suite for miRNA analysis which includes some utilities for putative function determination and pathway integration [14, 15]. More accurate predictors based on the neural network architecture and machine learning algorithms were also recently described such as MTar [16], Diana micro-T-ANN [17], and miRmap [18, 19]. Some other recent predictors such as miRTar Hunter [20] claimed to be more successful than the others by using proprietary methods based on a dynamic programming algorithm that incorporates more sequence-specific features and reflects the already known properties of various types of target sites.

High-throughput analysis of miRNA-target interactions based on selective immunoprecipitation of RISC complex components followed by deep-sequencing analysis showed the existence of a relative enrichment of targets comprising exact miRNA seed matches [21]. Moreover, recent studies showed that noncanonical miRNA-target recognition is very frequent, reaching up to 60 % of the total miRNA targets [21]. Target prediction methods frequently

Table 1
Characteristics, strengths, and drawbacks of the main individual miRNA target prediction applications

Target prediction algorithm	Characteristics		Quality of the prediction	
	Final score based on	Cross-species conservation	Robustness	Drawbacks
miRanda	Complementarity and free energy binding	Interspecies conservation filter	Beneficial for prediction sites with imperfect binding within seed region	Low precision, too many false positives. Medium performance in noncanonical targets
TargetScan	Seed match, 3′ complementarity local AU content and position contribution 5	The application scores each result by conservation	Many parameters included in target scoring. The final score correlates with protein downregulation	Sites with poor seed pairing are omitted. Low performance in noncanonical target prediction
TargetScanS	Seed match type	Only conservative sites are considered	Simple tool for search of conserved sites with stringent seed pairing	Underestimate miRNAs with multiple target sites
PicTar	Binding energy, complementarity, and conservation	Required pairing at conserved positions	MiRNAs with multiple alignments are favored	Does not predict non-conservative sites. Low performance in noncanonical sites
DIANA-microT	Free energy binding and seed complementarity	Dataset of conserved UTRs among human and mouse is used	Probability given for each target site. Possibility of using own miRNA sequence as an input	Some miRNAs with multiple target sites may be omitted
PITA	Target site accessibility energy	User-defined	The secondary structure of 3′UTR is considered for miRNA interaction	Low efficiency compared to other algorithms, mainly in noncanonical target prediction
Rna22	Pattern recognition and folding energy	Not available	Allows identifying sites targeted by yet-undiscovered miRNAs	Low efficiency compared to other algorithms

Adapted from Witkos et al. [31]

fail when predicting these noncanonical and compensatory targets. In consequence, more powerful algorithms are needed, probably including more complex biological rules that we are starting to understand from biological data. Other drawbacks from the currently available algorithms are related with the variations in the annotations of the reference databases used to map the miRNA targets. Recently, Ritchie and coworkers [22] analyzed the overlap in the number of predicted targets when the same algorithm was applied using two reference databases (UCSC and Ensembl) revealing that this overlap was less than 50 %. Moreover, the majority of the target predictors do not take into account the splicing pattern of the target gene and its cell-type specificity [22].

3 Multiple Predictors

One possible strategy to circumvent the intrinsic variability of the target prediction algorithms is to use what we like to call "multiple predictors." These applications are essentially web portals that will query a series of miRNA target predictors and represent the results in a user-friendly manner, showing which target is predicted by one or more algorithms. Among others, the most widely used multiple predictors are miRWalk [23], miRecords [24], and mirDIP [25]. MiRWalk and miRecords are integrated databases which include information about predicted and also validated miRNA targets. On the other hand, mirDIP is only a portal for target prediction analysis; however it offers a higher degree of customization when compared with miRWalk and miRecords. The user can filter the final results by deciding which applications to use as well as the minimum number of applications able to predict a particular miRNA target. Figure 2 shows the different number of algorithms interrogated by the multiple predictors miRWalk, miRDIP, and miRecords. All these multiple predictors are able to present the result in several formats, including tabular data format and Excel-friendly spreadsheets.

4 Protocol for miRNA Target Prediction Analysis in a Multiple Regulatory Scenario

Multiple regulatory scenarios are common in miRNA regulatory effect and characteristic of higher eukaryotes. They consist in a group of miRNAs which are differentially expressed in a particular cellular process, regulating different targets at the same time. In complex cellular networks, miRNAs can control the genomic output regulating genes in a coordinate or divergent manner. Unfortunately, we are far to understand which are the factors that control the quantitative contribution of each miRNA over a particular network or pathway. Indeed, scientists always have to face the dilemma of how to proceed

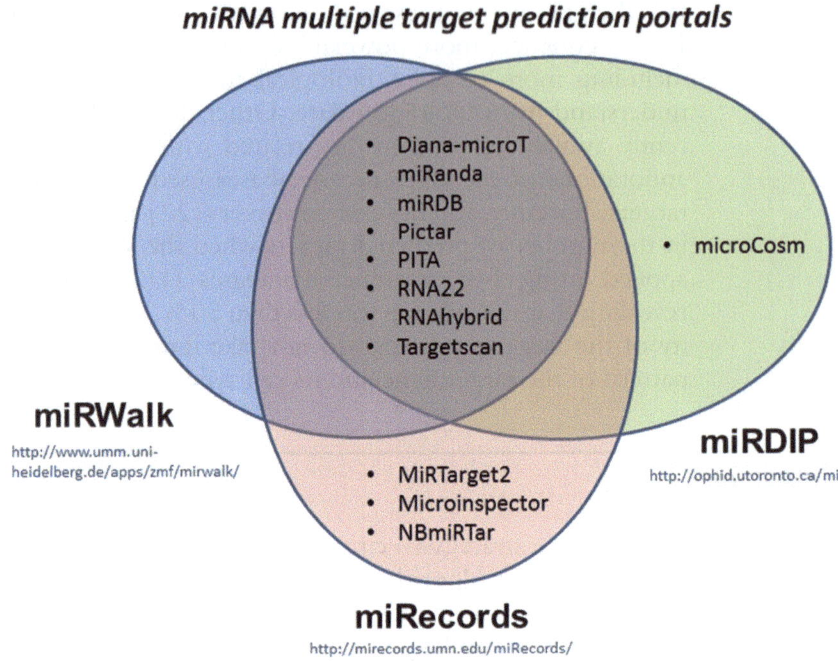

Fig. 2 Venn diagram representing the cohort of single target prediction algorithms used by the multiple predictors mirDIP, miRWalk, and miRecords

with laboratory experiments after determining the miRNA expression profile in a particular biological situation (*see* **Note 1**).

Following the identification of misregulated miRNAs using methods such as qPCR or small RNA-seq, a coordinated target analysis can be performed as described here. The following step-by-step protocol takes advantage of the mirDIP web server for multiple target prediction [25] and the Navigator software [26] for results analysis and representation, but can be easily adapted to other softwares such as Cytoscape (http://www.cytoscape.org/).

1. For target prediction, access mirDIP data integration portal at the web address: http://ophid.utoronto.ca/mirDIP/.
2. In the upper menu, select the option "search mirDIP."
3. Once open the search window, introduce the identification of the misregulated miRNAs following the rules of miRBase v.18.0 separated by "enter." It is possible to enter as many miRNAs as the user like; however, if the total number of predicted targets exceeds 50,000, the software will prompt an error message suggesting the use of filters to significantly reduce the number of predictions. Species identification should be added at the beginning of the miRNA name (i.e., "hsa" for human miR-122). MirDIP is exclusive for human miRNAs; miRWalk allows the user to search for human, mouse, and rat miRNAs; and miRecords contains information for human, mouse, rat, zebrafish, fly, chicken, and dog (*see* **Notes 2** and **3**).

a
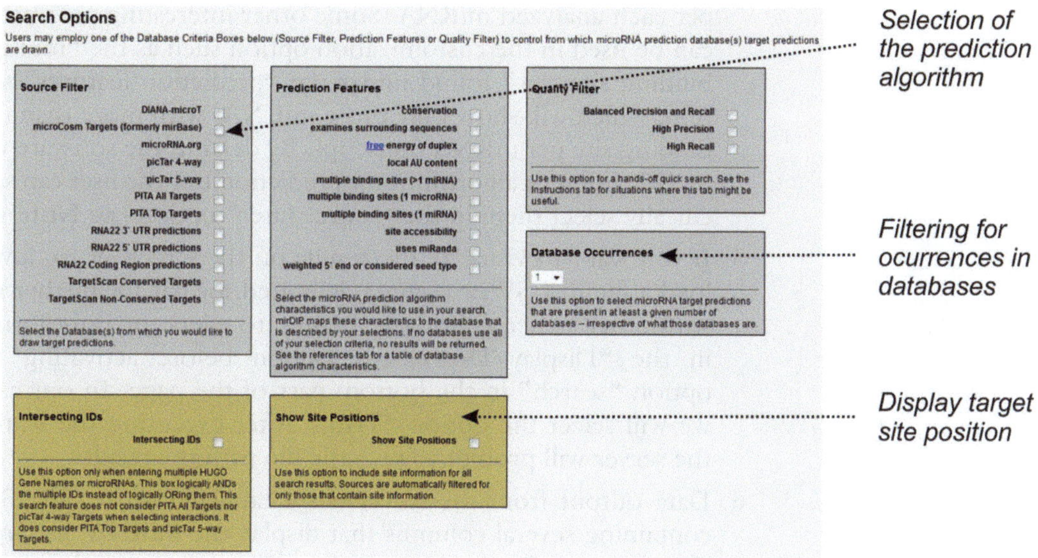

b **Results (14) :**

Gene Symbol	Links	microRNA	Source	Score (orig)	Score (std)	Rank
TRAF6	i2D GC	hsa-mir-146a	PITA Top Targets	-11.0	55.11	Top Third
TRAF6	i2D GC	hsa-mir-146a	DIANA-microT	8.0	2.03105	Top Third
TRAF6	i2D GC	hsa-mir-146a	microRNA.org	63.0	41.26984127	Bottom Third
TRAF6	i2D GC	hsa-mir-146a	microCosm Targets (formerly mirBase)	18.1423	50.89943614	Top Third
TRAF6	i2D GC	hsa-mir-146a	RNA22 3' UTR predictions	-29.9	16.89655172	Top Third
TRAF6	i2D GC	hsa-mir-146a	TargetScan Conserved Targets	-0.415	73.49382244	Top Third
TRAF6	i2D GC	hsa-mir-146a	TargetScan Non-Conserved Targets	-0.101	35.39042821	Mid Third
ZBTB2	i2D GC	hsa-mir-146a	PITA Top Targets	-12.7	58.17	Top Third
ZBTB2	i2D GC	hsa-mir-146a	DIANA-microT	5.0	1.1606	Top Third
ZBTB2	i2D GC	hsa-mir-146a	microRNA.org	66.0	46.03174603	Bottom Third
ZBTB2	i2D GC	hsa-mir-146a	microCosm Targets (formerly mirBase)	15.9983	27.42541195	Mid Third
ZBTB2	i2D GC	hsa-mir-146a	RNA22 3' UTR predictions	-29.200001	14.48276207	Top Third
ZBTB2	i2D GC	hsa-mir-146a	TargetScan Conserved Targets	-0.457	78.84345943	Top Third
ZBTB2	i2D GC	hsa-mir-146a	TargetScan Non-Conserved Targets	-0.222	50.62972292	Top Third

Fig. 3 mirDIP interface features. (**a**) Input screen with the most useful customization options (predictor selection, display of target positions, and filtering for the number of applications that predict the same target); (**b**) mirDIP tabular output for the miRNA target prediction. Data includes the direct score generated by each individual predictor and a rank classification to assess the quality of the prediction. Data can be exported in tab- and comma-delimited format. Other multiple predictors such as miRWalk and miRecords have similar outputs

4. In order to customize the search, several filters can be applied in the "Search option" of the query page (Fig. 3a). In our hands, mirDIP software showed good performance in target prediction using the default conditions; however we strongly advise the user to modify the "Database occurrence" parameter to restrict the number of targets which are predicted by a precise number of applications. As a rule of thumb, to get a reliable prediction at least 50 % of the applications must predict a particular target. The interpretation of the results could be very difficult and, therefore, further filtering is advised to reduce the number of targets to a maximum of 100–150 genes

per each analyzed miRNA. Some other interesting parameters can be used in the customization option such as the "multiple binding sites >1" found under the "prediction features" section, which will show only those miRNAs with more than one binding site per mRNA transcript. By default, the software will use 12 different algorithms for prediction but the user can specifically select them in the "source filter" option (*see* **Note 4**).

5. If you want to visualize the results on the screen or to download a file in CSV or comma delimited format for further use with other local applications, be sure to select the right format in the "Display/Download" option before activating the option "search" in the bottom part of the page. In our case, we will select the option "TAB." After executing the search, the server will prompt to save the file with the results.
6. Data output from the server has a tabular format (Fig. 3b), containing several columns that display the miRNA, its target genes, the application that predicts these targets, and two columns with original and standardized scores for each prediction. The last column of the table ranks the predictions in four groups depending on the standardized score: 1 % top, top third, medium third, and bottom third, being particularly useful in the assessment of each prediction. Tabular data can be imported in MS-Excel and filtered to select the targets according to a gene or a prediction score.
7. Graphical representation of the predictions can be achieved by specialized software such as Navigator (http://ophid.utoronto.ca/navigator/). At the time of the publication of this protocol, the Navigator version available was 2.2.1. The software must be downloaded and locally installed in a computer.
8. Once Navigator is installed in your operative system, execute the program and open the tabular data file generated by mirDIP selecting File>Open and the file name and format (*.txt—TAB delimited format) of the data file. The following screens will allow the user to select the header of the file and also each interaction node. The software will construct a network on the basis of the relationships between two different nodes defined by columns in the table (miRNAs and targets).
9. Interpretation of raw networks is very difficult as they are produced by Navigator. However this software permits a customization of the network visualization using several parameters in the "appearance filters" section. For the particular case of miRNA target prediction from multiple miRNAs, we found extremely useful to filter the appearance of the network using the number of connections between nodes: number of miRNAs regulating a target and vice versa. The software will adjust

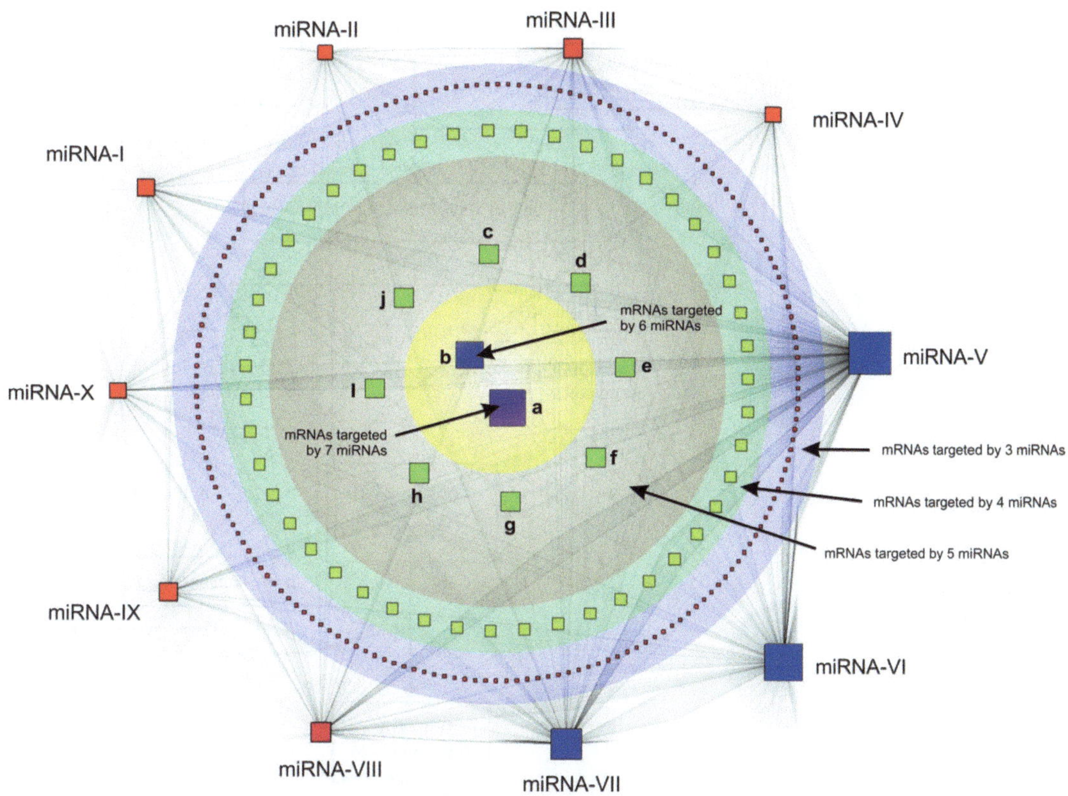

Fig. 4 Layered circular representation of target prediction for multiple miRNAs in an idealized scenario. The working scenario hypothesizes the existence of ten differentially expressed miRNA (numbered in the graph using roman numerals) in a biological process. Previously to the construction of the graph, the predicted targets of each miRNA must be calculated using mirDIP, and all the data imported into Navigator or other visualization software. Symbols representing miRNAs and targets are size and color-coded to visualize the number of relationships between them. Representation of targets in layers taking into account the number of miRNAs regulating each of the targets simultaneously allows the reader to clearly visualize which target genes could be more relevant in the analyzed biological process. For instance, target genes "a" and "b" are putatively regulated simultaneously by 7 and 6 miRNAs respectively, and could be subject of further laboratory studies. These representations can be produced by software for network analysis such as Cytoscape [30] or Navigator [25] and can be complemented with further information such as protein interaction networks or relative expression values of each miRNA

the color and size of the node symbols according to the number of connections (Fig. 4). Once filtering by the number of connections, the bigger symbols correspond to these genes or miRNAs with more connections in the network (*see* **Note 5**).

10. Other actions that may improve the observation of the overall relationships between miRNAs and predicted targets are the distribution of the nodes according to the levels of relationships. This can be achieved in different ways but we suggest using a circular distribution to generate regulatory layers (Fig. 4).

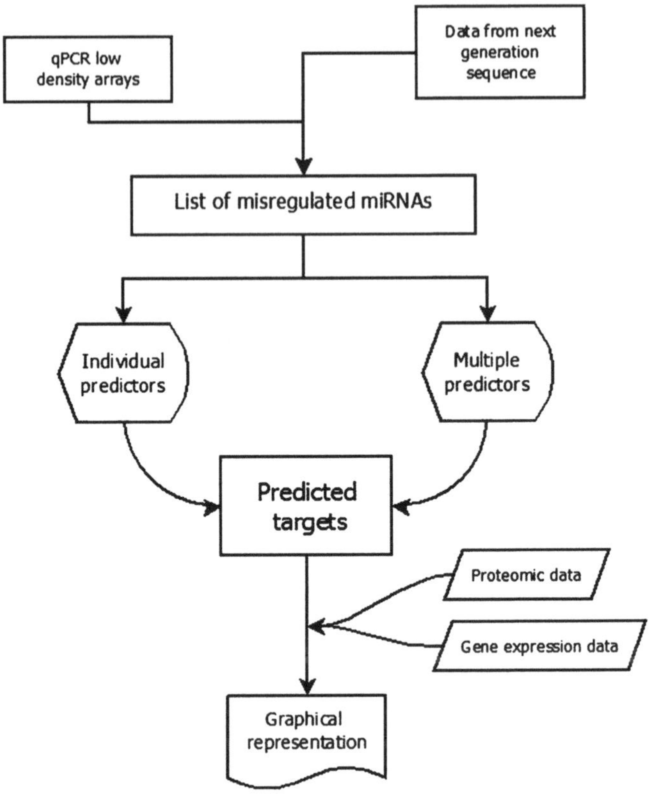

Fig. 5 Flowchart for target prediction analysis using web-based applications and graphical representation

In principle, the nodes with more connections will be more relevant for the process, and susceptible of laboratory validation. For instance, this strategy has been successfully employed to pinpoint the combinatorial effect of miRNAs involved in the Myc oncogenic pathway [27].

11. Graphical representations could be always considered as a skeleton. Additional complexity layers can be included in the graph by using information obtained from gene expression, proteomic data, or specific databases. For instance, functional information about the relationships between the proteins encoded by the targeted mRNAs is very easy to include in the Navigator graph by overlaying another data group using the File>Import>Node features option (*see* **Note 6**). This information can be found in specific databases such as STRING (http://string-db.org/) [28].

12. The proposed flowchart for this protocol (Fig. 5) can be adapted also to other multiple predictors as miRecords and miRWalk by producing the corresponding table of miRNA–mRNA relationships and importing the data into Navigator.

5 Notes

1. MiRNA regulation of genomic output is often a consequence of the action of many miRNAs working simultaneously. The analysis of miRNA regulatory activity over a particular biological problem must be always oriented from the point of view of the systems biology. In higher eukaryotes like humans, no master miRNA is expected; in consequence, a complex network of regulatory events may exist.
2. Target prediction algorithms are erratic, so the use of many applications in parallel is advised. Applications such as mirDIP, miRecords, or miRWalk, which employ many algorithms simultaneously (multiple predictors), are preferred over any single predictor.
3. Careful attention must be paid to the new miRNA nomenclature rules established from miRBase18.0 onwards. Current miRBase version 20 (released on June 2013) applies the same rules started in 2011 to identify the mature miRNAs coming from the same pre-miRNA by using their relative positions in the precursor transcript (5′ and 3′ ends) instead or their abundance within the cell [29]. Unfortunately, many of the target predictors run old versions of miRBase, a fact that should be taken into account.
4. It is advisable to use multiple predictors for target analysis in order to understand the effect of a group of miRNAs over a biological network. Those miRNAs putatively regulating a higher number of targets, which are regulated by more than one miRNA simultaneously, must be our focus for future laboratory experiments. This approach will direct our efforts in a more specific and intelligent way.
5. Graphical representation of miRNA-target interactions is always a plus to understand the complex regulatory networks. Initially, these graphical networks could be difficult to interpret but with proper filtering some interesting things may pop-out. The representation of targets in layers of complexity may help to the overall interpretation. Those targets putatively regulated by more miRNAs are probably the most relevant in our biological problem. This fact does not prevent the need for a biological validation and wet-lab studies.
6. Other layers of complexity can be added to the graphical representation of miRNA–mRNA interactions, including transcriptomic and/or proteomic data as well as functional data. In particular, information about protein–protein interactions will be extremely useful to understand the long-range effects of a miRNA over a metabolic pathway or network.

Acknowledgements

M.C.C. was supported by a postdoctoral fellowship from Fundação para a Ciência e Tecnologia, Portugal (Ref. SFRH/BPD/65131/2009). The authors would like to acknowledge Francisco Enguita Jr. for his friendship and excellent technical support in the preparation of the manuscript.

References

1. Ameres SL, Zamore PD (2013) Diversifying microRNA sequence and function. Nat Rev Mol Cell Biol. doi:10.1038/nrm3611
2. Holcik M, Pestova TV (2007) Translation mechanism and regulation: old players, new concepts. Meeting on translational control and non-coding RNA. EMBO Rep 8:639–643
3. Elkayam E, Kuhn CD, Tocilj A et al (2012) The structure of human argonaute-2 in complex with miR-20a. Cell 150:100–110
4. Turchinovich A, Burwinkel B (2012) Distinct AGO1 and AGO2 associated miRNA profiles in human cells and blood plasma. RNA Biol 9:1066–1075
5. Brennecke J, Stark A, Russell RB et al (2005) Principles of microRNA-target recognition. PLoS Biol 3:e85
6. Brodersen P, Voinnet O (2009) Revisiting the principles of microRNA target recognition and mode of action. Nat Rev Mol Cell Biol 10: 141–148
7. Smalheiser NR, Torvik VI (2006) Complications in mammalian microRNA target prediction. Methods Mol Biol 342:115–127
8. Zhang Y, Verbeek FJ (2010) Comparison and integration of target prediction algorithms for microRNA studies. J Integr Bioinform 7. doi:10.2390/biecoll-jib-2010-127
9. Thomas M, Lieberman J, Lal A (2010) Desperately seeking microRNA targets. Nat Struct Mol Biol 17:1169–1174
10. Friedman RC, Farh KK, Burge CB et al (2009) Most mammalian mRNAs are conserved targets of microRNAs. Genome Res 19:92–105
11. Krek A, Grun D, Poy MN et al (2005) Combinatorial microRNA target predictions. Nat Genet 37:495–500
12. Betel D, Wilson M, Gabow A et al (2008) The microRNA.org resource: targets and expression. Nucleic Acids Res 36:D149–153
13. Kertesz M, Iovino N, Unnerstall U et al (2007) The role of site accessibility in microRNA target recognition. Nat Genet 39:1278–1284
14. Maragkakis M, Vergoulis T, Alexiou P et al (2011) DIANA-microT Web server upgrade supports Fly and Worm miRNA target prediction and bibliographic miRNA to disease association. Nucleic Acids Res 39:W145–148
15. Maragkakis M, Reczko M, Simossis VA et al (2009) DIANA-microT web server: elucidating microRNA functions through target prediction. Nucleic Acids Res 37:W273–276
16. Chandra V, Girijadevi R, Nair AS et al (2010) MTar: a computational microRNA target prediction architecture for human transcriptome. BMC Bioinformatics 11(Suppl 1):S2
17. Reczko M, Maragkakis M, Alexiou P et al (2011) Accurate microRNA target prediction using detailed binding site accessibility and machine learning on proteomics data. Front Genet 2:103
18. Vejnar CE, Blum M, Zdobnov EM (2013) miRmap web: comprehensive microRNA target prediction online. Nucleic Acids Res 41: W165–W168
19. Vejnar CE, Zdobnov EM (2012) MiRmap: comprehensive prediction of microRNA target repression strength. Nucleic Acids Res 40: 11673–11683
20. Park K, Kim KB (2013) miRTar Hunter: a prediction system for identifying human microRNA target sites. Mol Cells 35: 195–201
21. Helwak A, Kudla G, Dudnakova T et al (2013) Mapping the human miRNA interactome by CLASH reveals frequent noncanonical binding. Cell 153:654–665
22. Ritchie W, Flamant S, Rasko JE (2009) Predicting microRNA targets and functions: traps for the unwary. Nat Methods 6:397–398

23. Dweep H, Sticht C, Pandey P et al (2011) miRWalk–database: prediction of possible miRNA binding sites by "walking" the genes of three genomes. J Biomed Inform 44:839–847
24. Xiao F, Zuo Z, Cai G et al (2009) miRecords: an integrated resource for microRNA-target interactions. Nucleic Acids Res 37:D105–110
25. Shirdel EA, Xie W, Mak TW et al (2011) NAViGaTing the micronome–using multiple microRNA prediction databases to identify signalling pathway-associated microRNAs. PLoS One 6:e17429
26. Brown KR, Otasek D, Ali M et al (2009) NAViGaTOR: Network Analysis, Visualization and Graphing Toronto. Bioinformatics 25: 3327–3329
27. Bueno MJ, Gomez de Cedron M, Gomez-Lopez G et al (2011) Combinatorial effects of microRNAs to suppress the Myc oncogenic pathway. Blood 117:6255–6266
28. Franceschini A, Szklarczyk D, Frankild S et al (2013) STRING v9.1: protein-protein interaction networks, with increased coverage and integration. Nucleic Acids Res 41: D808–815
29. Kozomara A, Griffiths-Jones S (2011) miRBase: integrating microRNA annotation and deep-sequencing data. Nucleic Acids Res 39: D152–157
30. Zinovyev A, Viara E, Calzone L et al (2008) BiNoM: a Cytoscape plugin for manipulating and analyzing biological networks. Bioinformatics 24:876–877
31. Witkos TM, Koscianska E, Krzyzosiak WJ (2011) Practical aspects of microRNA target prediction. Curr Mol Med 11:93–109

Chapter 24

Tapping MicroRNA Regulation Networks Through Integrated Analysis of MicroRNA–mRNA High-Throughput Profiles

Anthony D. Saleh and Hui Cheng

Abstract

Understanding the biological relevance and context of microRNA (miRNA) regulation of target mRNAs is difficult to ascertain because an individual miRNA aids simultaneously in the regulation of hundreds of mRNAs in a cell. With the increasing availability of large public datasets that profile both mRNA and miRNA expression levels from the same samples, it is possible to apply robust statistical methods to identify global negative correlations in miRNA and target mRNA expression. Using a dataset from The Cancer Genome Atlas as a case study, we show how to use linear regression analysis followed by permutation-based false discovery rate to assign high statistical power to pair-wise negative correlations of miRNA and mRNA expression. Used in conjunction with available prediction tools or other target databases, a high confidence dataset of global miRNA–mRNA interactions can be generated. We also describe further methods to prioritize identified interactions by integrating with mutation, copy number variation, methylation, or survival data to support observations and provide context. Finally, we discuss methods to experimentally validate selected novel targets.

Key words MicroRNA regulation, MicroRNA integrated analysis, MicroRNA expression profiling, MicroRNA statistical methods, MicroRNA correlation, The Cancer Genome Atlas (TCGA) microRNA analysis, MicroRNA target prediction, MicroRNA target prioritization

1 Introduction

An individual microRNA (miRNA) can affect simultaneously the expression of hundreds of mRNA targets in a cell [1]. To understand the role of miRNAs in a biological phenomenon, it is imperative to unravel the network of mRNAs that they target. Several in silico miRNA target prediction tools have been created to aid in this process [2–8]. These tools rely on complementary sequence information and empirical rules to score interactions, e.g., sequence alignment, free energy calculations of the miRNA–mRNA duplex, and conservation of binding sites between species. Although these tools are a necessary and valuable step in target identification, they

M. Lucrecia Alvarez and Mahtab Nourbakhsh (eds.), *RNA Mapping: Methods and Protocols*, Methods in Molecular Biology, vol. 1182, DOI 10.1007/978-1-4939-1062-5_24, © Springer Science+Business Media New York 2014

all suffer from high false-positive prediction rates [2]. Several studies have used integrated pair-wise expression analysis of miRNA and mRNA or protein levels in conjunction with prediction tools to provide a context for regulation and significantly reduce false-discovery rate [3]. However, miRNAs are only one of many factors that regulate mRNA expression levels [4], and miRNA expression changes often result in moderate changes in mRNA expression level [5]. Therefore, previous methods have limited sensitivity because they look for an inverse correlation between mRNA and miRNA expression levels using small datasets. Here we present a method that overcomes the sensitivity limitation of this approach by using linear regression analysis of large public datasets that profile mRNA and miRNA expression levels from the same samples, followed by permutation-based false discovery rate (FDR) to assign high statistical power to pair-wise negative correlations of miRNA and mRNA expression. Used in conjunction with available prediction tools or other target databases, a high-confidence dataset of global miRNA–mRNA interactions can be generated. We also describe methods to prioritize identified interactions by further integration with mutation, copy number variation, methylation, or survival data. Finally, we discuss methods to experimentally validate selected novel identified targets.

2 Methods

2.1 Data Selection

In order to obtain increased statistical sensitivity, large datasets with both miRNA and mRNA profiling results are required. We would suggest at least 100 samples, but the larger the better. The Cancer Genome Atlas (TCGA, http://cancergenome.nih.gov/) offers paired expression-profiling datasets of miRNA and mRNA for up to 500 samples from several human cancers. Other small datasets may be identified in profiling databases such as Arrayexpress from the European Bioinformatics Institute (http://www.ebi.ac.uk/arrayexpress/) [6] or gene expression omnibus from the National Center for Biotechnology (http://www.ncbi.nlm.nih.gov/geo/) [7]. If suitable datasets cannot be identified, one may consider meta-data analysis techniques for combining small datasets [8]. However, we would recommend that miRNA and target mRNA or protein profiling data be always collected at the same time from the same sample. In this chapter, we utilize matched miRNA and mRNA RNA-seq profiling data from 279 head and neck squamous cell carcinoma (HNSCC) specimens as a case study for this method.

2.2 Data Normalization

In the example presented in this chapter, sequencing of both miRNAs and mRNAs was performed using the Illumina HiSeq platform for the HNSCC dataset [9]. miRNA sequences were mapped to

1,222 miRBase 5p and 3p strands as the reference database of miRNA transcripts/genes [9]. For mRNA, aligned reads to the hg19 genome were assembled using Mapsplice (http://www.netlab.uky.edu/p/bioinfo/MapSplice) [10]. Gene expression was quantified with RNA-seq by expectation maximization (RSEM) [11].

For each RNA-seq profile, the copy number of RNA sequence reads can be used to estimate the expression level of each mRNA or miRNA. To compare RNA level in different samples, a normalization step is needed. Currently, miRNA-Seq normalization methods are largely adapted from the mRNA-Seq normalization methods.

The most widely used normalization methods for RNA-seq data are global scaling to adjust for varying sequencing depths and technical effects in different libraries [12–14]. The assumption is that the total RNA population or the upper-quartile value remains fairly constant between samples (without necessarily trying to make assumptions on the RNA subpopulation) [14] (*see* **Note 1**).

Normalize miRNAs to reads per million (RPM); that is, for each sample, estimate the frequency of miRNAs by computing the proportion of summed reads that mapped to a particular miRNA relative to the total number of aligned reads; transform the proportions by multiplying by one million, and then compare these values across all samples [9, 12]. Normalize the RSEM values of mRNA to a fixed upper quartile within each sample. Set the expression values of zeros with the overall minimum nonzero RSEM value.

To increase detection power for these high-throughput experiments, data filtering is a necessary step. Due to the large number of hypothesis tests that we perform on these mRNA and miRNA profiles, a multiple testing adjustment is warranted. However, when the number of tests is very large and the proportion of correlated mRNA and miRNA pairs is relatively low, the power to detect weakly correlated pairs would be reduced [15, 16]. Variance-based data filtering allows for a reduction in the number of tests and a corresponding increase in statistical power. Rank mRNAs and miRNAs by the variance of their abundance across the samples, and use the most variant 50 % miRNA and mRNA for integrated analysis. Log2 transform both miRNA and mRNA expression data.

2.3 Target Prediction by Correlation Analysis

Until high-throughput paired mRNA and miRNA expression profiling became available, many computational algorithms to predict miRNA targets were mainly based on perfect base-pairing of the 5′ seed of miRNA to a complementary site in the 3′ untranslated region (UTR) of the mRNA, conservation of the miRNA-binding site, and a high duplex stability with a positive balance of minimum free energy [17, 18]. However, only a limited number of miRNA-predicted targets have been experimentally validated, and accurate target predictions are still obstacles in miRNA research.

Simultaneous high-throughput profiling of miRNA and mRNA expression allows computational methods to be more accurate in predicting the strongest miRNA–mRNA interaction [19, 20]. Most of these methods assume that miRNAs cause the degradation of their targets; therefore, the expression profiles of the miRNA and mRNA pairs should reveal an inverse relationship (*see* **Note 2**). A simple and straightforward method to analyze miRNA–mRNA relationship is Pearson correlation [3], which assumes a linear relationship between pairs (*see* **Note 3**). Use miRNAs as independent variables and mRNAs as dependent variables in a linear regression analysis for each miRNA–mRNA pair in a large dataset analyzed with the Matrix eQTL R package. Using this method, we identified 586,337 interactions between miRNAs and mRNAs. To estimate FDR for the regression *P*-values, randomly permute sample labels and compare the observed *P*-values to the null distribution of *P*-values obtained by permutation. Use an FDR threshold of 0.05 to identify significant negative miRNA–mRNA associations. After adjustment, 369,559 significant miRNA–mRNA interactions were identified using the method described here.

2.4 Identifying Direct and Indirect Targets of miRNA

2.4.1 Filtering Predicted or Direct Targets

The list of inversely correlated miRNA and mRNA targets contains both direct and indirect targets. In order to identify directly targeted mRNAs, prediction algorithms can be utilized. Table 1 contains a list of miRNA target prediction tools and verified target databases. We recommend combining prediction from at least two databases. A high-quality resource like miRecords (http://mirecords.biolead.org/) allows individual searches of 11 prediction tools simultaneously [21] (*see* **Note 4**).

Table 1
MicroRNA target prediction tools and databases

Name	Access	Website
miRanda	Online search	http://www.microrna.org
Target Scan	Online search	http://www.targetscan.org
microT	Online search	http://diana.cslab.ece.ntua.gr/
PicTar	Online search	http://pictar.mdc-berlin.de/
RNAHybrid	Download	http://bibiserv.techfak.uni-bielefeld.de/rnahybrid
miRGen++	Mathlab Code	http://www.psi.toronto.edu/genmir
MiTarget	Online search	http://cbit.snu.ac.kr/~miTarget
MiRtaget2	Online search	http://mirdb.org
RNA22	Online search	http://cbcsrv.watson.ibm.com/rna22.html

2.4.2 Filtering Method

miRNA and mRNA pairs with statistically significant inverse expression interactions were filtered using experimentally validated miRNA and mRNA interactions from miRTarBase (http://mirtarbase.mbc.nctu.edu.tw/) [22]. In our example, 166 previously tested interactions were identified after filtering. To find interactions that may be important in carcinogenesis, identify differentially expressed miRNAs and mRNAs in tumors and normal tissues using a Mann–Whitney–Wilcoxon test. Identify pairs that are regulated in opposite directions, i.e., a downregulated miRNA and its up-regulated mRNA target, or vice versa. For each miRNA–mRNA pair, we identified samples that exhibit at least a twofold change for both miRNA and mRNA expression, which resulted in 119 inversely deregulated pairs. Table 2 shows the example of previously verified mRNA targets that are significantly inversely expressed to the repressed miRNA let-7c and the overexpressed miRNA miR-155-5p. An expression plot for let-7c and its target gene HMGA2 is shown in Fig. 1 (*see* **Note 5**).

Table 2
Inversely correlated microRNA and mRNA

MicroRNA	Target mRNA	*t*-stat[a]	FDR[b]
let-7c	E2F1	-3.40	6.78E-03
	HMGA2	-9.47	7.22E-19
	AURKB	-5.52	2.69E-06
	CCNA2	-4.39	2.72E-04
	CDK6	-8.31	8.01E-13
	IFNB1	-2.78	3.34E-02
	IGF2BP1	-4.21	5.31E-04
	IGF2BP2	-10.21	2.27E-18
	PXDN	-3.76	2.31E-03
	RRM2	-5.03	2.33E-05
	UHRF1	-3.77	2.28E-03
miR-155-5p	AGTR1	-4.02	1.03E-03
	CYR61	-3.56	4.35E-03
	EDN1	-2.62	4.82E-02
	KRT80	-3.81	2.03E-03

[a] *t-stat* represents *t*-statistic from linear regression analysis. This test statistic is a measure of the likelihood that the actual value of the β coefficient for the predictor variable microRNA is not zero

[b] *FDR* represents false discovery rate. It is defined as the proportion of false discoveries among the discoveries. FDR procedures are used to control the falsely rejected hypotheses in multiple hypothesis testing

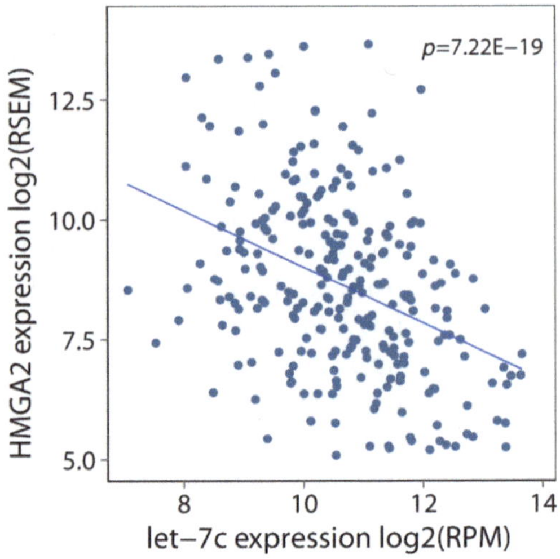

Fig. 1 Strong negative correlation between the level of expression of let-7c and its target gene, high mobility group AT-hook 2 (HMGA2). *P*-values from linear regression are annotated on *top* of the figure

2.5 Hit Prioritization

2.5.1 Overlap with Mutations

Once a large and valuable dataset has been generated, there is the difficult task of making sense of it. If interactions were identified using a TCGA dataset, integration with methylation or copy number variation (CNV) of miRNA genes can further support observations of expression changes and indicate importance in disease biology. Filtering of mRNA target lists for genes that are frequently mutated or have CNV is also useful in prioritization. Level 4 analysis files of TCGA datasets can be viewed or downloaded from the Broad Institute's GDAC firehose website (https://confluence.broadinstitute.org/display/GDAC/Home;jsessionid=AEE65AC8636BC707 38673257AFE14735) [23]. Genomic Identification of Significant Targets in Cancer (GISTIC) is a software that predicts focal somatic copy-number alteration in human cancers and can assign adjusted *P*-values for gain or loss of a particular genomic loci across the TCGA dataset for HNSC. Gene loci for miRNA identified in the correlation analysis are intersected with significant GISTIC peaks to identify miRNAs that were deleted or amplified. This analysis identified that the MIRLET7C gene resides in a significantly deleted GISTIC peak. As mentioned above, this miRNA is inversely correlated to several proven target oncogenes.

2.5.2 Transcriptional Regulation

Another method of prioritization consists in connecting selected miRNAs to transcription factors that are known to be important in the biological phenomenon under study. ChIP-seq data can be obtained from the Encode project for many transcription factors [24]. miRNA gene locations or known promoters can then be intersected with ChIP peaks to identify interactions. Alternatively, the Encode ChIP-seq data can be viewed in the UCSC genome browser (http://genome.ucsc.edu/cgi-bin/hgGateway) [25] or in the Integrated Genome Viewer (http://www.broadinstitute.org/igv/) [26] for querying of individual gene loci. A search of encode data in UCSC genome browser identifies a DNA-binding peak for the transcription factor PRDM1 adjacent to the known promoter of the MIRLET7C gene. PRDM1 has previously been shown to correlate with let-7 expression [27].

2.5.3 Survival Analysis

Many studies have found that miRNAs are deregulated in tumors, and their expression levels are more predictive of cancer subtypes than corresponding mRNA expression [28, 29]. The regulatory role of miRNAs has enabled them as either oncogenes or tumor suppressors [30]. To investigate whether let-7c is a predictive marker for survival and treatment response in head and neck cancer patients, analyses were performed using the R Survival statistical package [31] (Terry Therneau, http://CRAN.R-project.org/package=survival). Analyses of time to death due to HNSCC were performed using the product-limit procedure (Kaplan–Meier method), and differences between categories were estimated by the log-rank test. Patients who died of other causes were censored at the date of death. Let-7c expression values were subdivided into quartiles and assessed for their prognostic values as shown in Fig. 2. There is a statistically significant difference in the survival between the fourth quartile (highest let-7c subgroup) and the rest of the

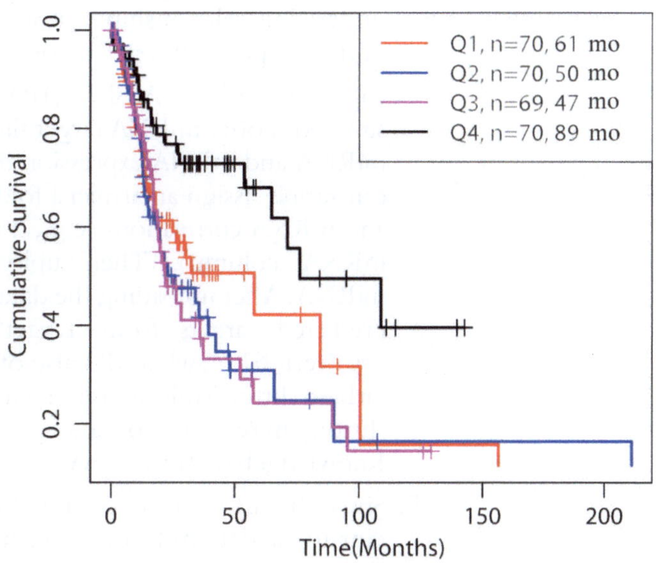

Fig. 2 High level of let-7c expression was associated with increased survival in head and neck cancer. Kaplan–Meier curves showed that head and neck squamous cell carcinoma (HNSCC) patients with high expression level of let-7c (Q4: 75 % percentile) had significantly longer overall survival (p=0.015, 0.0015, 0.0012 for Q1, Q2, and Q3, respectively, by log-rank test). *Mo* months

three quartiles (lower expression groups). These results suggest that let-7c acts as a tumor suppressor in head and neck carcinogenesis and predicts favorable prognosis for HNSCC. This finding is consistent with previous studies that have demonstrated let-7 functions as a tumor suppressor by regulating cell proliferation in several cancers [32–35], and low let-7 expression was identified as a marker for advanced cancer [30].

3 Notes

1. Many studies compared different normalization methods for mRNA-Seq data but only a few for miRNA-Seq data. Although global scaling normalizations are most widely used, some studies have found that non-scaling methods like Lowess normalization and quantile normalization perform better for miRNA-Seq data [13].
2. Most mathematical models using sequence-based expression profiles assume a negative correlation between miRNA and mRNA expression. However, some studies do consider positive correlations since some miRNAs may act as transcription factors [3, 36, 37].
3. In most cases, the assumption that miRNA and mRNA have a linear relationship would be sufficient for detection of their inverse relationship. In some situations, especially when the underlying relationship is not linear or with the presence of outliers, Spearman correlation is preferred [3].
4. Ingenuity Pathway Analysis (IPA) software and Partek genomics suite both offer miRNA target filtering tools for complementary miRNA and mRNA expression data. To utilize these tools, you can simply assign an artificial fold change value to your miRNA and mRNA correlations (e.g., 2 for miRNA column and -2 for mRNA column). Then upload the targets of a single miRNA. After uploading the data file, both IPA and Partek offer predicted targets from TargetScan (http://www.targetscan.org/vert_61/), while IPA also offers validated targets from mirtarbase (http://mirtarbase.mbc.nctu.edu.tw/) [22], miRecords (http://mirecords.biolead.org/) [21], and the Ingenuity Knowledgebase (http://www.ingenuity.com/products/ipa).
5. Since the information obtained is only correlative, it is important to validate miRNA target interaction in vitro once prioritized targets have been selected. The use of a cell line that expresses a relatively low level of a specific miRNA and one that expresses it at high level is recommended for these types of experiments. Both cell lines should have moderate expression of the target mRNAs. miRNA expression profiles for many cells lines can be obtained from miRanda (http://www.microrna.org)

or array databases such as GEO (http://www.ncbi.nlm.nih.gov/geo/) [7]. Matched miRNA and mRNA data is also available for the NCI60 cell lines [38]. Overexpression of the specific miRNA under study can be achieved using either vector that expresses the precursor miRNA (pre-miRNA) or by direct transfection of a miRNA mimic. Corresponding decreases in target protein level and usually target mRNA level should be observed, especially in the cell line that has low background expression. Short antisense oligonucleotides (anti-miRs) that are complementary to miRNAs are also commercially available and can knock down miRNA expression. Treatment of cells with anti-miR should increase target expression, especially in the cell line that has high miRNA background expression. Detection of target expression can confirm both direct and indirect targeting by miRNAs. To show direct targeting, reporter gene experiments or co-immunoprecipitation with Ago2 is usually employed [39].

References

1. Lim LP, Lau NC, Garrett-Engele P et al (2005) Microarray analysis shows that some microRNAs downregulate large numbers of target mRNAs. Nature 433:769–773
2. Witkos TM, Koscianska E, Krzyzosiak WJ (2011) Practical aspects of microRNA target prediction. Curr Mol Med 11:93–109
3. Muniategui A, Pey J, Planes FJ et al (2013) Joint analysis of miRNA and mRNA expression data. Brief Bioinform 14:263–278
4. Le TD, Liu L, Liu B et al (2013) Inferring microRNA and transcription factor regulatory networks in heterogeneous data. BMC Bioinformatics 14:92
5. Bartel DP (2009) MicroRNAs: target recognition and regulatory functions. Cell 136: 215–233
6. Brazma A, Parkinson H, Sarkans U et al (2003) ArrayExpress–a public repository for microarray gene expression data at the EBI. Nucleic Acids Res 31:68–71
7. Edgar R, Domrachev M, Lash AE (2002) Gene Expression Omnibus: NCBI gene expression and hybridization array data repository. Nucleic Acids Res 30:207–210
8. Kuo TY, Hsi E, Yang IP et al (2012) Computational analysis of mRNA expression profiles identifies microRNA-29a/c as predictor of colorectal cancer early recurrence. PLoS One 7:e31587
9. Network CGA (2012) Comprehensive molecular portraits of human breast tumours. Nature 490:61–70
10. Wang K, Singh D, Zeng Z et al (2010) MapSplice: accurate mapping of RNA-seq reads for splice junction discovery. Nucleic Acids Res 38:e178
11. Li B, Dewey C (2011) RSEM: accurate transcript quantification from RNA-Seq data with or without a reference genome. BMC Bioinformatics 12:323
12. Robinson MD, Oshlack A (2010) A scaling normalization method for differential expression analysis of RNA-seq data. Genome Biol 11:R25
13. Garmire LX, Subramaniam S (2012) Evaluation of normalization methods in mammalian microRNA-Seq data. RNA 18:1279–1288
14. Kadota K, Nishiyama T, Shimizu K (2012) A normalization strategy for comparing tag count data. Algorithms Mol Biol 7:5
15. Hackstadt AJ, Hess AM (2009) Filtering for increased power for microarray data analysis. BMC Bioinformatics 10:11
16. Bourgon R, Gentleman R, Huber W (2010) Independent filtering increases detection power for high-throughput experiments. Proc Natl Acad Sci U S A 107:9546–9551
17. Barbato C, Arisi I, Frizzo ME et al (2009) Computational challenges in miRNA target predictions: to be or not to be a true target? J Biomed Biotechnol 2009:803069
18. Wang X, Wang X (2006) Systematic identification of microRNA functions by combining target prediction and expression profiling. Nucleic Acids Res 34:1646–1652

19. Gennarino VA, Sardiello M, Avellino R et al (2009) MicroRNA target prediction by expression analysis of host genes. Genome Res 19: 481–490
20. Huang JC, Babak T, Corson TW et al (2007) Using expression profiling data to identify human microRNA targets. Nat Methods 4: 1045–1049
21. Xiao F, Zuo Z, Cai G et al (2009) miRecords: an integrated resource for microRNA-target interactions. Nucleic Acids Res 37(Database issue):D105–D110
22. Hsu SD, Lin FM, Wu WY et al (2011) miRTarBase: a database curates experimentally validated microRNA-target interactions. Nucleic Acids Res 39(Database issue):D163–D169
23. Broad Institute TCGA Genome Data Analysis Center (2013) Analysis overview for head and neck squamous cell carcinoma (Primary solid tumor cohort). Broad Institute of MIT and Harvard. doi: 10.7908/C1WJ66HQ3. Accessed on 21 Apr, 2013.
24. Birney E, Stamatoyannopoulos JA, Dutta A et al (2007) Identification and analysis of functional elements in 1 % of the human genome by the ENCODE pilot project. Nature 447:799–816
25. Kent WJ, Sugnet CW, Furey TS et al (2002) The human genome browser at UCSC. Genome Res 12:996–1006
26. Robinson JT, Thorvaldsdottir H, Winckler W et al (2011) Integrative genomics viewer. Nat Biotechnol 29:24–26
27. West JA, Viswanathan SR, Yabuuchi A et al (2009) A role for Lin28 in primordial germ-cell development and germ-cell malignancy. Nature 460:909–913
28. Lu J, Getz G, Miska EA et al (2005) MicroRNA expression profiles classify human cancers. Nature 435:834–838
29. Iorio MV, Ferracin M, Liu CG et al (2005) MicroRNA gene expression deregulation in human breast cancer. Cancer Res 65:7065–7070
30. Shell S, Park SM, Radjabi AR et al (2007) Let-7 expression defines two differentiation stages of cancer. Proc Natl Acad Sci U S A 104: 11400–11405
31. Team RC (2012) *R:* A language and environment for statistical computing. R Foundation for Statistical Computing, Vienna
32. Esquela-Kerscher A, Trang P, Wiggins JF et al (2008) The let-7 microRNA reduces tumor growth in mouse models of lung cancer. Cell Cycle 7:759–764
33. Boyerinas B, Park SM, Hau A et al (2010) The role of let-7 in cell differentiation and cancer. Endocr Relat Cancer 17:F19–F36
34. Akao Y, Nakagawa Y, Naoe T (2006) let-7 microRNA functions as a potential growth suppressor in human colon cancer cells. Biol Pharm Bull 29:903–906
35. Nadiminty N, Tummala R, Lou W et al (2012) MicroRNA let-7c is downregulated in prostate cancer and suppresses prostate cancer growth. PLoS One 7:e32832
36. Nunez-Iglesias J, Liu CC, Morgan TE et al (2010) Joint genome-wide profiling of miRNA and mRNA expression in Alzheimer's disease cortex reveals altered miRNA regulation. PLoS One 5:e8898
37. Wang YP, Li KB (2009) Correlation of expression profiles between microRNAs and mRNA targets using NCI-60 data. BMC Genomics 10:218
38. Liu H, D'Andrade P, Fulmer-Smentek S et al (2010) mRNA and microRNA expression profiles of the NCI-60 integrated with drug activities. Mol Cancer Ther 9:1080–1091
39. Hassan MQ, Gordon JA, Lian JB et al (2010) Ribonucleoprotein immunoprecipitation (RNP-IP): a direct in vivo analysis of microRNA-targets. J Cell Biochem 110:817–822

Chapter 25

miRWalk Database for miRNA–Target Interactions

Harsh Dweep, Norbert Gretz, and Carsten Sticht

Abstract

miRWalk (http://mirwalk.uni-hd.de/) is a publicly available comprehensive resource, hosting the predicted as well as the experimentally validated microRNA (miRNA)–target interaction pairs. This database allows obtaining the possible miRNA-binding site predictions within the complete sequence of all known genes of three genomes (human, mouse, and rat). Moreover, it also integrates many novel features such as a comparative platform of miRNA-binding sites resulting from ten different prediction datasets, a holistic view of genetic networks of miRNA–gene pathway, and miRNA–gene–Online Mendelian Inheritance in Man disorder interactions, and unique experimentally validated information (e.g., cell lines, diseases, miRNA processing proteins). In this chapter, we describe a schematic workflow on how one can access the stored information from miRWalk and subsequently summarize its applications.

Key words miRWalk, MicroRNA, Promoter, Prediction, 5′-UTR, CDS, 3′-UTR, Pathways, OMIM, Validated, Application

1 Introduction

MicroRNAs (miRNAs) comprise a class of small, noncoding RNA molecules of 21–25 nucleotides (nt) in length that regulate the expression of their target genes at posttranscriptional level [1]. In addition to their role in such posttranscriptional repression, miRNAs are now implicated in transcriptional gene silencing by targeting the promoter region [2]. Thus, miRNAs can inhibit gene expression via translational repression, mRNA degradation, or transcriptional inhibition. Recently, miRNAs have also been found in activating the expression of their target genes [3].

The miRNA gene family is one of the largest in higher eukaryotes: more than 1,900 miRNAs have been identified in the human genome [4]. In silico and experimental methods indicate that each of these miRNAs has the capability to anneal more than a hundred mRNAs. Furthermore, over 60 % of protein-coding genes are predicted to harbor the possible miRNA-binding sites within their 3′-UTR regions.

M. Lucrecia Alvarez and Mahtab Nourbakhsh (eds.), *RNA Mapping: Methods and Protocols*, Methods in Molecular Biology, vol. 1182, DOI 10.1007/978-1-4939-1062-5_25, © Springer Science+Business Media New York 2014

Since the breakthrough discovery of the very first miRNAs [5], many programs/databases such as Targetscan [6], miRanda [7], Pictar [8], PITA [9], RNA22 [10], DianaMT [11], miRDB [12], RNAhybrid [13], TarBase [14], miRTarBase [15], PhenomiR [16], and miR2Disease [17] have been developed to supply information on the putative and the experimentally verified miRNA–target interaction pairs for animal miRNAs (reviewed in ref. 18). These resources have been proven to be useful as centralized sources for collecting target information; however, there are a number of limitations: These databases are restricted to offer the putative miRNA–target interaction pairs within the mRNA 3′-UTR region, whereas accumulating studies demonstrated a novel action of miRNAs by which they may modulate the expression of the target genes by annealing in their promoter, 5′-UTR, and CDS regions [19–22]. Additionally, seed matches, conservation of the sequence, and thermodynamics of the miRNA–mRNA interactions are widely used in predicting potential targets. The main disadvantage in this approach is the enormous number of possible targets, and, therefore, results may contain many false-positive predictions. Moreover, comparative studies conducted with the earlier miRNA–target prediction algorithms found that no program was consistently superior to all others [23]. Demanding a target to be predicted by several algorithms can be helpful in reducing the high number of putative and maybe false-positive targets. Furthermore, the experimentally validated information on mammalian miRNAs is scattered in available databases such as TarBase, miRTarBase, miR2Disease, and PhenomiR. Moreover, these resources take about 6 months or a year (if at all) to update the documented information due to their surveying pertinent literature.

To eliminate these limitations, we provide a new comprehensive database: miRWalk [24], encompassing the predicted as well as the experimentally validated miRNA–target interactions. miRWalk is different from the existing resource as:

- A novel algorithm, i.e., miRWalk, is utilized for generating the possible miRNA-binding sites by walking the genes of three genomes (human, mouse, and rat) [24].
- miRWalk is so far the only database, offering the putative miRNA-binding sites within the complete sequence (promoter, CDS, 5′- and 3′-UTR) of all known genes of three species (human, mouse, and rat).
- One of the key features of this database is that it provides a comparative platform of the possible miRNA-binding sites within the mRNA 3′-UTR regions which are resulted from the miRWalk algorithm and the eight other prediction programs [DIANA-mT [11], miRanda [7], miRDB [12], RNAhybrid [13], PICTAR4 [8], PICTAR5 [8], PITA [9], RNA22 [10], and TargetScan [6], i.e., (ten different prediction datasets)].

- miRWalk also comprises a holistic view of genetic networks of miRNA-gene pathway, miRNA–gene–Online Mendelian Inheritance in Man (OMIM), and mitochondrial genome–miRNA interactions.
- miRWalk amalgamates the scattered data and subclassifies the collected information of miRNAs at one place. Furthermore, miRWalk hosts new and unique features on experimentally validated miRNAs.

2 Materials

Methods described in this chapter are based on Web-based functionality of miRWalk.

2.1 Web-Based Protocols

miRWalk is publicly accessible via a user Web interface based on Hyper Text Markup Language (HTML), Cascading Style Sheets (CSS), and Hypertext Preprocessor (PHP) scripts at the URL http://mirwalk.uni-hd.de/. In order to use this resource, users need to have:

1. A computer, with access to the Internet
2. Candidates of interest such as miRNAs, genes, pathways, OMIM disorders, chromosomes, diseases, organs, cell lines, or miRNA-processing proteins.

3 Methods

In this section, we provide a comprehensive overview about the content documented in miRWalk (Subheading 3.1). We then depict a schematic workflow on how one can employ the Web interface (different search methods) of miRWalk to obtain information on the predicted (Subheading 3.2) and the experimentally validated miRNA–target interactions (Subheading 3.3). Lastly, we briefly outline the applications of miRWalk with the help of recently published studies (Subheading 3.4).

3.1 Content of miRWalk

miRWalk is implemented as a relational database on a MySQL (Structure Query Language) database management system (DBMS). The data content of miRWalk is broadly classified into two modules: *Predicted Target* and *Validated Target*. Therefore, information on the putative and the experimentally validated miRNA–target interactions can be obtained via these two modules. Figure 1 displays the home page of miRWalk.

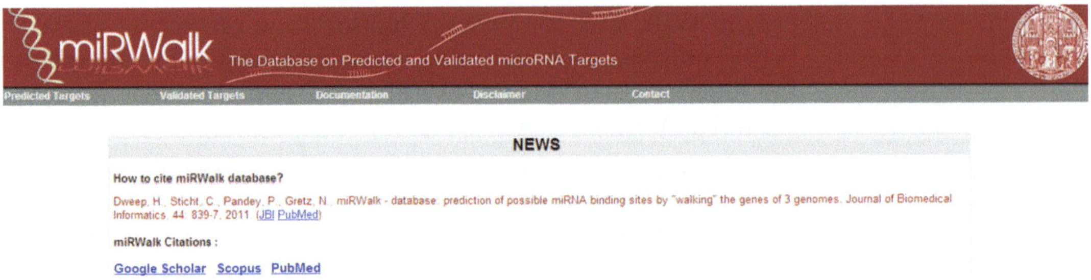

Fig. 1 Home page of miRWalk. The information of miRWalk is broadly categorized under the Predicted and the Validated Target modules. Moreover, the Documentation, Disclaimer, and Contact pages are also integrated for providing the help on Web interface and term of uses and for reporting the problems/bugs, respectively

3.2 The Predicted Target Module

The Predicted Target module offers possible miRNA-binding site predictions on the known genes of three species, resulting from miRWalk (novel predictions) and other eight algorithms. Moreover, it also supplies putative miRNA-binding site interactions on the representative members of human biological pathways and OMIM disorders.

The Predicted Target module is further divided into six different search pages: Target Gene, miRNA, Pathway, Chromosome, OMIM, and Mitochondrial Target. These search interfaces are restricted to some limits (*see* **Note 2**).

3.2.1 Search Interface: Target Gene

"Target Gene" search interface is broadly designed into four blocks (Fig. 2). Possible miRNA-binding site predictions within the complete sequence of genes of interest can be collected by initiating a default or a customized query using "Target Gene" interface in the following manner:

Block 1: Species and gene of interest

1. Select a species of interest from the given drop-down menu.
2. Identifiers or upload a file (*see* **Note 1**): A user can provide gene symbols or Entrez Gene ID either directly into the given textbox or by uploading a file (*see* **Note 2** for the limit of input identifiers).

Block 2: Regions of interest

3. Gene region: There are four options, i.e., promoter, 5′-UTR, CDS, and 3′-UTR. Choosing promoter, a user can restrict the query for the upstream gene flanking region by providing a value in the "Upstream flanked" textbox.

Block 3: Additional options for a customized query

4. Options: A user can select either the longest transcript or all transcripts encoded by the gene(s) of interest. Moreover, one can customize the search by providing a minimum seed length of miRNA and/or *p*-value (*see* **Note 4**).

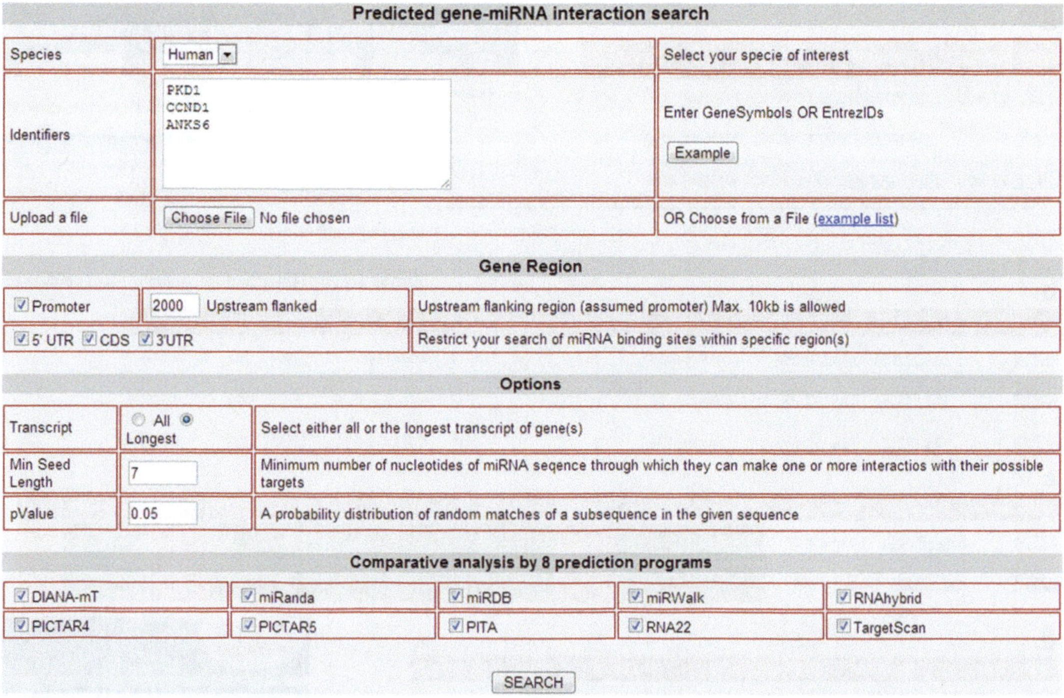

Fig. 2 Overview of "Target Gene" Search interface under the Predicted Target module. A default or a customized search for putative miRNA-binding sites can be submitted by (**a**) selecting a species, i.e., human, mouse, or rat; (**b**) providing identifiers; (**c**) choosing regions of interest; (**d**) clicking either the longest or all transcripts encoded by genes of interest; (**e**) inputting a minimum seed length of the miRNA through which they can bind with the genes of interest and/or the *p*-value; and (**f**) selecting at least miRNA–target prediction algorithms for comparing the results and submitting the query by pressing "SEARCH" button

Block 4: Comparative platform

5. Prediction programs: Choose at least two programs for comparing the putative miRNA-binding sites obtained from different prediction algorithms (*see* **Note 5**). Click on "Submit" button to initiate a search. Figure 2 schematically depicts the query procedure.

Figure 3 displays putative miRNA-binding site predictions on PKD1, CCND1, and ANKS6 genes. The result page contains information in four tables, i.e., Fig. 3a: Gene information table; Fig. 3b: Putative miRNA-binding sites generated by miRWalk according to mRNA selected regions; Fig. 3c: A comparative view of miRNA-binding sites identified by miRWalk and other eight established programs within the mRNA 3′-UTR region; and Fig. 3d: Predicted miRNA-binding sites within the promoter region. All four tables are hyperlinked with GenBank, Ensembl, and miRBase for more annotations and further information on genes, their mRNAs, and predicted miRNAs (*see* **Note 3**).

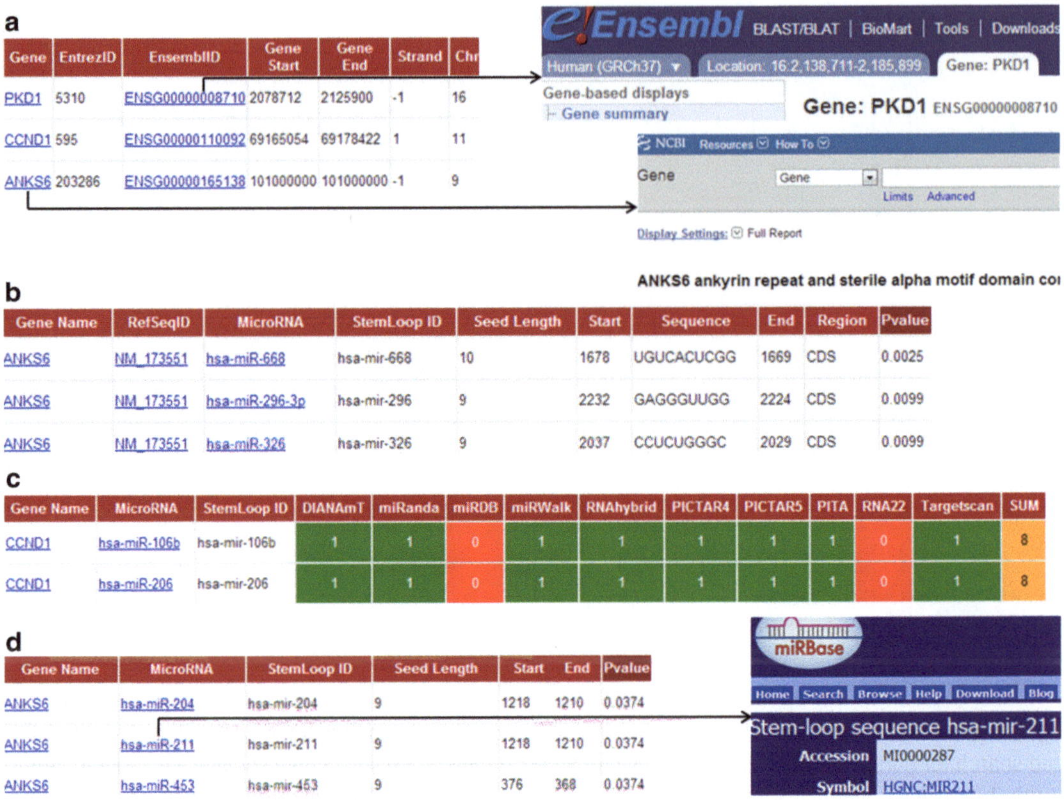

Fig. 3 Overview of the result page obtained after performing a query via Target Gene Search interface. The result panel shows four tables, i.e., (**a**) gene information table, (**b**) predicted miRNA-binding sites according to mRNA selected regions, (**c**) a comparative analysis of miRNA-binding sites within the 3′-UTR region (*green* and *red* colors indicate sites found and not found, respectively), and (**d**) putative miRNA-binding sites within the promoter regions. Links to external databases (EntrezGene and RefSeq at NCBI, Ensembl, and miRBase) are also integrated within all four tables (Color figure online)

3.2.2 Search Interface: MicroRNA Target

MicroRNA Target search (*see* **Note 2**) and result pages are organized similar to "Target Gene." The result of the query for miRNA "hsa-miR-31" is shown in Fig. 4. The output is displayed in four tables: Fig. 4a: MicroRNA information table; Fig. 4b: Putative targets of hsa-miR-31 according to mRNA regions (predicted by miRWalk); Fig. 4c: A comprehensive view of putative targets of hsa-miR-31 predicted by miRWalk and eight established programs within mRNA 3′-UTR region; and Fig. 4d: The last table displays putative target genes of hsa-miR-31 within the promoter region (default 2 kb, but up to 10 kb is possible). All four tables are hyperlinked with external databases for further information. Other information such as the host gene of miRNAs, miRNAs having similar seed sequences, regulatory binding sites (*see* **Note 6**), and target mining for miRNA names with asterisk (*see* **Note 7**) can be queried via this search interface.

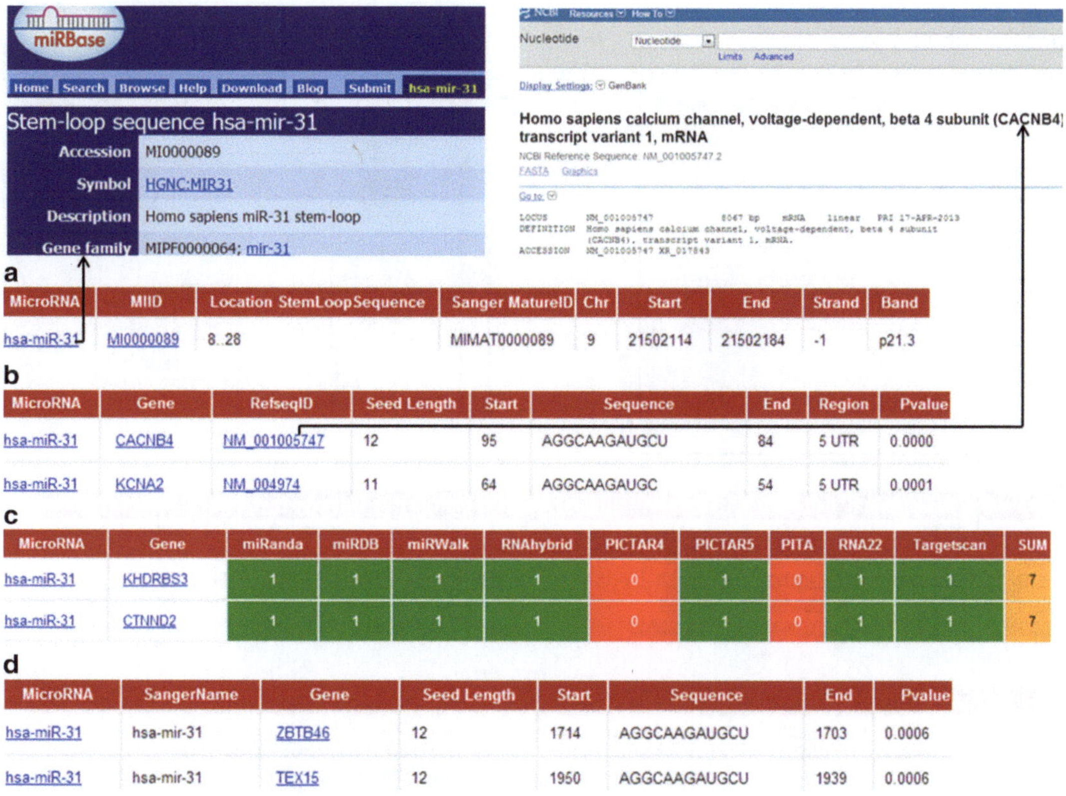

Fig. 4 Overview of results obtained from MicroRNA Target Search interface. The panel displays results in four tables, i.e., (**a**) miRNA information table, (**b**) putative target genes identified by miRWalk within mRNA selected regions, (**c**) a comparative analysis of miRNA–target interactions within the 3′-UTR region, and (**d**) possible targets within the promoter region. Links to external databases are also included

3.2.3 Search Interface: Pathway Target

Pathway Target search pages can be used to obtain a more holistic view of the genetic network of miRNA–target pathway interactions. For example, a user can view how many genes of a pathway can be targeted by similar or dissimilar miRNAs. This task can be easily achieved by selecting a pathway from the given drop-down menu (Fig. 5a). The result of the query is shown in Fig. 5b. Furthermore, it also offers putative miRNA-binding sites on the complete sequence of genes associated with 449 biological pathways. These pathways are hyperlinked to their respective resources [KEGG [25] or Biocarta (http://www.biocarta.com)] for more information.

3.2.4 Search Interface: Chromosome Target

Chromosome Target Search page is organized similar to “Pathway Target.” Users can gather putative binding site information on all miRNAs located within a chromosome against all genes (mRNAs) located on the same or different chromosome. As shown in Fig. 6a, such kind of data can be easily obtained by choosing the chromosomes where the miRNA and mRNA are located. Figure 6b shows an overview of the result obtained by selecting all miRNAs located on chromosome 1 against all genes located on chromosome 2 of humans.

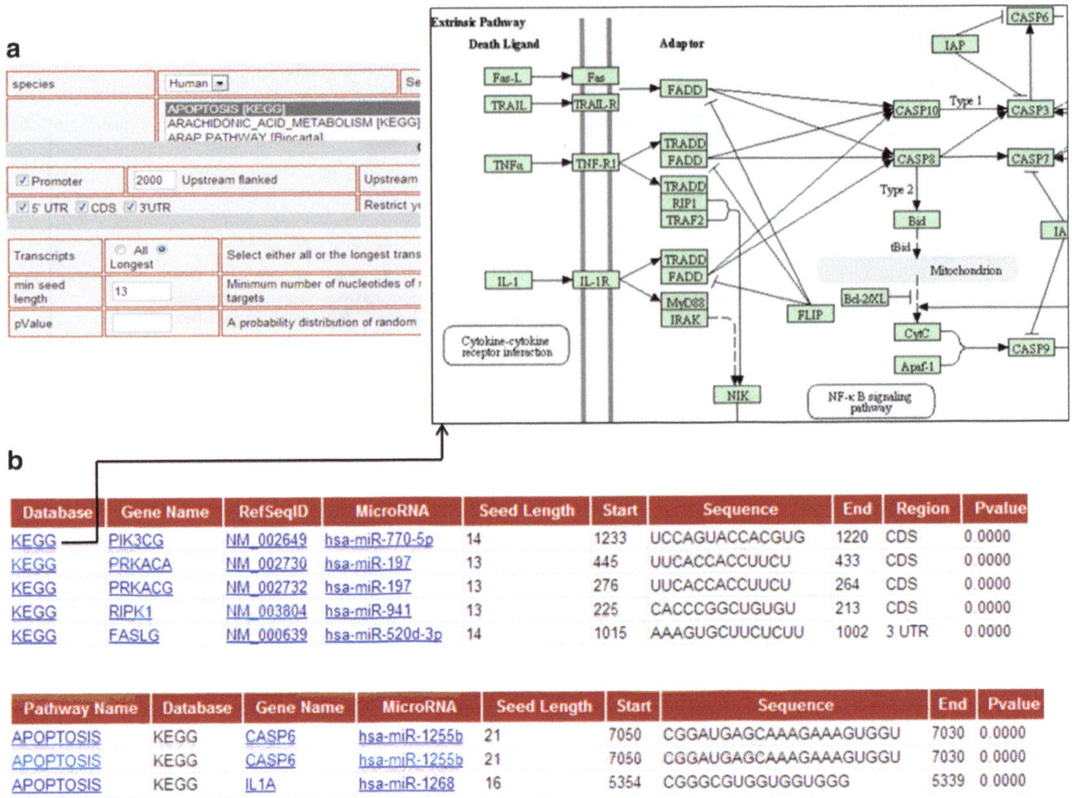

Fig. 5 Overview of query and result interfaces of Pathway Target. (**a**) Predicted Target Search page and (**b**) displays the result found on apoptosis pathway. Moreover, the pathway, gene, miRNA, and Refseqld are hyperlinked to their respective databases for further information

3.2.5 Search Interface: OMIM Target

OMIM Target search and result pages are organized similar to "Pathway Target." A user can fetch putative miRNA-binding sites on genes linked to 2,356 OMIM disorders (a maximum of ten disorders are allowed in a single query; *see* **Note 2**) by choosing one or more disorders from the given drop-down menu. For example, the possible miRNA-binding site predictions on genes linked with polycystic kidney and hepatic disease is described in Fig. 7a.

3.2.6 Search Interface: Mitochondrial Target

By using Mitochondrial Target Search page, all the possible miRNA-binding sites within the complete mitochondrial genome of a desired species can be obtained (Fig. 7b). Moreover, all the result tables of the Predicted Target module are linked with external databases such as NCBI, Ensembl, UCSC, miRBase, KEGG, Biocarta, and OMIM for further information (*see* **Note 3**).

3.3 The Validated Target Module

The "Validated Target" module of miRWalk increases the miRNA validated information more than threefold as compared to the existing resources and offers the experimentally verified interactions

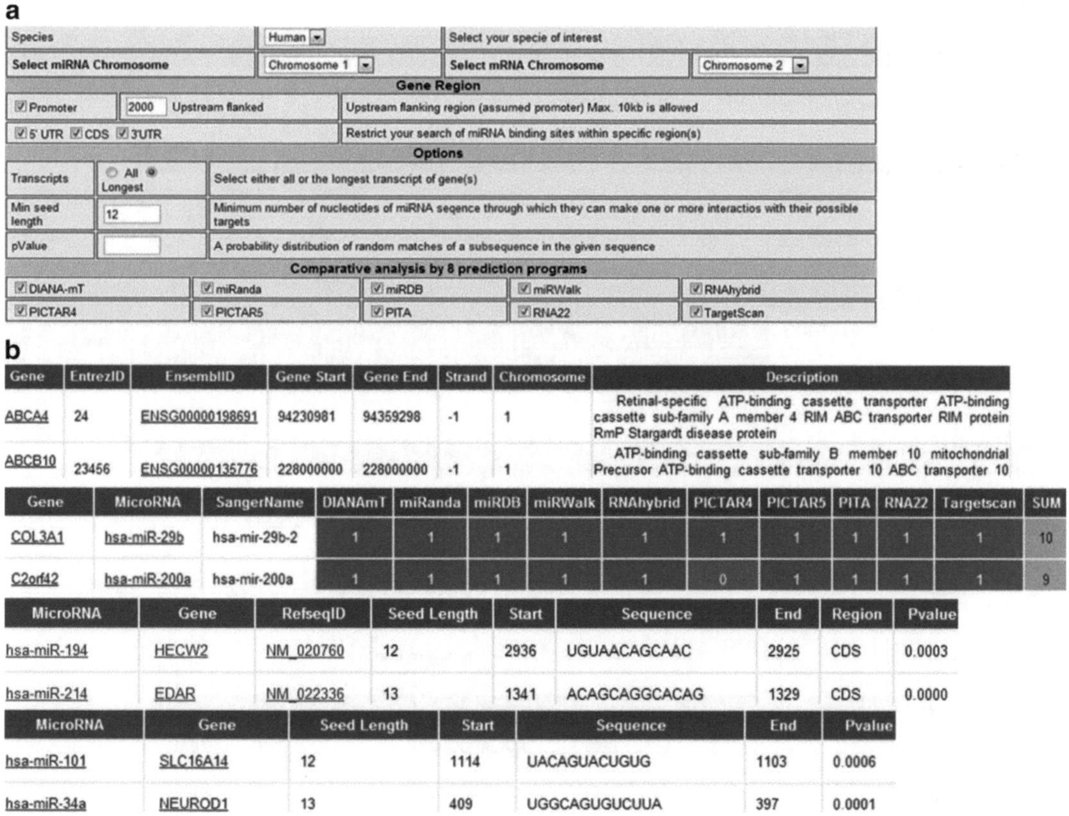

Fig. 6 Overview of Chromosome Target interfaces. (**a**) Search interface and (**b**) result page which shows the possible miRNA–mRNA interactions in four different tables along with external database links for further annotations

on 2,044 miRNAs linked to 6,436 genes, 850 diseases, 622 organs, 123 cell lines, 2,033 OMIM disorders, 375 pathways, and 70 miRNA-processing proteins. More than 9,560 scientific articles on mammalian miRNAs have been documented in the current release of miRWalk (*see* **Note 8**).

The Web interface of this module has different search pages. These are called Target Gene, miRNA, Pathway, Disease, Organ, Cell line, miRNA literature, OMIM disorder, and miRNA processing proteins (Fig. 8).

3.3.1 Search Interface: Target Gene, MicroRNA Target, and MicroRNA Literature

Using Target Gene, users can retrieve experimentally verified miRNAs on their genes of interest by supplying a list of gene symbols. The result page is presented in a tabulated form with all the available information (Fig. 9).

MicroRNA Target is organized similar to "Target Gene." All the validated miRNA–target information on human, mouse, and rat miRNAs can be easily obtained by providing miRNA names. The output of the query is displayed in a tabulated form together with links to get further information on the data (Fig. 9).

a

Disorder	MIM Number	Gene	EntrezID	EnsemblID	Gene Start	Gene End	Strand	Chr	Description
Polycystic kidney and hepatic disease	606702	PKHD1	5314	ENSG00000170927	51588104	52060382	-1	6	Fibrocystin Precursor Polycystic kidney and hepatic disease 1 protein Polyductin Tigmin

Gene	RefseqID	MicroRNA	Seed Length	Start	Sequence	End	Region	Pvalue
PKHD1	NM_138694	hsa-miR-103	8	210	AGCAGCAU	203	5 UTR	0.0042
PKHD1	NM_138694	hsa-miR-107	8	210	AGCAGCAU	203	5 UTR	0.0042

Gene	MicroRNA	SangerName	DIANAmT	miRanda	miRDB	miRWalk	RNAhybrid	PICTAR4	PICTAR5	PITA	RNA22	Targetscan	SUM
PKHD1	hsa-miR-520g	hsa-mir-520g	1	1	1	1	0	0	1	0	0	1	6
PKHD1	hsa-miR-17	hsa-mir-17	1	1	1	1	0	0	1	0	0	1	6

b

Gene	Gene Start	Gene End	Orientation	Descritpion
ATP6	8528	9208	-	ATP synthase 6 ATPase subunit 6
ATP8	8367	8573	-	ATP synthase 8 ATPase subunit 8
COX1	5905	7446	-	cytochrome c oxidase I
COX2	7587	8270	-	cytochrome c oxidase II
COX3	9208	9988	-	cytochrome c oxidase III
CYTB	14748	15882	-	protein coding

Gene	MicroRNA	Seed Length	Start	Sequence	End	Pvalue
COX1	hsa-miR-1279	12	6901	UCAUAUUGCUUC	6890	0.0001
COX3	hsa-miR-1825	13	9555	CCAGUGCCCUCCU	9543	0.0000

Fig. 7 Overview of results obtained by submitting queries to OMIM disorder and Mitochondrial Target Search interfaces. (**a**) Displays the result tables on polycystic kidney disease and (**b**) describes the putative miRNA-binding site predictions on human mitochondrial genome

Similarly, the literature information on one or more miRNAs can be enquired by entering miRNA names in the given textbox on the query page. The result of the query is depicted in Fig. 9.

3.3.2 Search Interface: Pathway, Disease, Organ, Cell Line, and OMIM Disorder Target

The validated miRNAs on the genes linked to one or more pathways can be interrogated by choosing a pathway (Biocarta or KEGG) from the drop-down menu (Fig. 8). Users can obtain a holistic view of gene–miRNA pathway network (Fig. 9) by using the Pathway Target query page. Currently, it hosts validated gene–miRNA interactions on 375 human biological pathways.

Disease Target Search encompasses several experimentally verified miRNAs on more than 150 human diseases. The interaction networks on different disease-miRNAs are available for interrogating on one or more diseases of interest (Fig. 9).

It has been shown that miRNAs play an important role during the development and functional maintenance of many organs [26, 27]. Thus, the information on miRNAs (Fig. 9) which have previously been reported in the scientific literature on various

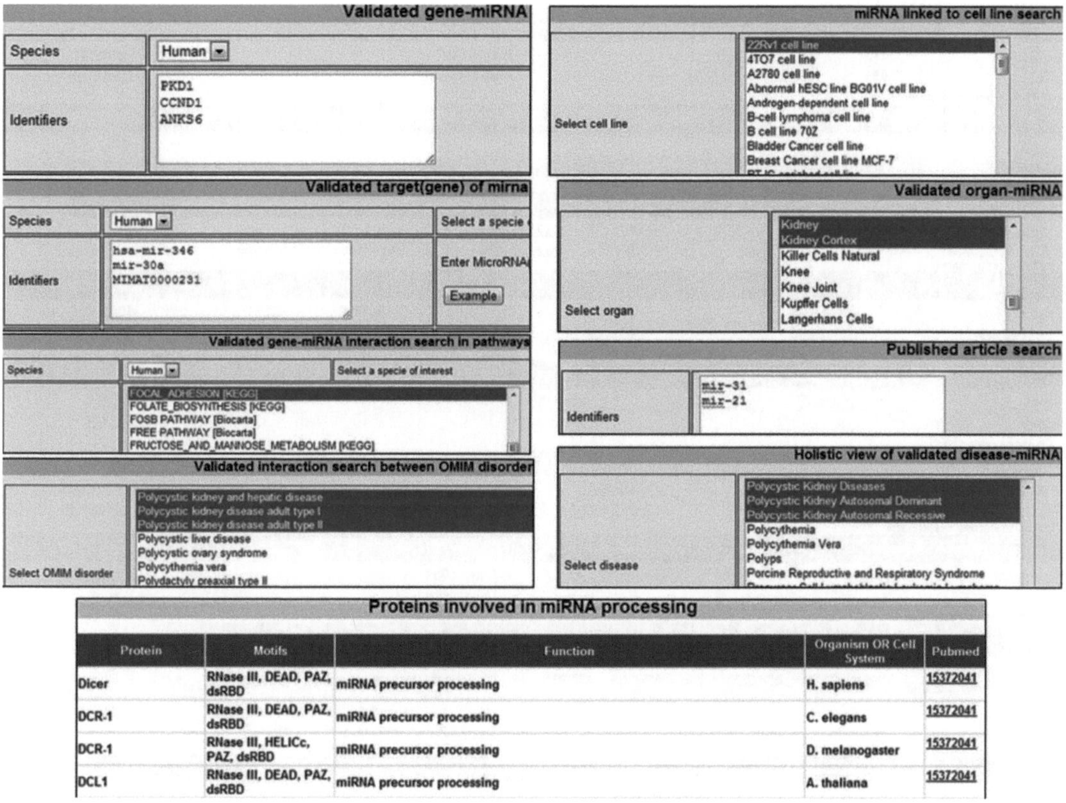

Fig. 8 Overview of the different search interfaces of the Validated Target module

organs under different conditions can be collected using "Organ Target" (Fig. 8).

Cell lines have been established in life sciences as easy-to-manipulate model systems for studying cellular processes. Cell lines and/or animals have been extensively used for the knock-down, knockout, or over-expression of several miRNAs to measure the expression level of predicted target genes under different pathological conditions [28]. Cell Line Target Search interface offers miRNA investigations on 139 cell lines. Such kinds of interactions (Fig. 9) can be accessed via Cell Line Target for one or more cell lines of interest (Fig. 8). In addition, the information on genes linked with 311 OMIM disorders is hosted under the "Validated Target" module. This information could possibly supply some helpful suggestions for treating OMIM disorders by miRNAs. The validated miRNA-binding sites on OMIM disorders (Fig. 9) can be queried by simply selecting the disorders of interest from the given drop-down menu (Fig. 8).

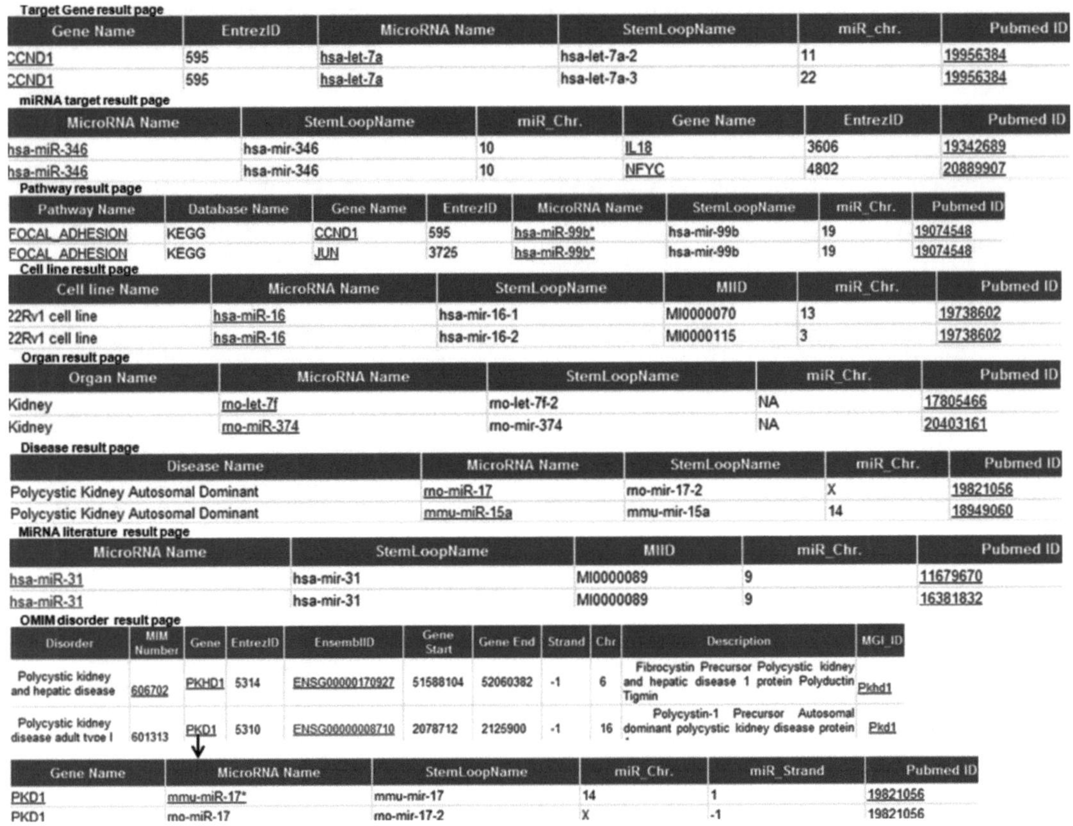

Target Gene result page

Gene Name	EntrezID	MicroRNA Name	StemLoopName	miR_chr.	Pubmed ID
CCND1	595	hsa-let-7a	hsa-let-7a-2	11	19956384
CCND1	595	hsa-let-7a	hsa-let-7a-3	22	19956384

miRNA target result page

MicroRNA Name	StemLoopName	miR_Chr.	Gene Name	EntrezID	Pubmed ID
hsa-miR-346	hsa-mir-346	10	IL18	3606	19342689
hsa-miR-346	hsa-mir-346	10	NFYC	4802	20889907

Pathway result page

Pathway Name	Database Name	Gene Name	EntrezID	MicroRNA Name	StemLoopName	miR_Chr.	Pubmed ID
FOCAL_ADHESION	KEGG	CCND1	595	hsa-miR-99b*	hsa-mir-99b	19	19074548
FOCAL_ADHESION	KEGG	JUN	3725	hsa-miR-99b*	hsa-mir-99b	19	19074548

Cell line result page

Cell line Name	MicroRNA Name	StemLoopName	MIID	miR_Chr.	Pubmed ID
22Rv1 cell line	hsa-miR-16	hsa-mir-16-1	MI0000070	13	19738602
22Rv1 cell line	hsa-miR-16	hsa-mir-16-2	MI0000115	3	19738602

Organ result page

Organ Name	MicroRNA Name	StemLoopName	miR_Chr.	Pubmed ID
Kidney	rno-let-7f	rno-let-7f-2	NA	17805466
Kidney	rno-miR-374	rno-mir-374	NA	20403161

Disease result page

Disease Name	MicroRNA Name	StemLoopName	miR_Chr.	Pubmed ID
Polycystic Kidney Autosomal Dominant	rno-miR-17	rno-mir-17-2	X	19821056
Polycystic Kidney Autosomal Dominant	mmu-miR-15a	mmu-mir-15a	14	18949060

MiRNA literature result page

MicroRNA Name	StemLoopName	MIID	miR_Chr.	Pubmed ID
hsa-miR-31	hsa-mir-31	MI0000089	9	11679670
hsa-miR-31	hsa-mir-31	MI0000089	9	16381832

OMIM disorder result page

Disorder	MIM Number	Gene	EntrezID	EnsemblID	Gene Start	Gene End	Strand	Chr	Description	MGI_ID
Polycystic kidney and hepatic disease	606702	PKHD1	5314	ENSG00000170927	51588104	52060382	-1	6	Fibrocystin Precursor Polycystic kidney and hepatic disease 1 protein Polyductin Tigmin	Pkhd1
Polycystic kidney disease adult type I	601313	PKD1	5310	ENSG00000008710	2078712	2125900	-1	16	Polycystin-1 Precursor Autosomal dominant polycystic kidney disease protein	Pkd1

Gene Name	MicroRNA Name	StemLoopName	miR_Chr.	miR_Strand	Pubmed ID
PKD1	mmu-miR-17*	mmu-mir-17	14	1	19821056
PKD1	rno-miR-17	rno-mir-17-2	X	-1	19821056

Fig. 9 Overview of the result pages of the Validated Target module

3.3.3 MicroRNA-Processing Proteins

Several studies have demonstrated a key role of Dicer [29, 30] and other proteins (e.g., p53) in generating mature miRNAs. The information on proteins known to be involved in miRNA processing is therefore hosted for users with the help of an HTML page under the "Validated Target" module of miRWalk (Fig. 8).

3.4 Applications of miRWalk

The information of miRWalk has been extensively employed in the field of miRNA research. A few applications are described below with the help of some recent studies.

- In the past few years, many researchers have been adapting a combinatorial approach, in which both mRNA and miRNA microarray profiling between healthy and diseased organ is performed for identifying significantly deregulated genes and miRNAs [31]. These candidates are then scanned for gathering the experimentally verified and the putative miRNA–target interactions [31]. Further, some potential miRNA–target interactions are selected and verified using cell lines and/or animals. Such information, i.e., both the putative and the experimentally validated miRNA–target interactions, can be fetched from miR-

Walk either by supplying identifiers or via uploading a file under the "Predicted Target" module (Subheading 3.2) and the "Validated Target" module (Subheading 3.3) [32]. In addition, the information on cell lines is hosted under the same module.

- A recent growing body of miRNA research suggests a novel mechanism by which miRNAs may direct the expression of the target genes by base-pairing not only within the mRNA 3′-UTR but also in other regions: promoter, 5′-UTR, and CDS. Such kinds of possible miRNA-binding sites within the complete sequence of known genes can be retrieved from miR-Walk (Fig. 2), and thereafter, these novel interactions can be confirmed on the other regions of known genes [33, 34].
- The comparative platform of miRWalk database has been widely utilized and referenced in many articles [31, 35], since it is very useful for carrying out a comprehensive analysis of miRNA-binding sites resulting from ten different prediction datasets (Fig. 3c) and also helpful in reducing the false-positive predictions.
- Recently, many studies have applied miRWalk in the diverse fields of miRNA research such as copy number variants [36], different splice variants of a gene [37], single-nucleotide polymorphisms (SNP) [35], combinatorial approach [31], and a cross talk between nucleus and mitochondria through miRNAs [38].
- Moreover, miRWalk encompasses verified miRNAs [39] and their interactions with genes [40], diseases, organs, and OMIM disorders (Subheading 3.3). This information may provide possible suggestions for treating several diseases by targeting miRNAs.

4 Notes

1. In order to avoid parameter processing errors, users need to provide their identifiers of interest (miRNA names, MIIDs, MIMATIDs, Genes, Entrez IDs, etc.) in a new line-separated format by either entering or uploading a text file. The other file formats, MS Excel (.xls or .xlxs), portable document format (.pdf), rich text format (.rtf), and MS Word document (.doc or .docx), are strictly prohibited. Therefore, these files should not be uploaded at any of the query pages.
2. Target Gene, MicroRNA Target, and OMIM Target Search interfaces under the Predicted Target module of miRWalk are restricted to 200 genes or Entrez IDs; 20 miRNAs, MIMATIDs, or MIIDs; and 10 OMIM disorder identifiers in a single query, respectively. Therefore, it is of utmost importance to supply the identifiers in accordance with their maximum limits.

However, users are allowed to upload any number of identifiers under the search interfaces of the Validated Target module, as these search pages are not restricted to any limit.

3. The result tables of the output pages under the Predicted Target module display only first 20 records; however, the complete interactions (records) can be viewed and/or downloaded by clicking on the given links, i.e., "View Complete Table," "Paging View," "Download Table," or "Click Here."

4. By default values are provided to minimum seed length and *p*-value textboxes to lower the burden of the miRWalk server. Due to these default values, the result pages display only those interactions which satisfy both conditions. In order to collect the complete information on the possible miRNA-binding site predictions, it is necessary to either remove the number from the *p*-value textbox or lower it, e.g., to 0.1.

5. At least two different miRNA–target prediction algorithms must be selected for generating a comprehensive (comparative) atlas of the possible miRNA-binding sites resulting from different programs. The comparative platform of miRWalk documents all the predictions of the third-party programs [DIANA-mT [11], miRanda [7], miRDB [12], RNAhybrid [13], PICTAR4 [8], PICTAR5 [8], PITA [9], RNA22 [10], and TargetScan [6]]; these binding sites are either matched or unmatched with miRWalk findings. The stored information in the current release of miRWalk is based on miRBase (version 14), DIANA-microT (version 3.0), miRanda (August 2010), miRDB (April 2009), PicTar (March 2007), PITA (August 2008), RNA22 (May 2008), TargetScan/TargetScanS (version 5.1), and RNAhybrid (version 2.1).

6. Of interest, just by clicking on the miRNA name, users can obtain basic information on miRNAs [i.e., mature and stem–loop sequence, identifiers, chromosome, strand and band (the q and p arms of each chromosome can be classified using special stains which result in different bands, also known as a banding pattern)] as well as other necessary data required for conducting a miRNA research, such as regulatory binding sites upstream and/or downstream of pre-miRNA, information on the host gene of miRNA, and a list of miRNAs that share a similar seed to those miRNAs in the list that the user provided (Fig. 10). Moreover, different links redirect users to NCBI, UCSC (for more annotations and information on miRNAs), and PubMed databases for a quick access to available publications on the user's miRNAs of interest such as their expression in different organs and cells lines and their association with diseases.

7. Due to security issues, MicroRNA Target Search interface does not allow direct searching for miRNAs with an asterisk (*) in

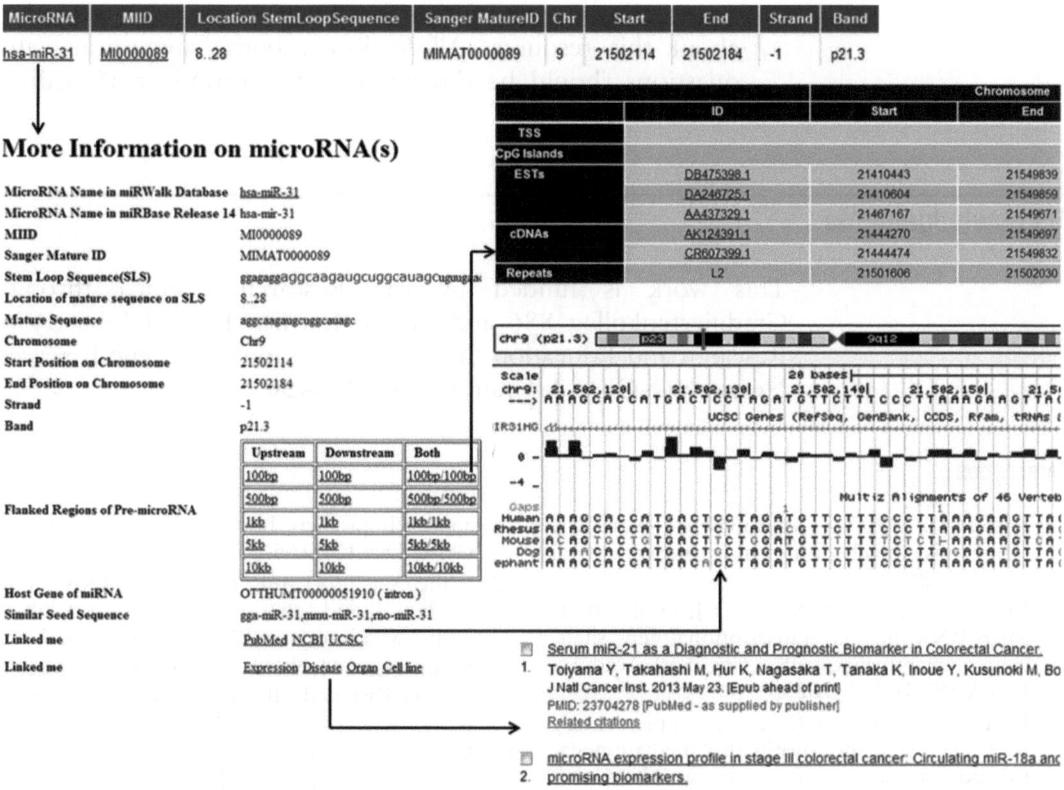

Fig. 10 Overview of the additional information on the user input miRNAs via MicroRNA Target Search interface of the Predicted Target module

its name. However, the putative target information on such type of miRNA names can be queried by their accession numbers (e.g., MIMAT0004494 is the accession number for hsa-miR-21*).

8. Information hosted under the Validated Target module is gathered via (1) accomplishing an automated text-mining survey in titles and abstracts of the PubMed articles against curated dictionaries using name entity recognition (NER) and (2) retrieving the experimentally verified miRNA information from the existing resources (TarBase, miRTarBase, MiRecords, miR2Disease, and PhenomiR). Each of these interactions of this module is hyperlinked to PubMed by using PubMed identifiers (PMIDs) and allowing the retrieval of abstracts. More information on miRWalk can be easily obtained by contacting the miRWalk team (*see* **Note 9**).

9. Users are welcome to report problems/bugs to the miRWalk's authors. Feedback and suggestions for improving the content as well as the Web interface of miRWalk are very welcome. Moreover, users are encouraged to contact the miRWalk team

for collaborations as well as for obtaining further information about this resource. All feedback, bugs, suggestions, and questions should be directly sent to mirwalkteam@medma.uni-heidelberg.de.

Acknowledgements

This work is funded by the Research Council through Graduiertenkolleg 886 and by the German Federal Ministry of Research and Education through the National Genome Research Network (NGFN-2, Grant no. 01GR 0450).

References

1. Bartel DP (2004) MicroRNAs: genomics, biogenesis, mechanism, and function. Cell 116:281–297
2. Kim DH, Saetrom P, Snove O Jr et al (2008) MicroRNA-directed transcriptional gene silencing in mammalian cells. Proc Natl Acad Sci U S A 105:16230–16235
3. Lai EC (2002) Micro RNAs are complementary to 3′ UTR sequence motifs that mediate negative post-transcriptional regulation. Nat Genet 30:363–364
4. Kozomara A, Griffiths-Jones S (2011) miRBase: integrating microRNA annotation and deep-sequencing data. Nucleic Acids Res 39: D152–D157
5. Lee RC, Feinbaum RL, Ambros V (1993) The C. elegans heterochronic gene lin-4 encodes small RNAs with antisense complementarity to lin-14. Cell 75:843–854
6. Lewis BP, Burge CB, Bartel DP (2005) Conserved seed pairing, often flanked by adenosines, indicates that thousands of human genes are microRNA targets. Cell 120:15–20
7. Enright AJ, John B, Gaul U et al (2003) MicroRNA targets in Drosophila. Genome Biol 5:R1
8. Krek A, Grun D, Poy MN et al (2005) Combinatorial microRNA target predictions. Nat Genet 37:495–500
9. Kertesz M, Iovino N, Unnerstall U et al (2007) The role of site accessibility in microRNA target recognition. Nat Genet 39:1278–1284
10. Miranda KC, Huynh T, Tay Y et al (2006) A pattern-based method for the identification of MicroRNA binding sites and their corresponding heteroduplexes. Cell 126:1203–1217
11. Kiriakidou M, Nelson PT, Kouranov A et al (2004) A combined computational-experimental approach predicts human microRNA targets. Genes Dev 18:1165–1178
12. Wang X (2008) miRDB: a microRNA target prediction and functional annotation database with a wiki interface. RNA 14:1012–1017
13. Rehmsmeier M, Steffen P, Hochsmann M et al (2004) Fast and effective prediction of micro RNA/target duplexes. RNA 10:1507–1517
14. Vergoulis T, Vlachos IS, Alexiou P et al (2012) TarBase 6.0: capturing the exponential growth of miRNA targets with experimental support. Nucleic Acids Res 40:D222–D229
15. Hsu SD, Lin FM, Wu WY et al (2011) miRTarBase: a database curates experimentally validated microRNA-target interactions. Nucleic Acids Res 39:D163–D169
16. Ruepp A, Kowarsch A, Schmidl D et al (2010) PhenomiR: a knowledgebase for microRNA expression in diseases and biological processes. Genome Biol 11: R6.
17. Jiang Q, Wang Y, Hao Y et al (2009) miR2Disease: a manually curated database for microRNA deregulation in human disease. Nucleic Acids Res 37:D98–D104
18. Dweep H, Sticht C, Gretz N (2013) In-silico algorithms for the screening of possible microRNA binding sites and their interactions. Curr Genomics 14:127–136
19. Guang S, Bochner AF, Pavelec DM et al (2008) An Argonaute transports siRNAs from the cytoplasm to the nucleus. Science 321:537–541
20. Lytle JR, Yario TA, Steitz JA (2007) Target mRNAs are repressed as efficiently by microRNA-binding sites in the 5′ UTR as in the 3′ UTR. Proc Natl Acad Sci U S A 104: 9667–9672
21. Place RF, Li LC, Pookot D et al (2008) MicroRNA-373 induces expression of genes

with complementary promoter sequences. Proc Natl Acad Sci U S A 105:1608–1613

22. Tay Y, Zhang J, Thomson AM et al (2008) MicroRNAs to Nanog, Oct4 and Sox2 coding regions modulate embryonic stem cell differentiation. Nature 455:1124–1128
23. Rajewsky N (2006) microRNA target predictions in animals. Nat Genet 38:S8–S13
24. Dweep H, Sticht C, Pandey P et al (2011) miRWalk – database: prediction of possible miRNA binding sites by "walking" the genes of three genomes. J Biomed Inform 44: 839–847
25. Kanehisa M, Goto S, Kawashima S et al (2002) The KEGG databases at GenomeNet. Nucleic Acids Res 30:42–46
26. Dong J, Jiang G, Asmann YW et al (2010) MicroRNA networks in mouse lung organogenesis. PLoS One 5:e10854
27. Ucar A, Vafaizadeh V, Jarry H et al (2010) miR-212 and miR-132 are required for epithelial stromal interactions necessary for mouse mammary gland development. Nat Genet 42: 1101–1108
28. Ikemura K, Yamamoto M, Miyazaki S et al (2013) MicroRNA-145 post-transcriptionally regulates the expression and function of P-glycoprotein in intestinal epithelial cells. Mol Pharmacol 83:399–405
29. Ho J, Ng KH, Rosen S et al (2008) Podocyte-specific loss of functional microRNAs leads to rapid glomerular and tubular injury. J Am Soc Nephrol 19:2069–2075
30. Shi S, Yu L, Chiu C et al (2008) Podocyte-selective deletion of dicer induces proteinuria and glomerulosclerosis. J Am Soc Nephrol 19: 2159–2169
31. Dweep H, Sticht C, Kharkar A et al (2013) Parallel analysis of mRNA and microRNA microarray profiles to explore functional regulatory patterns in polycystic kidney disease: using PKD/Mhm rat model. PLoS One 8: e53780
32. Cirera-Salinas D, Pauta M, Allen RM et al (2012) Mir-33 regulates cell proliferation and cell cycle progression. Cell Cycle 11:922–933
33. Wulfken LM, Moritz R, Ohlmann C et al (2011) MicroRNAs in renal cell carcinoma: diagnostic implications of serum miR-1233 levels. PLoS One 6:e25787
34. Zhao C, Huang C, Weng T et al (2012) Computational prediction of MicroRNAs targeting GABA receptors and experimental verification of miR-181, miR-216 and miR-203 targets in GABA-A receptor. BMC Res Notes 5:91
35. Papagregoriou G, Erguler K, Dweep H et al (2012) A miR-1207-5p binding site polymorphism abolishes regulation of HBEGF and is associated with disease severity in CFHR5 nephropathy. PLoS ONE 7:e31021
36. Felekkis K, Voskarides K, Dweep H et al (2011) Increased number of microRNA target sites in genes encoded in CNV regions. Evidence for an evolutionary genomic interaction. Mol Biol Evol 28:2421–2424
37. Durand C, Roeth R, Dweep H et al (2011) Alternative splicing and nonsense-mediated RNA decay contribute to the regulation of SHOX expression. PLoS One 6:e18115
38. Bandiera S, Ruberg S, Girard M et al (2011) Nuclear outsourcing of RNA interference components to human mitochondria. PLoS One 6:e20746
39. Xu LM, Li JR, Huang Y et al (2012) AutismKB: an evidence-based knowledgebase of autism genetics. Nucleic Acids Res 40:D1016–D1022
40. Santamaria C, Muntion S, Roson B et al (2012) Impaired expression of DICER, DROSHA, SBDS and some microRNAs in mesenchymal stromal cells from myelodysplastic syndrome patients. Haematologica 97:1218–1224

Chapter 26

A Schematic Workflow for Collecting Information About the Interaction Between Copy Number Variants and MicroRNAs Using Existing Resources

Harsh Dweep, Norbert Gretz, and Kyriakos Felekkis

Abstract

MicroRNAs (miRNAs) and copy number variations (CNVs) are two extensively studied genomic components in the field of modern biology—as they have been found to be associated with many disorders such as cancer, Alzheimer, pancreatitis, HIV susceptibility, beta-thalassemia, and glomerulonephritis. Several studies suggested that an alteration in CNV–miRNA interaction could result in some human diseases such as cancer. Therefore, the possible miRNA-binding site information within the CNV genes opens new avenues in understanding such disorders. In this chapter, we present a schematic approach for collecting the information on CNV–miRNA interactions using miRWalk and TargetScan databases.

Key words Copy number variant, MicroRNA, miRWalk, CNV–miRNA interaction predictions, 3′-UTR

1 Introduction

MicroRNAs (miRNAs) and copy number variations (CNVs) are two extensively studied genomic components in the field of modern biology—as they have been found to be associated with many disorders such as cancer [1, 2], Alzheimer [3, 4], pancreatitis [5, 6], HIV susceptibility [7, 8], and glomerulonephritis [9, 10].

miRNAs comprise a class of small, noncoding RNA molecules of 21–25 nucleotides (nt) in length that regulate the expression of their target genes at posttranscriptional level [11]. In addition to their role in such posttranscriptional repression, miRNAs are now implicated in transcriptional gene silencing by targeting the promoter region [12]. Thus, miRNAs can inhibit gene expression via translational repression, mRNA degradation, or transcriptional inhibition. Recently, miRNAs have also been found in activating

Electronic supplementary material–The online version of this chapter (doi:10.1007/978-1-4939-1062-5_26) contains supplementary material, which is available to authorized users.

M. Lucrecia Alvarez and Mahtab Nourbakhsh (eds.), *RNA Mapping: Methods and Protocols*, Methods in Molecular Biology, vol. 1182, DOI 10.1007/978-1-4939-1062-5_26, © Springer Science+Business Media New York 2014

the expression of their target genes [13]. The miRNA gene family is one of the largest in higher eukaryotes: more than 1,900 miRNAs have been identified in the human genome [14]. In silico and experimental methods indicate that each of these miRNAs has the capability to anneal more than a hundred mRNAs. Furthermore, over 60 % of protein-coding genes are predicted to harbor the possible miRNA-binding sites within their 3′-UTR regions.

CNVs are a newly discovered category of genomic elements which encompass segmental duplications of greater than 1 kb that are present at significant frequencies within a population and generally include protein-encoding genes [15, 16]. Many studies have linked CNVs with various biological processes and human diseases. In humans, more than 30,000 genomic regions with segmental duplications have been identified; however, exact boundaries and allele number of CNVs can be hardly estimated by current methods, adding to the complexity of their role in gene expression and disease [17, 18]. There is a significant evidence that a number of genes located within CNVs have increased expression level [18]. At the same time it should be noted that the different gene classes are not randomly distributed between CNV and non-CNV regions. CNV regions are enriched with dosage-insensitive genes [19] in contrast to the dosage-sensitive genes that are underrepresented in CNV regions [20].

The functional interaction between miRNAs and CNVs has been the focus of various recent studies. It is hypothesized that the coevolution of CNVs and miRNAs as well as their molecular interactions contributed to maintain a balance in gene expression during evolution. It was demonstrated that both the number of miRNAs and their binding sites are significantly higher in target genes found in CNV than in non-CNV regions [21, 22]. These recent data suggest a novel function of miRNAs contributing to the increasing regulatory complexity and the evolvability of genomes. As a result, deciphering the functional significance of miRNA–CNV interactions will help in the understanding of both the evolutionary impact of CNVs and the pathophysiologic mechanisms attributed to CNVs in various species. Therefore, the possible miRNA-binding site information within the CNV genes opens new avenues in understanding such mechanisms. In this chapter, we present a schematic approach for collecting the information on CNV–miRNA interactions using miRWalk [23] and TargetScan [24] databases.

2 Materials

1. A computer with access to the Internet.
2. Information on all known genes of human genome can be easily obtained from NCBI ftp site at ftp://ftp.ncbi.nlm.nih.

gov/gene/DATA/GENE_INFO/Mammalia/Homo_sapiens.gene_info.gz (*see* Subheading 3.1).

3. List of all known human genes that belong to a CNV region can be downloaded from the Database of Genomic Variants (DGV) [16] at http://dgv.tcag.ca/dgv/docs/GRCh37_hg19_variants_2013-05-31.txt (*see* Subheading 3.1).
4. miRNA–target prediction datasets resulting from miRWalk (http://www.umm.uni-heidelberg.de/apps/zmf/mirwalk/) [23] and TargetScan (http://www.targetscan.org/) [24] within the 3′-UTR of genes in the entire human genome.
5. The miRNA-binding sites can be located in CNV and non-CNV regions. Therefore, Perl (http://www.perl.org) or any other programming languages [such as Python (http://www.python.org/), Java (http://www.java.com/), Ruby (https://www.ruby-lang.org/), and R (http://www.r-project.org/)] are required for categorizing these possible miRNA-binding sites (obtained from miRWalk and TargetScan databases) into two classes: CNV and non-CNV regions.

3 Methods

In this section, we provide a schematic three-step approach for collecting the information on CNV–miRNA interactions using miRWalk and TargetScan (Fig. 1). First, we describe the procedure to download the information on all known genes and the ones that come under a CNV region of human genome using NCBI and DGV [16], respectively (Subheading 3.1; Fig. 1a). Next, we briefly outline the searching algorithm of miRWalk [23] and Targetscan [24], followed by the steps required for obtaining the possible miRNA-binding site prediction datasets (Subheading 3.2; Fig. 1b). Lastly, we supply information on how the miRNA–target interactions can be classified into two categories, CNVs and non-CNVs (Subheading 3.3; Fig. 1c) as well as a precise summary of a statistical method that has been applied in a recent study [21] for comparing the means of binding sites between these two classes.

3.1 Acquisition of Information on All Known Genes and CNV Regions of the Human Genome

Information on all known human genes can be downloaded via the ftp site of NCBI at ftp://ftp.ncbi.nlm.nih.gov/gene/DATA/GENE_INFO/Mammalia/Homo_sapiens.gene_info.gz. After downloading it, the archived file can be extracted using any file-archiving software such as Filzip (http://www.filzip.com/en/index.html), 7-ZIP (http://www.7-zip.org/), or WinRAR (http://www.win-rar.com/download.html). Then, the unnecessary data (*see* **Note 1**) can be removed from the text file for achieving the information of interest; that is, only one column ("*Symbol*") is required ("*humangene.txt*"; Fig. 1a). The "*humangene.txt*" is the final file containing the list of human genes.

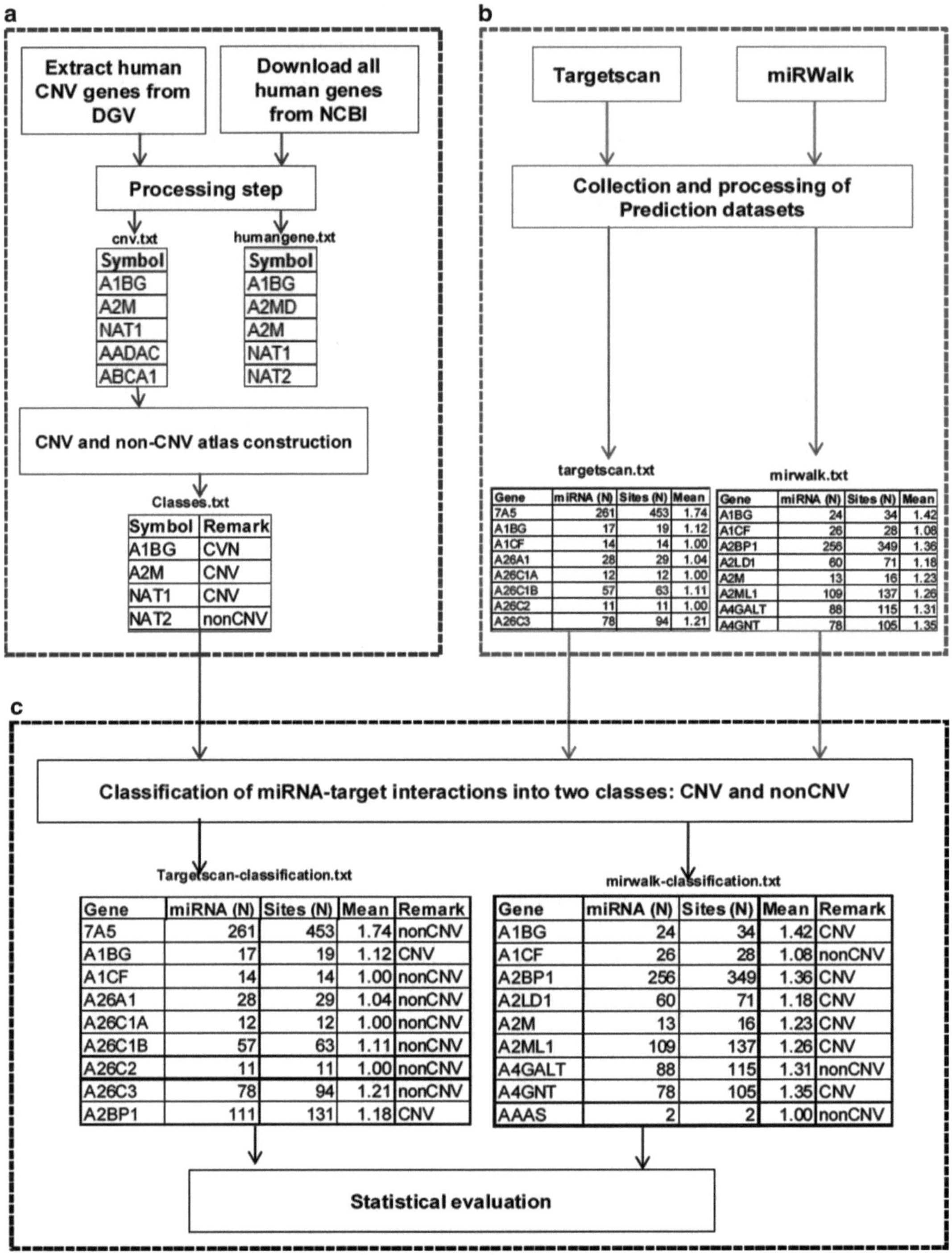

Fig. 1 A schematic workflow for collecting CNV–miRNA interactions. This workflow is broadly divided into three steps. (**a**) Illustrates the steps required to download and to process the data (resulting from NCBI and DGV) in a file; (**b**) depicts the procedures needed for extracting, processing, and compiling the prediction datasets obtained from miRWalk and TargetScan; and (**c**) supplies the steps used to classify the output results into two different classes and their statistical evaluation using Student's *t*-test

The human genes that belong to a CNV region can be gathered from the DGV [16] at http://dgv.tcag.ca/dgv/docs/GRCh37_hg19_variants_2013-05-31.txt. This file needs to be processed because it contains some additional information (rows and columns; *see* **Note 2**) which is not required for the current protocol. Therefore, such data should be removed before proceeding to the next step. Figure 1a depicts an example of the final CNV file ("*cnv.txt*").

A comprehensive atlas of CNV and non-CNV classes can be constructed by matching the content ("*Symbol*") of the above files ("*humangenes.txt*" and "*cnv.txt*"). The final file (e.g., "*classes.txt*"; Fig. 1a) should only contain two columns ("*Gene*" and "*Remark*").

3.2 Collecting miRNA–Target Interactions from miRWalk and TargetScan

miRWalk [23] and TargetScan [24] databases are chosen to retrieve the miRNA–target prediction datasets. These two databases are selected due to their popularity as well as for their user-friendly search interfaces. The incorporation of two independent and different algorithms has previously been shown to improve the reliability of the predictions identified [10, 21, 23, 25, 26]. The workflow described in this chapter is solely based on the use of miRWalk and TargetScan databases. However, other existing resources (reviewed in ref. 25) can also be considered to extract the information on miRNA–target interactions.

3.2.1 miRWalk

The miRWalk algorithm [23] is a newly constructed computational approach for the screening of perfect complementary binding site interactions between miRNA and gene sequences. Briefly, this algorithm scans for heptamers (7 nt from the 5′ end of miRNA sequences) or longer seeds based on Watson–Crick complementarity, walking on the complete sequence (promoter, CDS, 5′- and 3′-UTR) of a target gene. Next, all the identified miRNA-binding sites with 7 nt or longer matches are assigned into five parts, according to promoter region, 5′-UTR, CDS, and 3′-UTR and mitochondrial genes. These prediction results are then stored into a relational database named "miRWalk database" on a MySQL (Structure Query Language) database management system (DBMS). The data content of miRWalk database is broadly classified into two modules: *Predicted Target* and *Validated Target*. This chapter is focused on the predicted miRNA–target interactions within the mRNA 3′-UTR regions. Therefore, the possible miRNA-binding sites within the 3′-UTR of all known genes of humans can be easily collected via the Predicted Target module of miRWalk in the following manner:

1. Visit the user interface "*Target Gene*" search page of miRWalk [23] at http://mirwalk.uni-hd.de/predictedmirnagene.html (implemented under the Predicted Target module).
2. Select "human" from the drop-down menu.

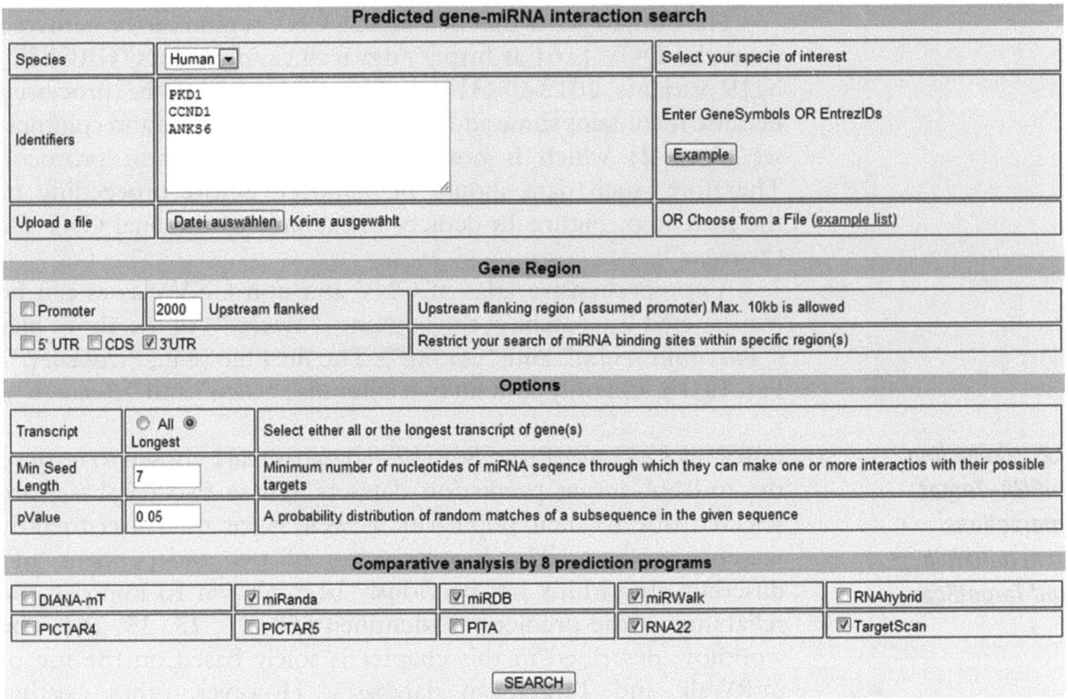

Fig. 2 Overview of "Target Gene" search interface under the Predicted Target module. A customized search for collecting the putative miRNA-binding sites within the known human genes can be submitted by (**a**) selecting human from the drop-down menu, (**b**) providing identifiers, (**c**) choosing 3′-UTR as the region of interest, and (**d**) keeping the remaining parameters default and submitting the query by pressing the "SEARCH" button

3. Provide identifiers or upload a file (*see* **Note 3**): Users can provide gene symbols or Entrez Gene ID either directly into the given textbox or by uploading a file (*see* **Note 4**).
4. Users can proceed with the default values for the remaining parameters such as "gene region," additional options ("minimum seed length" and/or "*p*-value"), and prediction programs (Fig. 2 schematically depicts the query procedure). Click on "Submit" button to initiate a search.
5. The result page displays the possible miRNA-binding site information in three tables. The first table provides gene information (Fig. 3a); the second table shows the putative miRNA-binding sites generated by miRWalk within the mRNA 3′-UTR region (Fig. 3b); and the third table depicts a comparative view of miRNA-binding sites predicted by different programs (Fig. 3c). The information documented under Fig. 3b is needed for this workflow. Therefore, download the prediction datasets on all known human genes by clicking on "Download Table" link (Fig. 3b).

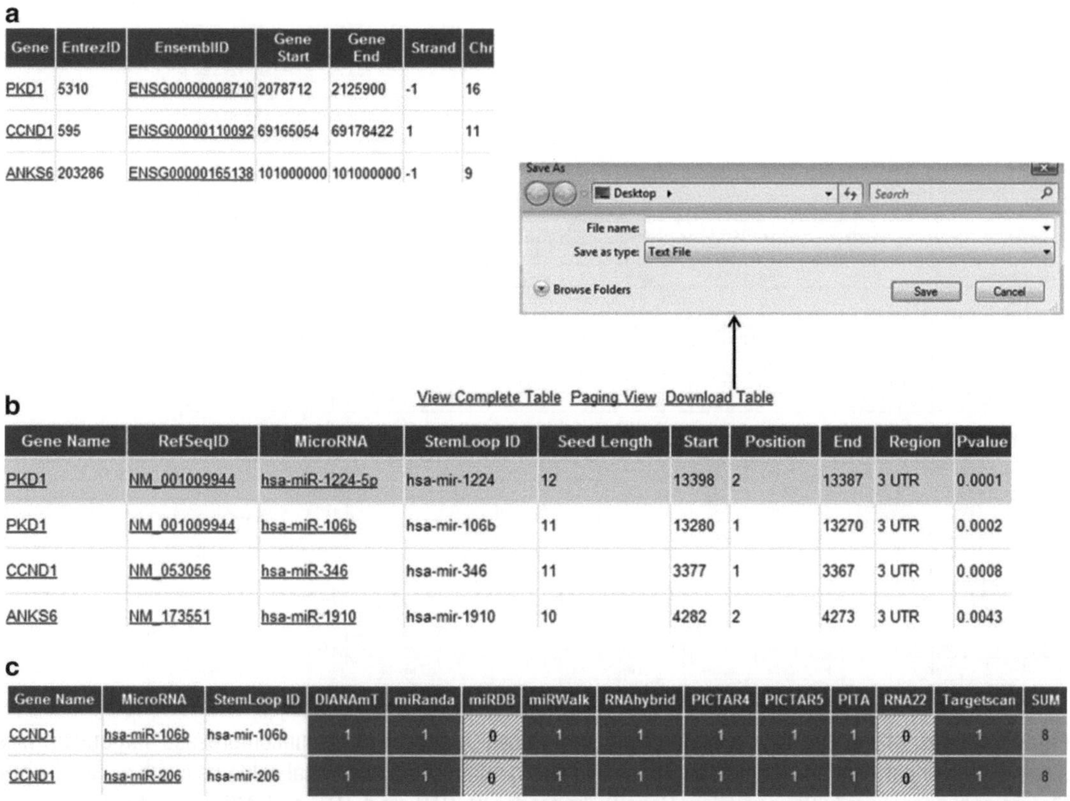

Gene	EntrezID	EnsemblID	Gene Start	Gene End	Strand	Chr
PKD1	5310	ENSG00000008710	2078712	2125900	-1	16
CCND1	595	ENSG00000110092	69165054	69178422	1	11
ANKS6	203286	ENSG00000165138	101000000	101000000	-1	9

Gene Name	RefSeqID	MicroRNA	StemLoop ID	Seed Length	Start	Position	End	Region	Pvalue
PKD1	NM_001009944	hsa-miR-1224-5p	hsa-mir-1224	12	13398	2	13387	3 UTR	0.0001
PKD1	NM_001009944	hsa-miR-106b	hsa-mir-106b	11	13280	1	13270	3 UTR	0.0002
CCND1	NM_053056	hsa-miR-346	hsa-mir-346	11	3377	1	3367	3 UTR	0.0008
ANKS6	NM_173551	hsa-miR-1910	hsa-mir-1910	10	4282	2	4273	3 UTR	0.0043

Gene Name	MicroRNA	StemLoop ID	DIANAmT	miRanda	miRDB	miRWalk	RNAhybrid	PICTAR4	PICTAR5	PITA	RNA22	Targetscan	SUM
CCND1	hsa-miR-106b	hsa-mir-106b	1	1	0	1	1	1	1	1	0	1	8
CCND1	hsa-miR-206	hsa-mir-206	1	1	0	1	1	1	1	1	0	1	8

Fig. 3 Overview for downloading the prediction data from miRWalk. The result panel shows three tables providing information about (**a**) genes, (**b**) predicted miRNA-binding sites according to 3′-UTR regions, and (**c**) a comparative analysis of miRNA-binding sites within the 3′-UTR region (*green* and *red* colors indicate miRNA-binding sites found and not found, respectively). The possible miRNA-binding site prediction files can be downloaded by clicking on "Download Table" link (Color figure online)

6. Only two columns ("Gene name" and "MicroRNA") should be compiled in a file, whereas the remaining data should be removed from all the downloaded files. Thereafter, the number of miRNAs as well as their target sites should be calculated for each human gene, and then, this information can be stored into a tab-delimited text file (for instance, "*mirwalk.txt*"; Fig. 1b).

3.2.2 TargetScan

TargetScan [24] predicts a perfect Watson–Crick complementary subsequence between 7 nt long seed of miRNA (position 2–8 nt) and the analyzed 3′-UTR sequence. It also extends each seed match with additional base pairings to the miRNA. Furthermore, it evaluates the thermodynamic free energy of the identified miRNA–mRNA interactions and conducts a comparative sequence analysis to find out the evolutionary conserved miRNA-binding

miR family.txt

miR family	Seed+m8	Species ID	MiRBase ID	Mature sequence	Family Conservation?	MiRBase Accession
let-7/98/4458/4500	GAGGUAG	9606	hsa-let-7a	UGAGGUAGUAGGUUGUAUAGUU	2	MIMAT0000062
let-7/98/4458/4500	GAGGUAG	9606	hsa-let-7b	UGAGGUAGUAGGUUGUGUGGUU	2	MIMAT0000063
let-7/98/4458/4500	GAGGUAG	10090	mmu-let-7g	UGAGGUAGUAGUUUGUACAGUU	2	MIMAT0000121
let-7/98/4458/4500	GAGGUAG	10116	rno-let-7d	AGAGGUAGUAGGUUGCAUAGUU	2	MIMAT0000562
let-7/98/4458/4500	GAGGUAG	9031	gga-let-7i	UGAGGUAGUAGUUUGUGCUGU	2	MIMAT0001098

Predicted conserved target info.txt

miR Family	Gene ID	Gene Symbol	Transcript ID	Species ID	UTR start	UTR end	MSA start	MSA end	Seed match	PCT
miR-22/22-3p	1	A1BG	NM_130786	9615	76	82	86	92	7mer-1a	0
miR-23abc/23b-3p	1	A1BG	NM_130786	9606	143	150	167	174	8mer	0
miR-23abc/23b-3p	1	A1BG	NM_130786	9615	144	151	167	174	8mer	0

Conserved family info.txt

miR Family	Gene ID	Gene Symbol	Transcript ID	Species ID	UTR start	UTR end	MSA start	MSA end	Seed match	PCT
let-7/98/4458/4500	29974	A1CF	NM_001198820	9598	3283	3290	4796	4809	8mer	0
let-7/98/4458/4500	1	A1BG	NM_130786	9606	3292	3299	4796	4809	8mer	0
let-7/98/4458/4500	29974	A1CF	NM_001198819	9606	3292	3299	4796	4809	8mer	0
let-7/98/4458/4500	29974	A1CF	NM_001198819	9785	3203	3210	4796	4809	8mer	0

Summary counts.txt

Transcript ID	Gene Symbol	miRNA family	Species ID	Total n	Numbe	Number of c	Number of	Total nu	Numbe	Number	Number of	Representative miRNA	Total context scor	Aggregate PCT
NM_000014	A2M	AAAAAAA	9913	0	0	0	0	2	0	2	0	bta-miR-2391	-0.04	NULL
NM_130786	A1BG	AAAAGUG	9606	0	0	0	0	1	0	1	0	hsa-miR-548t	-0.128	NULL
NM_000014	A2M	AAAGAAU	10090	0	0	0	0	1	0	1	0	mmu-miR-186	-0.078	NULL
NM_000014	A2M	AAGUCUU	9598	0	0	0	0	1	0	1	0	ptr-miR-1264	-0.198	NULL
NM_000014	A2M	AAGUCUU	9606	0	0	0	0	1	0	1	0	hsa-miR-1264	-0.198	NULL

Nonconserved family info.txt

miR Family	Gene ID	Gene Symbol	Transcript ID	Species ID	UTR start	UTR end	MSA start	MSA end	Seed match	PCT
miR-2391	178	AGL	NM_000028	9606	1071	1077	2029	2134	7mer-1a	NULL
miR-2391	178	AGL	NM_000028	9615	1048	1054	2026	2032	7mer-1a	NULL
miR-2391	178	AGL	NM_000028	42254	539	545	826	832	7mer-1a	NULL
miR-2391	178	AGL	NM_000028	9544	1094	1100	2147	2157	7mer-1a	NULL

Fig. 4 Overview of the five files obtained from TargetScan. A subset of the uncompressed TargetScan files along with their corresponding columns. These five files contain some additional columns (such as Seed + m8, Mature sequence, Family Conservation, GeneID, Transcript ID, UTR start, UTR end, MSA start, MSA end, Seed match, PCT) which are not required for this protocol. Moreover, these files also hold the putative miRNA-binding site information on ten different species. Therefore, the possible miRNA-binding site predictions on the human genome but only the necessary columns (*see* Fig. 5) should be fetched from these files

sites among various genomes (different species). The prediction datasets from TargetScan can be gathered in the following manner:

1. Go to http://www.targetscan.org, the home page of TargetScan.
2. Scroll down the page, and click on "Download all data or code" link. This clicking redirects the users to a new web page (http://www.targetscan.org/cgi-bin/targetscan/data_download.cgi?db=vert_61), which contains several archived files. Download only five files (Fig. 4): "miR family," "Predicted conserved targets info," "conserved family info," "Nonconserved family info," and "Summary counts" to a desired location (e.g., raw-data).
3. Unzip these five files under the same folder (or any other location) to extract the tab-delimited text files.
4. Process these five files (*see* **Note 5**) to acquire the predicted miRNA-binding site information (such as miR family, MiRBase ID, MiRBase Accession, and Gene Symbol) which is only identified within human genome (*see* **Note 6**), combine the extracted information in a file, and delete all the duplicate rows

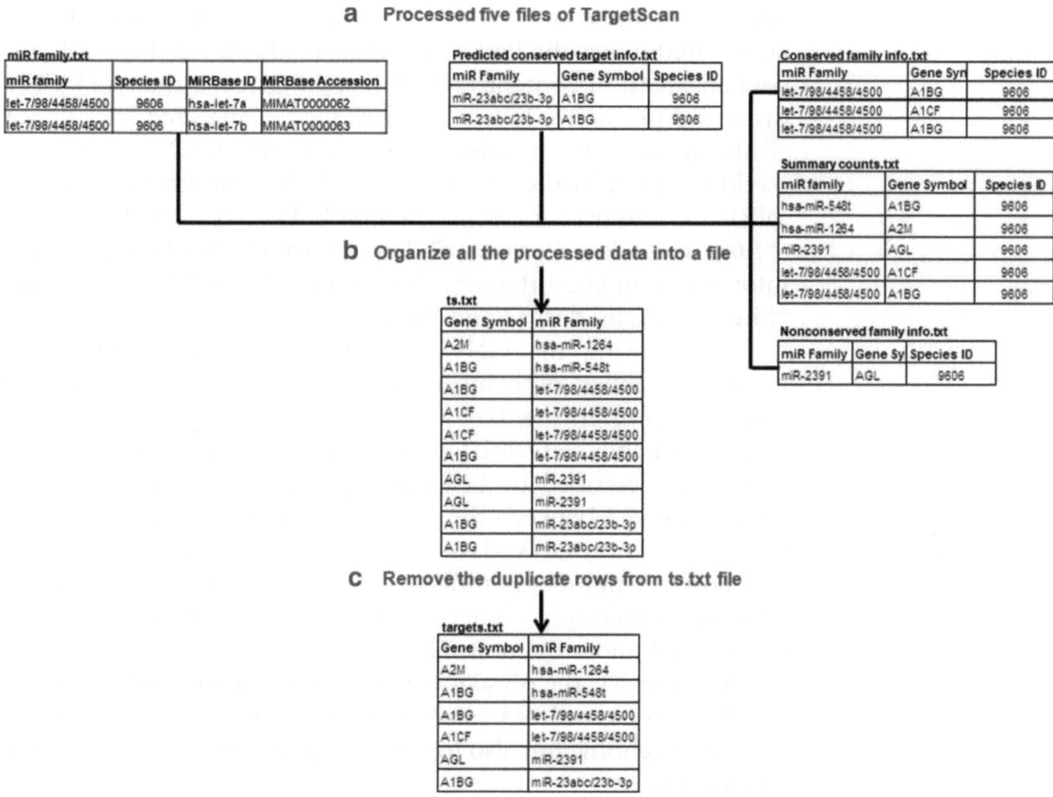

Fig. 5 Overview of the steps required to process and remove duplicate information. (**a**) All the five processed files only containing the relevant information (columns). (**b**) All the processed information is organized into a single file (ts.txt). (**c**) Final step: Remove all the duplicate rows from the "ts.txt" file

from it (Fig. 5). Organize collated information in a file having two columns, i.e., "Gene Symbol" and "miRNA name" (MiRBase ID).

5. In the final step, the number of miRNAs and their predicted binding sites should be determined for each gene and must be documented in such a manner that the final file should have four columns named, i.e., "Gene," "miRNA (N)," "Sites (N)," and "Mean" (*see* "*targetscan.txt*" in Fig. 1b).

3.3 Classification of miRNA–Target Interactions into CNV and Non-CNV

After attaining the processed information (prediction datasets as described in Subheading 3.2), the miRNA–target interactions ("*mirwalk.txt*" and "*targetscan.txt*") can then be categorized into two different classes (CNV and non-CNV) by comparing against the content of "*classes.txt*" file using a customized "*Perl script*" (Supplementary Material). Concisely (*see* **Note 7** for more details), the first step is to load "*mirwalk.txt*" file (all lines) into a Perl hash (the first column, i.e., "*Gene*" should be used as the key, whereas the entire line, i.e., all columns, should be stored as the value for a given key). Afterwards, the "*classes.txt*" file should be read line by line and its first column ("*Gene*") should be compared against the keys stored under the Perl hash—if both (the key and the column) contain identical content (100 % matching text), then add CNV; otherwise put non-CNV as a remark for the analyzed line. Repeat all the above steps in order to categorize the miRNA–target predictions stored in the "*targetscan.txt*" file. The final output files miRWalk ("*mirwalk-classification.txt*"; Fig. 1c) and Targetscan ("*targetscan-classification.txt*"; Fig. 1c) should hold the processed information in five different columns, i.e., "Gene," "miRNA (N)," "Sites (N)," and "Mean and Remark."

After having the CNV–miRNA interaction information in hand, a number of comparisons can be carried out between "*mirwalk-classification.txt*" and "*targetscan-classification.txt*" files using a suitable statistical method (such as Student's *t*-test)—as previously described by Felekkis et al. [21] in which the means were compared between the genes of human genome that belong to CNV and non-CNV classes. Moreover, the authors have shown that both the number of miRNAs and the number of their binding sites are significantly higher in the CNV genes than in the ones categorized under the non-CNV genes.

In summary, the present workflow is useful not only in obtaining the possible miRNA-binding site predictions within the genes of human genome but also for other genomes such as chimpanzee, mouse, and rat.

4 Notes

1. The "*Homo_sapiens.gene_info.txt*" tab-separated text file contains several additional rows (e.g., MIR4464) and columns (such as "*tax_id*," "*GeneID*," "*Locus Tag*," "*Synonyms*," and "*Chromosome*") which are not required, and, therefore, they should be deleted from the aforementioned file. Only the column entitled "*Symbol*" is necessary for the current protocol.
2. Similarly, the text file downloaded from DGV website also contains various additional columns (e.g., "*variant subtype*," "*reference*," "*pubmed id*," "*method*," "*platform*," "*sample size*," and "*samples*") which should be removed before proceeding further. Moreover, all the rows that hold information on the miRNA genes that belong to a CNV region should also be deleted because this chapter presents the workflow for extracting the possible miRNA-binding site predictions within the genes that do not encode for miRNAs.
3. In order to avoid parameter processing errors at the search interface of "*Target Gene*" at the *Predicted Target module* of miRWalk, users need to provide their identifiers (gene symbols or Entrez gene IDs) of interest in a new line-separated format

by either entering or uploading a text file. Other file formats such as MS Excel (.xls or .xlxs), portable document format (.pdf), rich text format (.rtf), and MS Word document (.doc or .docx) are strictly prohibited. Therefore, these files should not be uploaded.

4. The search interface of "*Target Gene*" is restricted to 200 genes in a single query. Therefore, it is of utmost importance to supply the genes in accordance with their maximum limits.
5. Process the five files downloaded from TargetScan (Fig. 4) as follows:
   (a) Open the five files into MS Excel and apply a sorting command (Sort A–Z) on the column, i.e., "*Species ID*."
   (b) Delete all the rows that do not contain "*9606*" under "*Species ID*" from all the five files.
   (c) Extract the following information from the five files (Fig. 5):
   "*miR family, miRBaseID, MiRBase Accession,*" "*miR family, Gene Symbol,*" and "*Gene Symbol, Representative miRNA*" from "*miR family.txt,*" "*Predicted conserved target info.txt,*" "*Conserved family info.txt,*" "*Nonconserved family info.txt,*" and "*Summary counts.txt.*"
   (d) The extracted columns need to be organized into a file.
   (e) In the final step, the duplicate information (rows) should be removed and the file saved (i.e., "*targets.txt*").
6. The information documented under the five files that were downloaded from TargetScan is not only confined to human genome but also supply the possible miRNA-binding site predictions within the 3′-UTR of other genomes such as mouse, rat, dog, chicken, chimpanzee, monkey, opossum, and frog. Therefore, users need to apply some filters (such as taxid, i.e., 9606) to extract the information that is only confined to human genome. This task can be easily accomplished by applying a sorting (Sort A–Z using MS Excel) on "Species ID" column (*see* **Note 5**).
7. The prediction datasets (mirwalk.txt and targetscan.txt) can be categorized into CNV and non-CNV regions in the following manner:
   (a) Install Perl (http://www.perl.org/get.html) programming language on the local computer (skip this step, if Perl is already installed).
   (b) Save the supplementary ("classification.txt") as well as "classes.txt" files (Subheading 3.1 and Fig. 1a) into a folder (e.g., "*D:/CNV*"), and rename the file "*classification.txt*" to "*classification.pl.*"

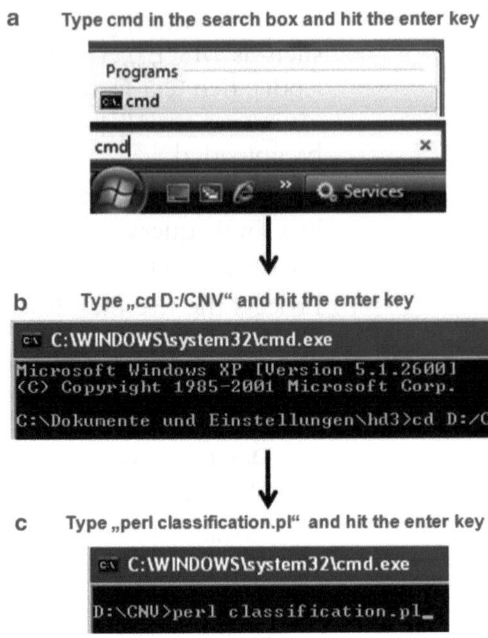

Fig. 6 Overview of the steps required to execute the Perl script from the DOS command prompt window. (**a**) Start Windows menu, (**b**) DOS command prompt, or cmd.exe window, (**c**) executing Perl script

(c) Create two new folders, i.e., "*input*" and "*output*," at the same location ("*D:/CNV*"), and move both files "*mirwalk.txt*" (Subheading 3.2.1 and Fig. 1b) and "*targetscan.txt*" (Subheading 3.2.2 and Fig. 1b) into the newly created "input" folder.

(d) Go to Start Windows menu, type "cmd" in the search box, and hit the Enter key. This opens a new window (DOS command prompt, i.e., "cmd.exe"); now type "cd D:/CNV"command in the "cmd.exe" window, and hit the "Enter" key (Fig. 6).

(e) The last step is to execute the Perl script by typing "perl classification.pl" in the "cmd.exe" window (Fig. 6) and hit the "Enter" key. This script, first, starts reading the text files (but one by one) stored in the "input" folder and loads all the lines into a Perl hash. This script then reads the "classes.txt" file and, at the same time, compares the first columns of both files, "classes.txt" and "mirwalk.txt or targetscan.txt," and stores the final files in the "output" folder.

Acknowledgements

This work was funded by the Research Council through Graduiertenkolleg 886 and by the German Federal Ministry of Research and Education through the National Genome Research Network (NGFN-2, Grant no. 01GR 0450).

References

1. Hastings PJ, Ira G, Lupski JR (2009) A microhomology-mediated break-induced replication model for the origin of human copy number variation. PLoS Genet 5:e1000327
2. Zaman MS, Shahryari V, Deng G et al (2012) Up-regulation of microRNA-21 correlates with lower kidney cancer survival. PLoS One 7:e31060
3. Rovelet-Lecrux A, Hannequin D, Raux G et al (2006) APP locus duplication causes autosomal dominant early-onset alzheimer disease with cerebral amyloid angiopathy. Nat Genet 38:24–26
4. Cheng C, Li W, Zhang Z et al (2013) MicroRNA-144 is regulated by activator protein-1 (AP-1) and decreases expression of Alzheimer disease-related a disintegrin and metalloprotease 10 (ADAM10). J Biol Chem 288:13748–13761
5. Le Marechal C, Masson E, Chen JM et al (2006) Hereditary pancreatitis caused by triplication of the trypsinogen locus. Nat Genet 38:1372–1374
6. Li A, Yu J, Kim H et al (2013) MicroRNA array analysis finds elevated serum miR-1290 accurately distinguishes patients with low-stage pancreatic cancer from healthy and disease controls. Clin Cancer Res 19:3600–3610
7. Townson JR, Barcellos LF, Nibbs RJ (2002) Gene copy number regulates the production of the human chemokine CCL3-L1. Eur J Immunol 32:3016–3026
8. Narayanan A, Iordanskiy S, Das R et al (2013) Exosomes derived from HIV-1-infected cells contain trans-activation response element RNA. J Biol Chem 288:20014–20033
9. Aitman TJ, Dong R, Vyse TJ et al (2006) Copy number polymorphism in Fcgr3 predisposes to glomerulonephritis in rats and humans. Nature 439:851–855
10. Papagregoriou G, Erguler K, Dweep H et al (2012) A miR-1207-5p binding site polymorphism abolishes regulation of HBEGF and is associated with disease severity in CFHR5 nephropathy. PLoS One 7:e31021
11. Bartel DP (2004) MicroRNAs: genomics, biogenesis, mechanism, and function. Cell 116:281–297
12. Kim DH, Saetrom P, Snove O Jr et al (2008) MicroRNA-directed transcriptional gene silencing in mammalian cells. Proc Natl Acad Sci U S A 105:16230–16235
13. Lai EC (2002) Micro RNAs are complementary to 3′ UTR sequence motifs that mediate negative post-transcriptional regulation. Nat Genet 30:363–364
14. Kozomara A, Griffiths-Jones S (2011) miRBase: integrating microRNA annotation and deep-sequencing data. Nucleic Acids Res 39: D152–157
15. Conrad DF, Pinto D, Redon R et al (2010) Origins and functional impact of copy number variation in the human genome. Nature 464: 704–712
16. Iafrate AJ, Feuk L, Rivera MN et al (2004) Detection of large-scale variation in the human genome. Nat Genet 36:949–951
17. Dear PH (2009) Copy-number variation: the end of the human genome? Trends Biotechnol 27:448–454
18. Henrichsen CN, Chaignat E, Reymond A (2009) Copy number variants, diseases and gene expression. Hum Mol Genet 18:R1–8
19. Nozawa M, Kawahara Y, Nei M (2007) Genomic drift and copy number variation of sensory receptor genes in humans. Proc Natl Acad Sci U S A 104:20421–20426
20. Schuster-Bockler B, Conrad D, Bateman A (2010) Dosage sensitivity shapes the evolution of copy-number varied regions. PLoS One 5:e9474
21. Felekkis K, Voskarides K, Dweep H et al (2011) Increased number of microRNA target sites in genes encoded in CNV regions. Evidence for an evolutionary genomic interaction. Mol Biol Evol 28:2421–2424
22. Woodwark C, Bateman A (2011) The characterisation of three types of genes that overlie copy number variable regions. PLoS One 6: e14814

23. Dweep H, Sticht C, Pandey P et al (2011) miRWalk - database: prediction of possible miRNA binding sites by "walking" the genes of three genomes. J Biomed Inform 44: 839–837
24. Lewis BP, Burge CB, Bartel DP (2005) Conserved seed pairing, often flanked by adenosines, indicates that thousands of human genes are microRNA targets. Cell 120:15–20
25. Dweep H, Sticht C, Gretz N (2013) In-silico algorithms for the screening of possible microRNA binding sites and their interactions. Curr Genomics 14:127–136
26. Dweep H, Sticht C, Kharkar A et al (2013) Parallel analysis of mRNA and microRNA microarray profiles to explore functional regulatory patterns in polycystic kidney disease: using PKD/Mhm rat model. PLoS One 8:e53780

Chapter 27

SYBR® Green and TaqMan® Quantitative PCR Arrays: Expression Profile of Genes Relevant to a Pathway or a Disease State

M. Lucrecia Alvarez and Stefania Cotta Doné

Abstract

Quantitative PCR arrays are the most reliable and accurate tool for analyzing the expression of a focused panel of genes relevant to a pathway or a disease state. PCR arrays allow gene expression analysis with the sensitivity, dynamic range, and specificity of a real-time PCR as well as the multi-gene profiling capability of a microarray. Differences among real-time PCR kits used in PCR arrays are largely restricted to the DNA polymerases and the detection methods used. In this chapter, we provide a step-by-step protocol for the two detection methods most commonly used in PCR arrays, known as SYBR® Green and TaqMan®, which are based on two different approaches to detect PCR products. While SYBR® Green uses a binding dye that intercalates nonspecifically into double-stranded DNA, the TaqMan® approach relies on a fluorogenic oligonucleotide probe that binds only the DNA sequence between the two PCR primers. Therefore, only specific PCR product can generate a fluorescent signal in TaqMan® PCR. Here we also provide a comparison of the SYBR® Green and TaqMan® approaches and highlight their advantages and disadvantages to help the user to choose the best platform.

Key words Quantitative PCR arrays, qPCR arrays, PCR arrays, TaqMan, SYBR Green, Gene expression, Expression profile, Real-time PCR, Quantitative PCR

1 Introduction

The advent of real-time quantitative PCR (qPCR) and real-time reverse transcription PCR (RT-PCR) has dramatically improved quantification of gene expression. qPCR allows data collection throughout the PCR process as it occurs, combining amplification and detection into a single step [1]. This is achieved using different fluorescent chemistries such as SYBR® Green and TaqMan®, which correlate PCR product concentration to fluorescence intensity [2]. The PCR cycle at which the amplification of the target is first detected is usually referred to as cycle threshold (Ct) or time at which fluorescence intensity is greater than background fluorescence. Therefore, the greater the amount of target DNA in the

M. Lucrecia Alvarez and Mahtab Nourbakhsh (eds.), *RNA Mapping: Methods and Protocols*, Methods in Molecular Biology, vol. 1182, DOI 10.1007/978-1-4939-1062-5_27, © Springer Science+Business Media New York 2014

sample, the faster a significant increase in fluorescence will appear and the lower the Ct value [3].

qPCR offers many benefits over other methods to quantify gene expression because it can produce quantitative data with an accurate dynamic range of 7–8 log orders of magnitude and does not require post-amplification manipulation. The sensitivity of qPCR assays is 10,000- to 100,000-fold higher than RNase protection assay [4], 1,000-fold higher than dot blot hybridization [5], and high enough to be able to detect a single copy of transcript [6]. In addition, qPCR can reliably detect as little as 23 % differences in gene expression between samples and discriminates between mRNA with almost identical sequences [1].

The two most commonly used PCR arrays are based on SYBR® Green or TaqMan® qPCR assays. SYBR® Green qPCR is widely used because of its ease assay design and its relatively low setup and running costs. The SYBR® Green dye intercalates into double-stranded DNA to detect the amplification of the target gene specifically initiated by gene-specific primers [7]. However, the dye is no specific for a DNA sequence and it can detect nonspecific amplification products. Unlike SYBR® Green, only specific PCR product can produce a fluorescent signal in TaqMan® qPCR because this method uses a fluorogenic single-stranded oligonucleotide probe that binds only to the DNA sequence between the two PCR primers, which significantly increases the specificity of the detection [8]. The TaqMan® probe consists of 18–22 bp oligonucleotide labeled with a fluorescent dye on the 5′ end, which is quenched by a minor groove binder (MGB)-nonfluorescent quencher on the 3′ end. During the DNA extension step of the qPCR in each cycle, Taq polymerase synthesizes a complementary DNA strand using the unlabeled primers and template as a guide. When the polymerase reaches the TaqMan® probe annealed to the DNA template, its 5′ nuclease cleaves the probe separating dye from the quencher and making the dye fluoresce [9]. More dye molecules are released in each cycle of the PCR increasing the fluorescence intensity proportionally to the amount of amplicon synthesized. The main characteristics, advantages, and disadvantages of the TaqMan® and SYBR® Green qPCR systems are compared in Table 1.

Figure 1 shows the general qPCR array procedure for both SYBR® Green and TaqMan® systems. The protocol starts by extracting RNA and assessing its quality and concentration. Only high-quality RNA samples are converted into PCR template (cDNA) using a reverse transcriptase enzyme. The template is then combined with a ready-to-use qPCR Master Mix, and equal aliquots of this mixture are added to each well of the same PCR array plate already containing dried gene-specific primer sets (SYBR® Green and TaqMan®) and probe (TaqMan®). The thermal cycling is performed, and the instrument's software is used to calculate the

Table 1
Comparison of SYBR® Green and TaqMan® qPCR systems

Characteristics	SYBR® Green qPCR arrays	TaqMan® qPCR arrays
qPCR chemistry	Uses a dsDNA-binding dye (SYBR® Green) to detect PCR product as it accumulates during qPCR	Uses a fluorogenic probe specific to target gene to detect PCR product as it accumulates during qPCR
Specificity	Medium[a]	High
Sensitivity	Variable[a]	1–10 gene copies
Reproducibility	Medium[a]	High
Cost	Medium (more economical than TaqMan® because it does not require a probe)	High
Simplicity of procedure	Low (requires extra controls and tests to compensate for lower specificity)	Medium
Effect of genomic DNA contamination in RNA samples	High (extra controls and steps are required to eliminate DNA contamination)	Low (most of the probes span an exon–intron junction, and only those TaqMan® assays with "s" and "g" suffix, which are designed to a single exon will detect genomic DNA)
Method used to analyze results	Free PCR Array Data Analysis Web portal (http://pcrdataanalysis.sabiosciences.com/pcr/arrayanalysis.php)	Free DataAssist software (Life Technologies)
Main advantages	Lower cost	Higher specificity and sensitivity
Main disadvantages	Lower specificity because SYBR Green binds to all double-stranded DNA independently of its sequence (SYBR Green qPCRs requires extra controls and dissociation curves)	Higher cost

[a]Depends on template quality and primer/design optimization

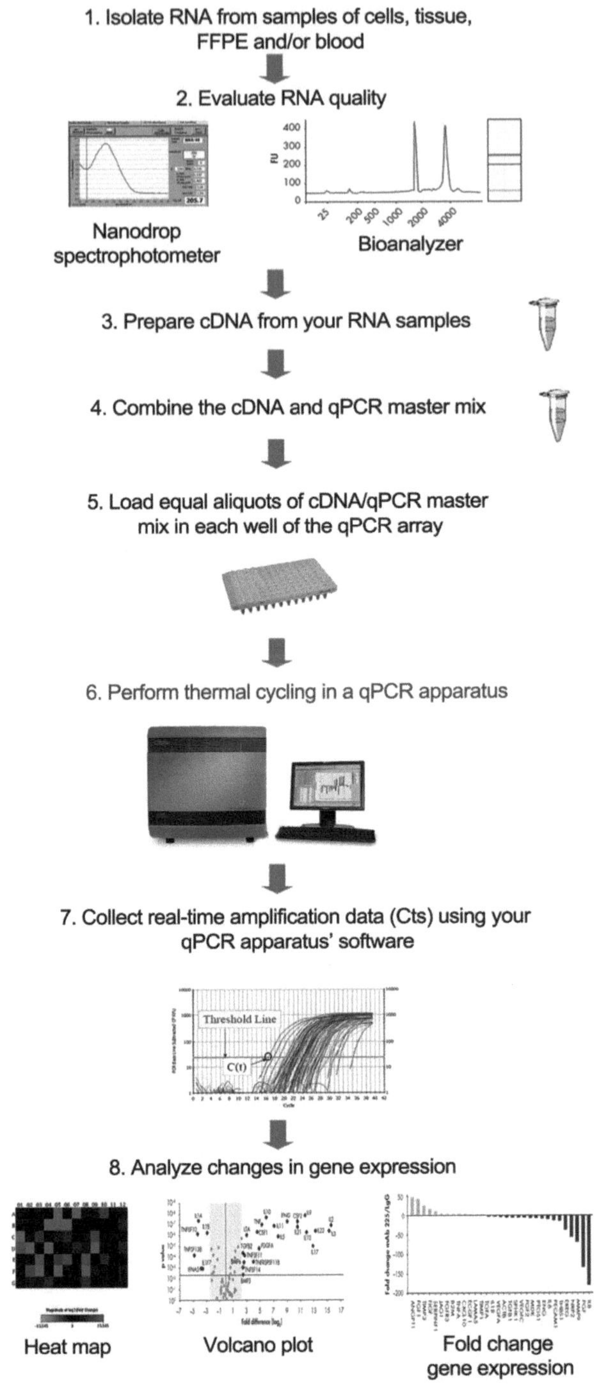

Fig. 1 General qPCR array procedure for both SYBR® Green and TaqMan® systems. High-quality RNA samples are required for analyzing gene expression using qPCR arrays. The quantity and quality (i.e., purity and integrity) of the RNA sample should be assessed preferably using both the Nanodrop spectrophotometer and the Bioanalyzer apparatus (*see* **Note 6**). Each high-quality RNA sample

threshold cycle (Ct) values for all the genes on each PCR array. Finally, the fold changes in gene expression for pair-wise comparison are calculated using the ΔΔCt method [10].

Here we provide a step-by-step protocol for a SYBR® Green PCR array using as example a commercial 384-well plate array that profiles the expression of 84 pathway-focused key genes involved in lipoprotein transport and cholesterol metabolism. This SYBR® Green PCR array was used to determine the effect of miR-27a over-expression on the cholesterol metabolism in HepG2 cells. In addition, we describe a step-by-step protocol for a TaqMan® PCR array using as example a custom-made 96-well plate array that profiles the expression of 29 key genes associated with extracellular matrix accumulation in kidney cells. This TaqMan® PCR array was used to determine the effect of miR-1207-5p on the expression of extracellular matrix-related genes in kidney mesangial cells.

2 Materials

Common materials for both SYBR® Green and TaqMan® qPCR arrays:

1. High-quality, nuclease-free water (*see* **Notes 1** and **2**).
2. Calibrated single- and multi-channel pipettes.
3. RNase/DNase-free pipette tips and tubes.
4. ABI 7900HT thermocycler (Life Technologies) and appropriate blocks for 384- and 96-well PCR array plates (*see* **Notes 3–5**).
5. Nanodrop Spectrophotometer to assess RNA purity and concentration (Thermo Fisher).
6. Bioanalyzer 2100 to determine RNA quality and integrity (Agilent).
7. RNaseZap for cleaning work surfaces, pipettes, and equipment that must be RNase free (Life Technologies).
8. RNeasy Mini kit for total RNA extraction (Qiagen) (*see* **Notes 6** and 7).
9. Gloves (*see* **Note 8**).

Fig. 1 (continued) is copied into cDNA, combined with the appropriate qPCR Master Mix, and loaded in the PCR array plate. After performing the thermal cycling in a qPCR apparatus, the results are initially analyzed using the qPCR apparatus' software to obtain the Cts (cycle thresholds) and then further processed with a different software for qPCR arrays or relative quantification of gene expression such as DataAssist (Life Technologies) or RT^2 Profiler PCR Array Data Analysis version 3.5 (http://sabiosciences.com/pcr/arrayanalysis.php). *FFPE* formalin-fixed paraffin-embedded, *Cts* cycle thresholds

Recommended loading direction →

	1	2	3	4	5	6	7	8	9	10	11	12
A	18S	COL1A1	COL1A2	COL3A1	18S	COL1A1	COL1A2	COL3A1	18S	COL1A1	COL1A2	COL3A1
B	COL4A1	COL4A2	COL4A3	COL4A4	COL4A1	COL4A2	COL4A3	COL4A4	COL4A1	COL4A2	COL4A3	COL4A4
C	COL4A5	FN1	LAMA1	LAMA4	COL4A5	FN1	LAMA1	LAMA4	COL4A5	FN1	LAMA1	LAMA4
D	LAMA5	LAMB1	LAMB2	LAMC1	LAMA5	LAMB1	LAMB2	LAMC1	LAMA5	LAMB1	LAMB2	LAMC1
E	ITGA1	ITGA3	ITGA6	ITGAV	ITGA1	ITGA3	ITGA6	ITGAV	ITGA1	ITGA3	ITGA6	ITGAV
F	ITGB1	ITGB3	MMP2	MMP9	ITGB1	ITGB3	MMP2	MMP9	ITGB1	ITGB3	MMP2	MMP9
G	TIMP1	TIMP2	TGFB1	TGFB2	TIMP1	TIMP2	TGFB1	TGFB2	TIMP1	TIMP2	TGFB1	TGFB2
H	CTGF	PAI1	UBC	PPIA	CTGF	PAI1	UBC	PPIA	CTGF	PAI1	UBC	PPIA

N=1 N=2 N=3

Fig. 2 TaqMan® custom-made array to quantify the expression of extracellular matrix-related genes. We selected 29 extracellular matrix-related genes as well as three endogenous controls (in *blue*:18S, PPIA, and UBC) for normalization. Three biological or technical replicates can be analyzed per plate (*see* **Note 48**) (Color figure online)

2.1 SYBR® Green qPCR Arrays

1. RT^2 First Strand kit (Qiagen) for reverse transcription of RNA (*see* **Note 9**).
2. 384-well SYBR® Green RT^2 Profiler PCR Array plate Human lipoprotein signaling and cholesterol metabolism (format E) (*see* **Notes 10** and **11**).
3. 384EZLoad Covers to load the samples into the PCR array plate (they come with the 384-well PCR array).
4. 2× RT^2 SYBR® Green RT^2 Profiler PCR Array Green/ROX qPCR Master Mix (Qiagen) (*see* **Note 12**).
5. RT^2 RNA QC PCR Array (optional, *see* **Note 13**).
6. RT^2 PCR Array Loading Reservoir (Qiagen).

2.2 TaqMan® qPCR Arrays

1. 96-well TaqMan® Gene Expression custom-made or predefined gene signature array (Life Technologies) (*see* **Notes 10** and **11**). Figure 2 shows the target genes included in the TaqMan® custom-made array that we designed, which includes, in triplicate, 29 extracellular matrix-related genes as well as three endogenous controls (in blue:18S, PPIA, and UBC) (*see* **Note 14**).
2. SuperScript® VILO™ Master Mix (Life Technologies) (*see* **Note 9**).
3. TaqMan® Universal PCR Master Mix 2× (Life technologies).
4. MicroAmp® Optical Adhesive Film (Life Technologies).
5. CD provided by Life Technology, containing the information about the assay, which will be loaded into the computer when setting up the run.

3 Methods

DNA contamination will artificially inflate the results of the qPCR, particularly SYBR® Green signal, yielding skewed gene expression profiles and false-positive signals. The amplification products from previous experiments spread into the air of your working environment and become a common source of DNA contamination. Before starting your PCR array experiments, please make sure to set up and maintain a working environment free of DNA contamination following the recommendations stated in **Notes 15–19**.

3.1 Total RNA Extraction

1. Spray RNaseZap onto work surfaces, pipettes, equipment, and the gloves that you are wearing before starting with the RNA extraction. Even trace quantities of RNase can lead to degradation during RNA purification protocols, lower yields from in vitro transcription reactions, and variable results in the PCR arrays.
2. Rinse the RNaseZap off with nuclease-free water.
3. Prepare RNA samples (at least three biological replicates, *see* **Note 20**) using RNeasy mini kit (Qiagen) according to the manufacturer's instructions and following methods specific for your biological samples (*see* **Note 5**). Other total RNA extraction kits from other suppliers can also be used as long as they are able to produce high-quality RNA according to the criteria specified in **Note 6**.
4. Determine RNA quantity and quality using the Nanodrop Spectrophotometer (Thermo Fisher) and Bioanalyzer. Only high-quality RNA should be used for qPCR arrays (*see* **Note 6**).

3.2 SYBR® Green qPCR Arrays

3.2.1 Reverse Transcription

Eliminating genomic DNA contamination is essential for obtaining optimal real-time gene expression profiling results using PCR arrays. We strongly recommend performing the on-column DNase treatment step in the RNeasy Mini Kit followed by using the RT^2 First Strand kit, which includes the genomic DNA elimination mixture, to remove any and all residual contamination from your RNA samples.

Before reverse transcribing your RNA sample, perform the following steps to eliminate residual genomic DNA:

1. Briefly (10–15 s) spin down all reagents from the RT^2 First Strand kit (Qiagen).
2. For each RNA sample, combine the following reagents in an RNase-free PCR tube:
   (a) 25.0 ng to 5.0 μg of total RNA (*see* **Notes 21–24**).
   (b) 2.0 μl of 5× genomic DNA elimination buffer.
   (c) Up to 10 μl with nuclease-free water.

3. Mix the genomic DNA elimination mixture (GEM) gently with a pipette followed by a brief centrifugation.
4. Incubate at 42 °C for 5 min.
5. Chill on ice immediately for at least 1 min.
6. Prepare the RT cocktail (amounts are for one reaction, final volume 10 μl):
   (a) 4 μl BC3 (5× RT buffer 3).
   (b) 1 μl P2 (primer and external control mix).
   (c) 2 μl RE3 (RT enzyme mix 3).
   (d) 3 μl Nuclease-free water.
7. Add 10 μl of RT cocktail to each 10 μl GEM from **step 5** to obtain 20 μl of first-strand cDNA synthesis reaction.
8. Mix well but gently with a pipette.
9. Incubate at 42 °C for *exactly* 15 min, and then *immediately* stop the reaction by heating at 95 °C for 5 min.
10. Add 91 μl of nuclease-free water to each 20 μl of cDNA synthesis reaction. Mix well.
11. Hold the obtained cDNA (first-strand cDNA synthesis reaction) on ice until the next step or store overnight at −20 °C.

3.2.2 Real-Time PCR

1. Briefly (10–15 s) spin down all reagents.
2. For a 384-well SYBR® Green RT² Profiler PCR array plate (format E or G), prepare the *experimental cocktail* by mixing the following reagents in a 5 ml tube (*see* **Notes 25–28**):
   (a) 550 μl 2× SABiosciences RT2 qPCR Master Mix (*see* **Notes 29** and **30**).
   (b) 102 μl Diluted first-strand cDNA synthesis reaction (from **step 10**, Subheading 3.2.1).
   (c) 448 μl Nuclease-free water.

   Final volume: 1,100 μl.
3. Carefully remove the PCR array plate from its sealed bag.
4. Dispense *experimental cocktail* (a new one per each sample) to the RT² PCR Array Loading Reservoir to assist in loading.
5. Load sample cocktails from a reservoir to the appropriate wells of the 384-well PCR array (format E or G) using a multichannel pipette with 8 tips only and the 384EZLoad Covers that come with the 384-well PCR array kit as follows (*see* **Note 31**):
   (a) Place cover #1 (white) on the plate. Add 10 μl of "sample 1 cocktail" to the open wells with odd numbers of rows A, C, E, G, I, K, M, and O. Remove and discard the cover.
   (b) Place cover #2 (yellow) on the plate. Add 10 μl of "sample 2 cocktail" to the open wells with even numbers of rows A, C, E, G, I, K, M, and O. Remove and discard the cover.

(c) Place cover #3 (black) on the plate. Add 10 μl of "sample 3 cocktail" to the open wells with odd numbers of rows B, D, F, H, J, L, N, and P. Remove and discard the cover.

(d) Place cover #4 (red) on the plate. Add 10 μl of "sample 4 cocktail" to the open wells with even numbers of rows B, D, F, H, J, L, N, and P. Remove and discard the cover.

6. Carefully but tightly seal the PCR array with the optical adhesive film (formats C, E, F, and G) (*see* **Note 32**).
7. Centrifuge the plate for 1 full minute at room temperature at 1,000 × *g* to remove bubbles (*see* **Note 33**).
8. Place the plate on ice while setting up the PCR cycling program (*see* **Note 34**).
9. In an ABI 7900HT qPCR apparatus (Life Technologies), start the SDS software and select "*file*" and then "*new*." For "assay," select "*standard curve* (*AQ*)"; for "container," select "*384-well clear plate*"; for "template," choose "*blank template*." Finally, click "OK."
10. Under the Setup tab, select "*add detector*" and enter "*SYBR*" in the "name" field. Select "*SYBR*" for the reporter, and click "OK."
11. Click once on the newly added detector, and select "*copy to plate document*." Click "*done*" button.
12. Choose *ROX* for the *Passive Reference* box.
13. Highlight all wells, and check the "*Use*" box in the setup tab.
14. Select the "Instrument" tab and then the "Thermal" profile. Use a two-step cycling program as follows (*see* **Note 35**):

    (a) 1 cycle: 10 min at 95 °C (*see* **Note 36**).

    (b) 40 cycles: 15 s at 95 °C and 1 min at 60 °C.

15. Click the "*add dissociation stage*" button; the preset dissociation stage will be added as *Stage 3* after the 40 cycles.
16. Select "standard mode" and 10 μl for 384-well plates.
17. Select "file" and then "*save as*" to save the template file as *SDS Template* (*.sdt) with the filename "*RT^{2}Profiler*™ *PCR Array Protocol Template*" (*see* **Note 37**).
18. Select the "*Instrument*" tab. Click "*Connect*" to link the computer to the thermal cycler. Open the plate tray, and place your plate in the precision plate holder with A1 in the top left corner.
19. Click "*Close*" to load the plate and then "*Start*" to begin the PCR run. The estimated run time will then appear on the screen.
20. *Dissociation (melting) curve*: A melting curve program should run immediately (*see* **Note 38**) after the qPCR cycling program, and generate a first derivative dissociation curve for each well in the entire plate using your instrument's software. No more than one peak should appear in each reaction at temperatures greater than 80 °C.

3.2.3 Analysis of qPCR Results

1. When the qPCR run is completed, a small dialog box stating "The run completed successfully" will appear on the screen. Click "*OK*"; this will close the box.
2. In the command tab, select "*Analysis*" and then "*Analysis Settings*" to open a dialogue box. Select "*Manual Baseline*" and "*Manual Ct*"; click "*OK*."
3. Click the "*Result*" tab. Display data as "*ΔRn vs. Cycle.*"
4. Click the square button in the upper left corner of the diagram of the 384-well plate (between the letter "*A*" and "*1*") to select all wells. The selected wells will be highlighted in yellow in the lower left panel.
5. Follow the procedures below to calculate the threshold cycle (Ct) for each well (Fig. 3):
   (a) *To define the baseline*: Use the linear view of the amplification plots. Double click on *y*-axis and the window for "*Display Settings*" will appear. For amplification plot properties, select "*Auto Scale*" for both the *y*- and *x*-axes. Select "*Linear view*" for *y*-axis, and then click "*OK*." With the linear plots, determine the cycle number at which the earliest amplification can be seen. Use the red sliding bars on the *x*-axis to set the manual baseline to start from cycle number 2 through two cycle values before the earliest visible amplification. The earliest amplification will usually be visible between cycles 14 and 18.
   (b) *To define the threshold value*: Use the log view of the amplification plots. Double click on y-axis and the window for "*Display Settings*" will appear. For amplification plot properties, select "*Auto Scale*" for both the *y*- and *x*-axes. Select "*Log view*" for *y*-axis, and click "*OK*." With the log plots, place the threshold line above the background signal but within the lower one-third to lower one-half of the linear phase of the amplification plot (*see* **Note 39**).
6. For the SDS software to analyze data after the run, click on the green arrow on the lower command tab or select "*Analysis*" and then "*Analyze.*".
7. The values for Ct will be displayed in the lower left panel for each well.
8. Analyze the Cts from the different types of controls included in the PCR array plate (Fig. 4):
   (a) *Genomic DNA control* (*GDC*): Specifically detects non-transcribed genomic DNA contamination in each sample with a high level of sensitivity. If the value is greater than 35, then the level of genomic DNA contamination is too low to affect gene expression profiling results. No action is needed. However, if the value is less than 35, then genomic DNA contamination is evident (*see* **Notes 40** and **41**).

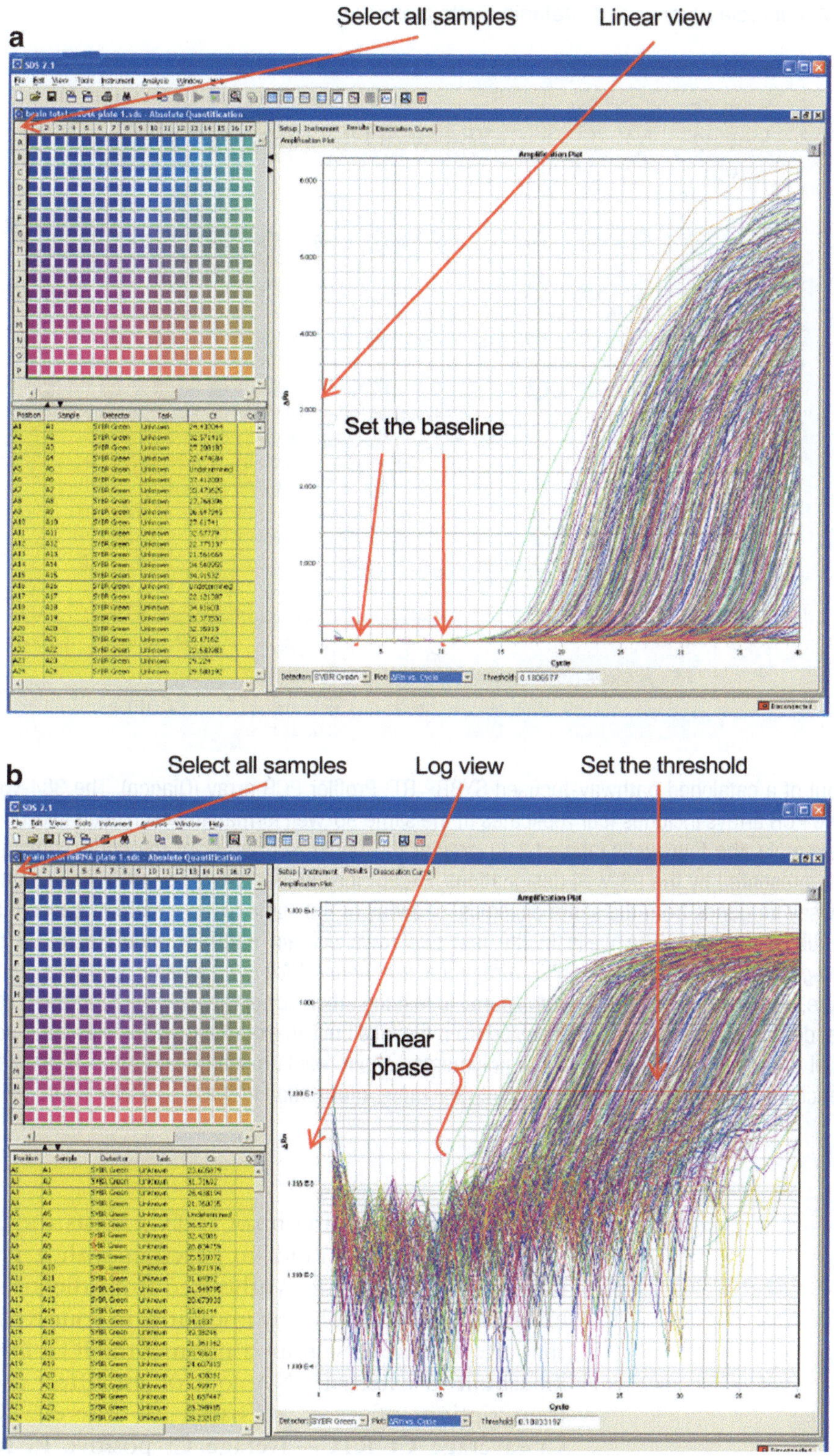

Fig. 3 Procedures for manually setting the baseline and threshold cycle (Ct) using the SDS software. (**a**) To define the baseline, double click on *y*-axis and select "*Linear*" view. Using the red sliding bars on *x*-axis, set the manual baseline to start from cycle number 2 through two cycle values before the earliest visible amplification. (**b**) To define the threshold value, double click on *y*-axis and select "*Log*" view for *y*-axis. Place the threshold line above the background signal but within the lower third of the linear phase of the amplification plot. Modified from SABiosciences' Technical Note "ABI 7900HT: for SDS software 2.3. Instrument set up instructions for RT2 Profiler PCR arrays"

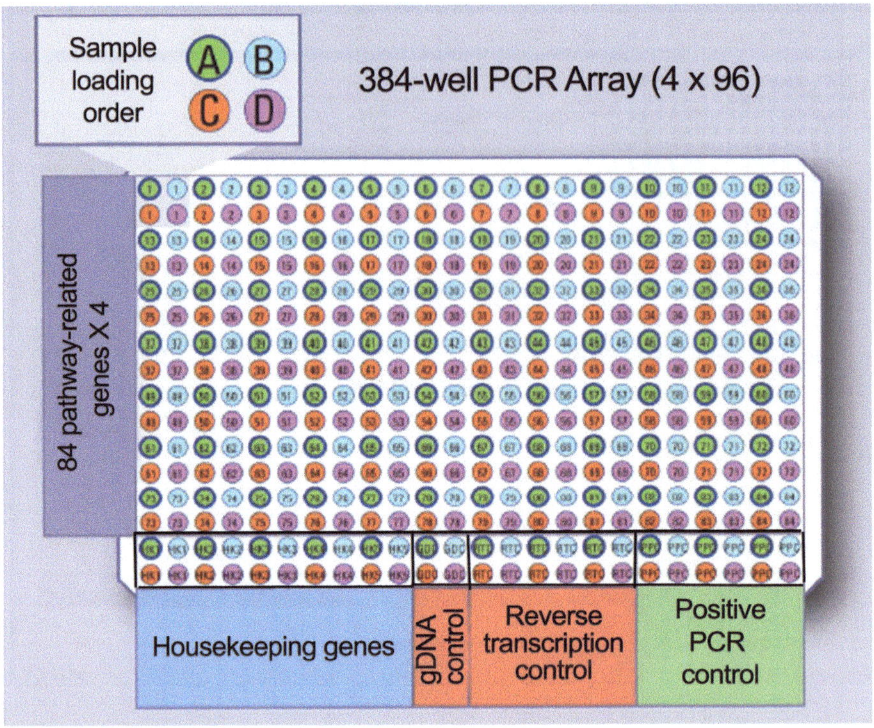

Fig. 4 Layout of a cataloged pathway-focused SYBR® RT[2] Profiler PCR array (Qiagen). The 384-well (4 × 96) format of the PCR arrays includes four replicates of the same 96-well format, in which each two-by-two set of wells (for example, the four wells labeled 1 in *grey* at the *upper left corner of the plate*) contain the same primer set represented by the 96-well designations. Except for the last two rows, each well contains a real-time PCR assay for genes from the same biological pathway or the same disease state or genes that are otherwise functionally related. The wells in the last two rows contain different panels of qPCR controls: a *housekeeping gene* panel to normalize PCR array data, a *genomic DNA control* panel to assess contamination of the sample, a *reverse transcription control* panel to test the efficiency of the RT[2] First Strand kit reaction with a primer set detecting the template synthesized from the kit's built-in external RNA control, and a *positive PCR control* panel to assess the efficiency of the polymerase chain reaction itself using a pre-dispensed artificial DNA sequence and the primer set that detects it (*see* more information in the text) (Color figure online)

(b) Reverse transcription control (RTC): Tests the efficiency of the RT2 First Strand kit reaction with a primer set detecting the template synthesized from the kit's built-in external RNA control. Any impurities in your RNA sample that affect the reverse transcription of the RT2 First Strand kit's built-in external RNA control also affect the reverse transcription of your messages of interest. Calculate Ct = average Ct RTC—average Ct positive PCR control (see below). If this value is less than 5, then no inhibition is apparent. If this value is greater than 5, then evidence of impurities that inhibited the reverse transcription phase of the procedure is evident (*see* **Note 42**).

(c) Positive PCR control (PPC): Tests the efficiency of the polymerase chain reaction itself using a pre-dispensed artificial DNA sequence and the primer set that detects it. Any impurities in your RNA sample that affect the PCR amplification of the positive control also affect the PCR amplification for your messages of interest. The average Ct PPC value should be 20±2 on each PCR array and should not vary by more than two cycles between PCR arrays being compared. Larger differences in average Ct PPC values between samples indicate the presence of different amounts of PCR amplification inhibitors in each sample and that all of the RNA samples require further purification. An average value of Ct PPC that is consistently greater than 22 for all of your samples may indicate a problem with the cycling conditions or may simply be indicative of the relative sensitivity of your instrument (*see* **Note 43**).

9. To export the results of the qPCR array from the SDS software (ABI 7900HT qPCR apparatus, Life Technologies) to an Excel spreadsheet, select "*File*," then "*Export*," and finally "*Results Table*." Save the file as "*Tab-delimited Text file*" (**.txt*).
10. The RT^2 Profiler PCR Array Data Analysis Web portal (http://pcrdataanalysis.sabiosciences.com/pcr/arrayanalysis.php) analyzes Ct values to calculate changes in gene expression. However, for the RT^2 Profiler PCR Array Data Analysis Web portal to correctly read/import your data, the Ct values must be first formatted into a new Excel spreadsheet. The RT^2 Profiler PCR Array Data Analysis Web portal only accepts Excel spreadsheets for analysis. All of the plates/wells must be copied into a single worksheet. Please use the templates below as guides.
11. Using Microsoft Excel Program, open the txt file previously generated by the qPCR apparatus.
12. From http://www.sabiosciences.com/pcrarraydataanalysis.php, download the patch for conversion of 384-well data into equivalent 96-well format for 96×4 PCR (an Excel file named "pcrarraydataanalysis384patch_4samples_96×4"). The same website provides patches for other PCR array formats.
13. From the Excel file that you obtained in **step 8**, copy the Ct values from your qPCR instrument's run.
14. Paste the Cts into the yellow cells starting at the Excel cell B11 (position A1 of the 384-well PCR array). The patch automatically converts the 384-well format of the data into four sets of 96-well formats (columns E–H), each representing your four different samples.
15. Copy the Gene Symbol column from your approved Gene List file and paste on Cell D11.
16. For Web-based data analysis: Copy cells C10 through H106 and paste (select "paste special" as "values") to a new worksheet

starting on Cell A1. Add the name or the treatment of each sample as title of each column (*see* Table 2). Save the Excel file as "results ready for analysis webportal.xls" (*see* **Note 44**).

17. Select "Standard RT² PCR array" and the catalogue of your PCR array ("PAHS-080" for the ver. 3 or "PAHS-080Z" for ver. 4 of 384-well PCR array plate human lipoprotein signaling and cholesterol metabolism).
18. At the RT² Profiler PCR Array Data Analysis version 3.5 (http://sabiosciences.com/pcr/arrayanalysis.php), browse the Excel file obtained in the previous step and click "upload." You will be asked to create a free account to analyze your results.
19. In the "Basic Setup" section of the "Analysis setup," assign samples to different group numbers (biological replicates should be in the same group number). At least two groups are needed, where one of those groups must be the "control group." In addition, you may exclude samples from the analysis by selecting "Exclude" on the drop-down menu. Click the "Update" bottom when finished.
20. Review the "Data QC" (quality control) section to assess each group's PCR reproducibility, reverse transcription efficiency, and presence of genomic DNA contamination.
21. In the "Select Housekeeping Genes" section, choose "manual selection" as preferred method of analysis and "geometric mean" as normalization method. Five different housekeeping genes (HG) are available in each PCR array plate. Only select HG with small changes in their expression across different sample groups (differences in Ct values less than 1). The expression level of the housekeeping genes chosen for normalization in the ΔΔCt method must not be influenced by your experimental conditions. Remove or add preferred housekeeping genes for data normalization by clicking the appropriate checkboxes. Click "perform normalization" when finished.
22. Review the "Data Overview" section to see each group's distribution of threshold cycle values and the average of the raw data in each group.
23. In the "Analysis" section, the average (AVG) ΔCt, $2^{-\Delta Ct}$, fold change, *p*-value, and fold regulation results are provided by the software from your data. The software uses the fold change and *p*-value results in subsequent graphical analyses (*see* **Note 45**).
24. In the "Plots & charts" section, you can choose between the following options to represent your results:
    (a) *Scatter plot*: Compares the normalized expression of every gene on the array between two groups by plotting them against one another to quickly visualize large gene expression changes. The central line indicates unchanged gene expression (Fig. 5a).

Table 2
Example of a Microsoft Excel table with Ct value results ready for Web-based data analysis

Position	Sample 1	Sample 2	Sample 3	Sample 4
A01	22.947102	23.203945	24.397852	24.240885
A02	32.64058	31.605745	39.514068	39.714066
A03	19.954372	19.855955	20.279177	20.966599
A04	24.866692	24.808746	23.617691	23.814632
A05	28.145184	26.973978	26.898994	37.534798
A06	24.8832	32.86193	24.577545	25.183624
A07	20.553404	20.005669	18.360966	18.818506
A08	15.33497	15.93494	16.21135	15.824698
A09	30.562838	30.947708	29.555754	30.167124
A10	16.697287	17.264198	17.55929	17.650913
A11	23.930744	30.32056	29.748398	31.202936
A12	19.658525	20.32671	19.662186	19.454039
B01	25.20977	25.773933	25.713951	25.810675
B02	21.636423	21.311886	23.146988	23.27905
B03	22.941557	22.91391	25.51358	24.61549
B04	35.116425	Undetermined	Undetermined	Undetermined
B05	Undetermined	35.95377	34.089092	35.964565
B06	25.824116	25.911173	25.502832	25.25561
B07	24.870405	24.954279	26.104229	25.145397
B08	17.567822	17.881706	17.56085	17.685776
B09	22.928915	23.481422	27.137724	27.263731
B10	25.280937	25.141603	25.50175	25.571594
B11	19.985317	19.461784	19.646757	20.094988
B12	28.76117	28.613745	30.369272	30.164816
C01	28.110167	27.849783	29.553253	29.673254
C02	Undetermined	Undetermined	37.735123	Undetermined
C03	20.548738	20.778555	19.835114	20.343662
C04	Undetermined	Undetermined	30.349026	30.314428
C05	28.620623	29.723993	29.344774	28.671759
C06	21.374655	21.618105	21.635664	20.721338
C07	21.649908	21.566235	20.715597	20.664833
C08	27.356163	27.359152	29.13274	29.372458

(continued)

Table 2 (continued)

Position	Sample 1	Sample 2	Sample 3	Sample 4
C09	31.651373	32.10653	32.289425	33.841404
C10	19.41836	19.563114	18.208511	18.259987
C11	23.596302	23.284672	23.088047	22.959435
C12	20.380278	20.590984	20.870806	20.56529
D01	22.313938	22.381845	22.210728	22.120174
D02	22.008171	21.962492	20.578918	20.79123
D03	32.7428	33.847244	31.627157	31.779516
D04	20.720102	20.945763	20.236698	20.21922
D05	25.784595	25.90658	26.262827	26.747755
D06	Undetermined	Undetermined	Undetermined	Undetermined
D07	35.117702	34.258812	31.01169	31.629148
D08	20.73037	20.920885	20.939959	21.070967
D09	25.247293	25.177979	25.472387	25.318346
D10	18.975967	18.954731	19.803394	19.92512
D11	28.455778	27.227055	28.80814	28.886637
D12	Undetermined	Undetermined	Undetermined	Undetermined
E01	38.25936	34.172813	33.426086	34.340023
E02	21.870417	21.662039	23.578161	23.198595
E03	25.71855	25.872715	26.393307	26.763252
E04	Undetermined	Undetermined	34.091473	Undetermined
E05	23.08461	23.106352	23.865164	23.505028
E06	19.573034	19.908016	19.862162	20.698463
E07	21.326319	21.410328	21.907417	21.864113
E08	22.372438	22.864193	21.411388	21.905571
E09	28.217123	28.026003	27.349588	27.522041
E10	25.928137	26.393015	27.019493	25.754826
E11	24.5588	24.469772	26.31139	25.919462
E12	26.877563	28.939928	25.161022	25.03504
F01	24.838243	24.971394	25.206577	24.9394
F02	33.87163	Undetermined	Undetermined	34.009136
F03	30.407818	30.821981	30.34593	31.687702
F04	Undetermined	Undetermined	34.541042	34.010612

(continued)

Table 2 (continued)

Position	Sample 1	Sample 2	Sample 3	Sample 4
F05	28.261038	28.0711	25.729658	26.645685
F06	26.237175	26.487864	27.468538	27.060848
F07	23.287455	24.181545	24.807642	25.30101
F08	21.359547	21.462915	22.27058	22.478079
F09	21.499184	21.735744	21.98313	22.298594
F10	24.541235	24.527061	25.179726	25.176617
F11	24.510782	24.68363	24.844604	24.749487
F12	27.945993	27.81604	29.821291	30.110443
G01	21.616755	21.316362	21.529284	21.350845
G02	23.184181	23.877026	23.260048	23.332691
G03	22.710815	22.781141	23.684061	24.329382
G04	25.189348	25.350182	25.127048	25.341005
G05	23.965261	23.924213	23.300932	23.384636
G06	30.02957	30.819132	30.97721	30.880545
G07	30.264639	31.262903	30.709425	31.360056
G08	25.256203	25.166363	25.924135	25.937561
G09	22.549492	22.774189	21.78226	21.943815
G10	30.418106	30.981133	30.768251	30.678495
G11	27.479012	27.789549	29.43417	30.29682
G12	34.184216	34.006542	30.128984	20.83217
H01	18.56581	19.301252	18.11646	18.497372
H02	20.361687	20.342052	20.254856	20.337875
H03	17.627934	17.703598	17.729275	17.865719
H04	15.948761	16.100574	16.66678	16.883123
H05	15.237252	15.264711	16.196733	16.349558
H06	Undetermined	33.969994	Undetermined	Undetermined
H07	22.44331	22.687386	22.91791	23.243818
H08	22.360973	22.625776	22.806044	23.201609
H09	22.366451	22.616217	22.804466	23.199217
H10	18.50138	18.797602	18.60998	18.682615
H11	18.610035	18.708664	18.7784	18.576456
H12	18.833483	18.841806	18.917484	18.706877

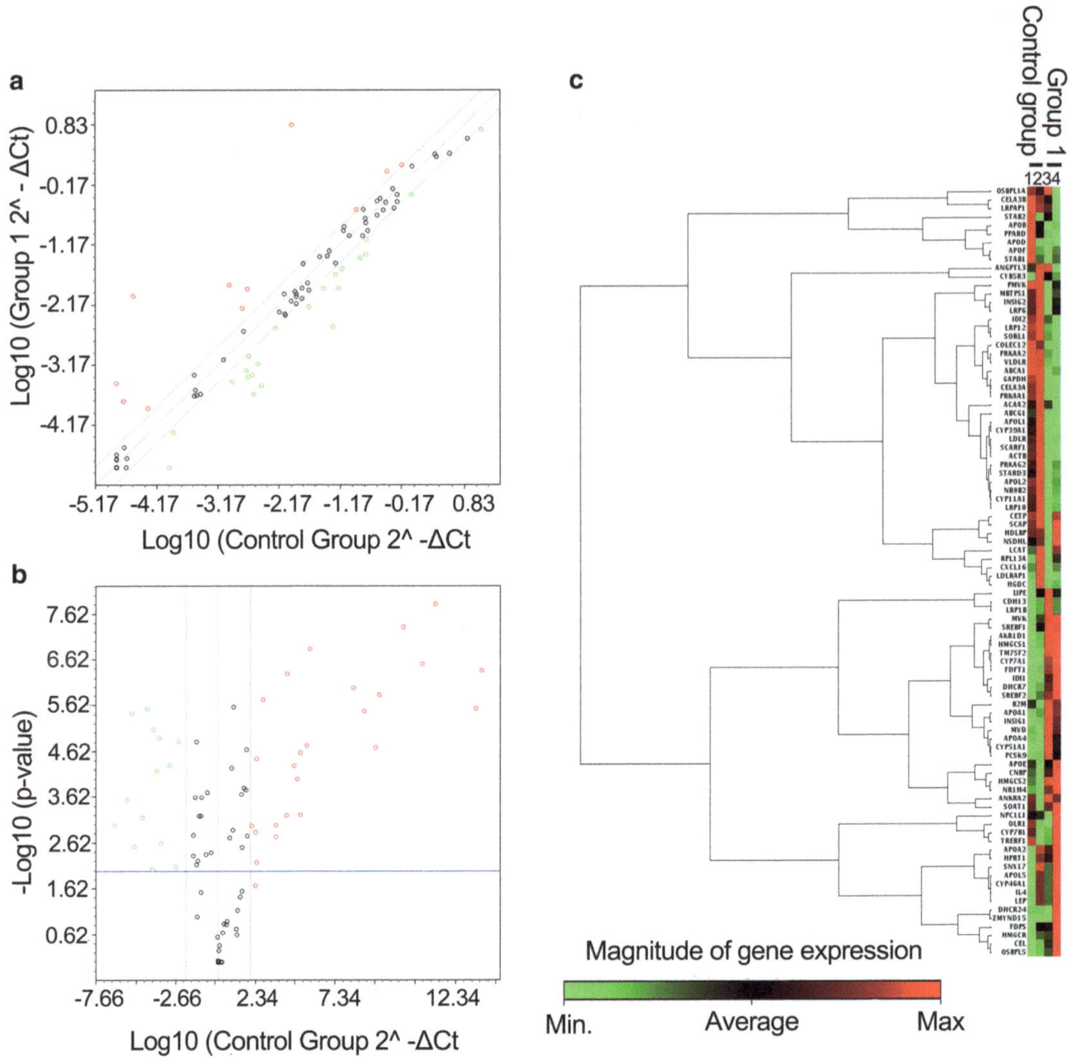

Fig. 5 Examples of scatterplot, volcano plot, and clustergram, three different ways to represent qPCR array results. To determine the effect of miR-27a over-expression on cholesterol homeostasis in HepG2 cells, HepG2 cells were transfected with 30 nM miR-27a mimics or negative control (NC), and total RNA was extracted 48 h after transfection. RNA was reverse transcribed and analyzed using a commercial 384-well plate SYBR® PCR array that profiles the expression of 84 pathway-focused key genes involved in lipoprotein transport and cholesterol metabolism (Qiagen). The RT[2] Profiler PCR Array Data Analysis version 3.5 (http://sabiosciences.com/pcr/arrayanalysis.php) was used to analyze and do a graphic of the results obtained. (**a**) *Scatterplot*: Compares the normalized expression of every gene on the array between two groups by plotting them against one another. The *central line* indicates unchanged gene expression; *red dots* represent upregulated genes, while *green dots* are downregulated genes. (**b**) *Volcano plot*: Combines a *p*-value statistical test with the fold regulation change enabling identification of genes with both large and small expression changes that are statistically significant. (**c**) *Clustergram*: A heat map with dendrograms (tree diagrams) indicating co-regulated genes across groups or individual samples

(b) *Volcano plot*: Displays statistical significance versus fold change on the y- and x-axes, respectively. The volcano plot combines a p-value statistical test with the fold regulation change enabling identification of genes with both large and small expression changes that are statistically significant. However, this plot requires three or more replicates in each group. Biological replicates are recommended instead of technical replicates (*see* **Note 20**) (Fig. 5b).

(c) *Clustergram*: Performs non-supervised hierarchical clustering of the entire dataset to display a heat map with dendrograms indicating co-regulated genes across groups or individual samples (Fig. 5c).

(d) *Heat map*: Provides a graphical representation of fold regulation expression data between two groups overlaid onto the PCR array plate layout. Simply select the experimental and control groups to display (Fig. 6).

(e) *Multi-group plot*: Provides both a line graph (Fig. 7a) and a bar chart (Fig. 7b) representation (with optional error bars) and is commonly used to examine the expression of a selected set of genes. Alternatively, follow **step 24**.

25. Click "Export Data" to download an MS Excel file containing all raw and processed data from the "Analysis Result" section (*see* **Note 46**). Construct a bar chart with your genes of interest that have been up- or downregulated (Fig. 8).

3.3 TaqMan® qPCR Arrays

3.3.1 Reverse Transcription

1. Briefly spin down the SuperScript VILO Master Mix to collect all the components at the bottom of the tube.
2. Prepare the RT mix as follows (amounts are for a 20 μl final reaction volume):
   (a) 4 μl SuperScript VILO Master Mix.
   (b) Up to 2.5 μg of high-quality RNA (*see* **Notes 21** and **22**).
   (c) Nuclease-free water up to 20 μl.
3. Gently mix the tubes and spin down using a microcentrifuge.
4. Incubate the tubes at 25 °C for 10 min followed by 60 min at 42 °C.
5. Incubate the tubes at 85 °C for 5 min to terminate the RT reaction.
6. Proceed to the qPCR step, or store cDNA at −20 °C until use.

3.3.2 Real-Time PCR

1. For a 96-well FAST TaqMan® array plate, prepare the *experimental cocktail* by mixing the following reagents in a 1.5 ml tube (*see* **Notes 26–28**):
   (a) 540 μl 2× TaqMan® Universal PCR Master Mix.
   (b) 540 μL cDNA + nuclease-free water (*see* **Note 47**).
   (c) Final volume: 1,080 μl.

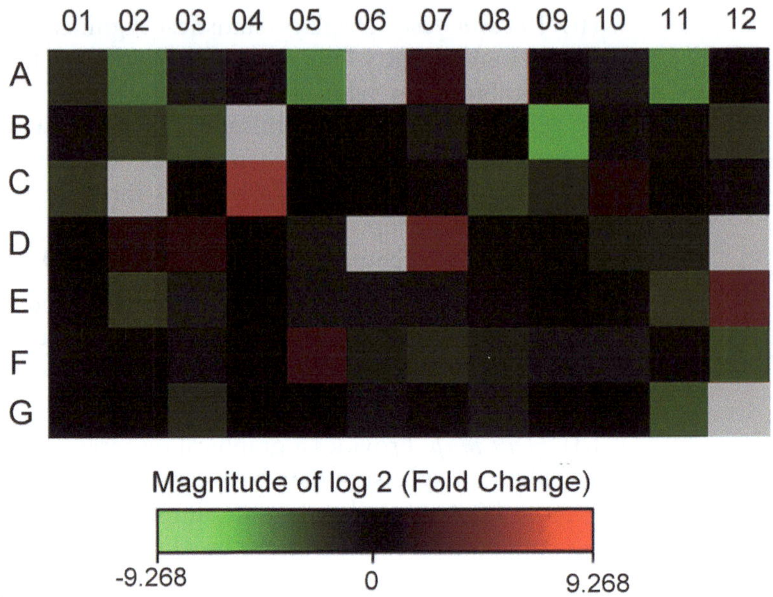

Layaout	01	02	03	04	05	06	07	08	09	10	11	12
A	ABCA1 -2.69 OKAY	ABCG1 -8.33 OKAY	ACAA2 -1.86 OKAY	AKR1D1 1.92 OKAY	ANGPTL3 -11.89 A	ANKRA2 -1.06 OKAY	APOA1 2.84 OKAY	APOA2 -1.13 OKAY	APOA4 1.64 A	APOB -1.75 OKAY	APOD -11.56 A	APOE 1.19 OKAY
B	APOF -1.37 OKAY	APOL1 -3.79 OKAY	APOL2 -4.99 OKAY	APOL5 -1.13 C	CDH13 1.21 B	CEL 1.24 OKAY	CETP -1.86 OKAY	CNBP -1.06 OKAY	COLEC12 -18.09 OKAY	CXCL16 -1.42 OKAY	CYB5R3 -1.26 OKAY	CYP11A1 -3.39 A
C	CYP39A1 -3.52 OKAY	CYP46A1 -1.13 OKAY	CYP51A1 1.31 OKAY	CYP7A1 22.42 B	CYP7B1 -1.01 OKAY	DHCR24 1.10 OKAY	DHCR7 1.67 OKAY	CELA3A -4.22 OKAY	CELA3B -2.58 B	FDFT1 2.11 OKAY	FDPS 1.18 OKAY	HDLBP -1.33 OKAY
D	HMGCR 1.00 OKAY	HMGCS1 2.17 OKAY	HMGCS2 2.66 B	IDI1 1.34 OKAY	IDI2 -1.79 OKAY	IL4 -1.13	INSIG1 8.74 B	INSIG2 -1.28 OKAY	LCAT -1.29 OKAY	LDLR -2.11 OKAY	LDLRAP1 -2.28 OKAY	LEP -1.13 C
E	LIPE 1.44 B	LRP10 -3.49 OKAY	LRP12 -1.95 OKAY	LRP1B 1.21 B	LRP6 -1.71 OKAY	LRPAP1 -1.65 OKAY	MBTPS1 -1.62 OKAY	MVD 1.72 OKAY	MVK 1.42 OKAY	NPC1L1 -1.33 OKAY	NR0B2 -3.44 OKAY	NR1H4 6.19 OKAY
F	NSDHL -1.27 OKAY	OLR1 -1.19 B	OSBPL1A -1.50 B	OSBPL5 1.46 B	PCSK9 3.47 OKAY	PMVK -2.12 OKAY	PPARD -2.83 OKAY	PRKAA1 -2.21 OKAY	PRKAA2 -1.63 OKAY	PRKAG2 -1.77 OKAY	SCAP -1.30 OKAY	SCARF1 -4.81 OKAY
G	SNX17 -1.11 OKAY	SOAT1 1.04 OKAY	SORL1 -2.72 OKAY	SREBF1 -1.11 OKAY	SREBF2 1.34 OKAY	STAB1 -1.61 B	STAB2 -1.37 B	STARD3 -1.87 OKAY	TM7SF2 1.53 OKAY	TRERF1 -1.15 B	VLDLR -5.32 OKAY	ZMYND15 -1.13 A

Fig. 6 Example of the use of heat map to represent SYBR® Green PCR array results. The same experiment described in the legend of Fig. 5 was analyzed with the RT[2] Profiler PCR Array Data Analysis version 3.5 (http://sabiosciences.com/pcr/arrayanalysis.php) and represented using a heat map, which provides fold regulation expression data between two groups overlaid onto the PCR array plate layout

2. Mix the contents well, and briefly spin the tubes to collect the contents at the bottom.
3. Carefully remove the PCR array plate from the box.
4. Before removing the cover, spin the plate briefly (1,000 rmp, for 1 min).

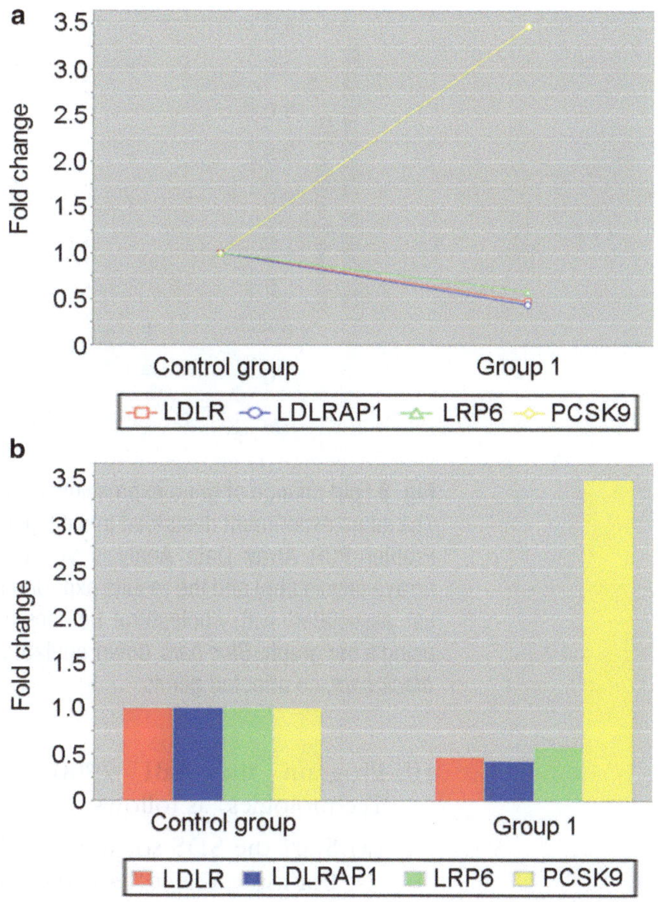

Fig. 7 Example of the use of a multi-group plot to represent SYBR® Green PCR array results. The same experiment described in the legend of Fig. 5 was analyzed with the RT² Profiler PCR Array Data Analysis version 3.5 (http://sabiosciences.com/pcr/arrayanalysis.php) and represented as a multi-group plot, which provides both a line graph (**a**) and a bar chart (**b**) representation (with optional error bars) and is commonly used to examine the expression of a selected set of genes

5. Remove the cover from the plates, one by one, as they are being used. *Never* leave an unused plate open, because the primer–probe mix can be degraded or lost.
6. Load 10 μl of experimental cocktail to the appropriate wells of the 96-well PCR array. If using a multichannel pipette with 8 tips only, make sure to load in the direction of the columns of the plate (A–H) (*see* **Note 48**).
7. Cover the plate with MicroAmp optical adhesive film.
8. Briefly centrifuge the plate to collect the reaction mix at the bottom and to remove any bubbles trapped in the well.
9. Place the plate on ice while setting up the PCR cycling program (*see* **Note 34**).

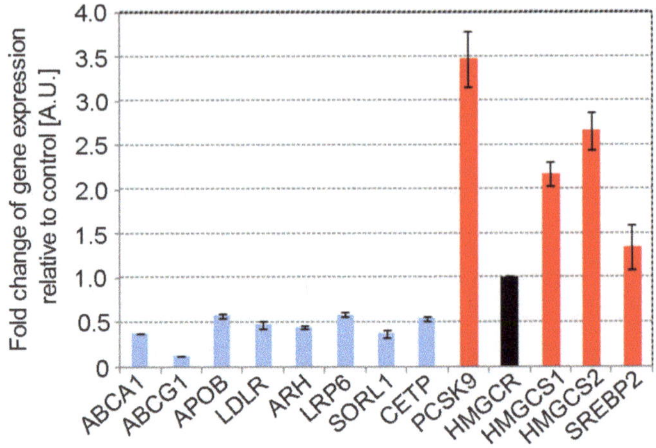

Fig. 8 Fold change of gene expression in HepG2 cells over-expressing miR-27a. The same experiment described in the legend of Fig. 5 was analyzed with the RT^2 Profiler PCR Array Data Analysis version 3.5 (http://sabiosciences.com/pcr/arrayanalysis.php) and the results exported to MS Excel. The main genes of interest associated with cholesterol homeostasis were selected and represented using a bar graph. *Blue bars*, downregulated genes; *red bars*, upregulated genes; *black bars*, no affected genes

10. Program the ABI 7900HT qPCR thermocycler (Life Technologies) as follows:
    (a) Start the SDS software, and select "*file*" and then "*new*." For assay, select "*relative quantification* (*RQ*)" or "*ΔΔCt*" [9] depending on the version of the software. For container, select "*96-well clear plate*"; for template, select "*blank template*." Finally click "*OK*."
    (b) Import the assay information that is in the CD that accompanies all the assays from Life Technologies. Place the CD in the appropriate drive. In the setup frame, select "file" in command tab, and then click on "import." Select the CD drive. From the list, select the file named "*ProdNum_7900_* SDS.txt." This file contains all the information necessary for the run. Import the file into the SDS program (*see* **Note 49**).
11. In the setup frame, highlight all wells and check the "Use" box in the setup tab.
12. Select the "Instrument" tab on the right and then the "Thermal" profile. Use a two-step cycling program as follows (*see* **Note 35**):
    (a) 1 cycle: 10 min at 95 °C.
    (b) 40 cycles: 15 s at 95 °C and 1 min at 60 °C.
13. Select "standard mode" and 10 μl for 96-well FAST plates.

14. Select "file" and then "*save as*" to save the template file as *SDS Template* (**.sdt*) with the filename "*TaqMan PCR Array Protocol Template.*"
15. Select the "*Instrument*" tab. Click "*Connect*" to link the computer to the thermal cycler. Open the plate tray, and place your plate in the precision plate holder with A1 in the top left corner.
16. Click "*Close*" to load the plate and then "*Start*" to begin the PCR run. The estimated run time will then appear on the screen.

3.3.3 Analysis of the Results

1. When your run is complete, a small dialog box stating "The run completed successfully" will appear on the screen. Click "*OK*"; this will close the box.
2. Click on the green arrow on the lower command tab. The program will then process the data and generate the file to be analyzed. Save this file, and close the SDS program.
3. Open the SDS RQ Manager program.
4. Click on "*File*," and select "*New study*" and then "*Add plate.*"
5. Browse the SDS file generated in **step 2**, and then click "*Add.*" This step can be repeated to add up to ten plates to the same study.
6. In the Amplification Tab (the upper right section of the computer screen), there are three different menu bar drop-down lists: "*Table Orientation,*" "*Calibrator,*" and "*Data.*" You can select different formats to view the study information (*see* **Note 50**).
7. In the "*Table Orientation*" drop-down list, select "*Detector.*" The list of detectors (target genes in the TaqMan® array plate) will appear in the left upper panel.
8. In the "*Data*" drop-down list, select "*ΔRn vs. cycle.*"
9. Before analyzing the resulting study data, specify parameter values for the analysis. Select "*Analysis*" and then "*Analysis Settings*" to open a dialogue box. Select "*Automatic Ct*" and "*Automatic Baseline*", and click "*Apply*" and then "*OK*" (*see* **Note 51**).
10. Click the square button "#" in the upper left corner of the lower left panel to select all the samples. The selected samples will be highlighted in yellow in the lower left panel.
11. For the SDS RQ Manager software to analyze data, click on the green arrow on the lower command tab or select "*Analysis*" and then "*Analyze all.*"
12. In the "*Table Orientation*" drop-down list, select "*Plate Centric.*" The list of plates included in the study will appear in the left upper panel, and the Ct values per well will be displayed in the left lower panel.

13. To export the results of the qPCR array from the SDS RQ Manager software, select "*File*," then "*Export*," and finally "*Results Data*." Save the file as "*Tab-delimited Text file*" (*.txt).
14. Download the free software DataAssist from Life Technologies website at http://www.invitrogen.com/site/us/en/home/Global/forms/dataassist-software-registration.html (a free registration is required to download the software).
15. Install and open DataAssist software.
16. Go to "*File*" and select "*New*" and a dialogue box will appear. Provide a "study name" and a "description," and select "*Add Files*." Browse for the txt file exported from SDS RQ Manager software in **step 13**, and click "*Open*." Follow the same procedure to add if you wish to add more plates in the same study, and then click "*OK*."
17. The list of samples used in your study will appear in the upper left panel (you can omit samples from this panel). The name given to each sample by the user is predefined in and imported with the results files. Data points in a study with the same sample and assay name are considered technical replicates. A name for a biological replicate group (e.g., normal, disease or time point 1, time point 2) can be assigned by clicking the "*Group*" box and manually entering a group name.
18. The "Assay Design" panel will show in the upper center and allows you to omit detectors and select endogenous controls. In "*Type*," select at least two detectors as "*candidate control*." The rest of the detectors are selected as "target" by default.

    In the "*Analysis Stetting*" (right upper panel), choose "*Endogenous Control*" as normalization method and click the bottom with the binocular. A graphic will display Ct values of candidate and selected controls for all samples. Select as endogenous controls those detectors that do not change more than 1 Ct across all your samples, and omit the rest. A score will display for each candidate control (*see* **Note 52**); the best endogenous controls are those with the lowest score. If you choose more than one gene for normalization, the mean CT value of the controls will be used for normalization.
19. Click next to the score in the "*Control Selection*" upper right panel to select the endogenous control that you are choosing for this study, and click "*OK*" at the right bottom of the screen.
20. In the "*Analysis Stetting*" (right upper panel), select a reference sample or calibrator (usually a negative control or no-treatment sample). Then click "*Perform Analysis*."
21. The lower panel displays the "Analysis Results" as "Average CT," "ΔCt," "$2^{-\Delta Ct}$," and "Fold Change." For each sample,

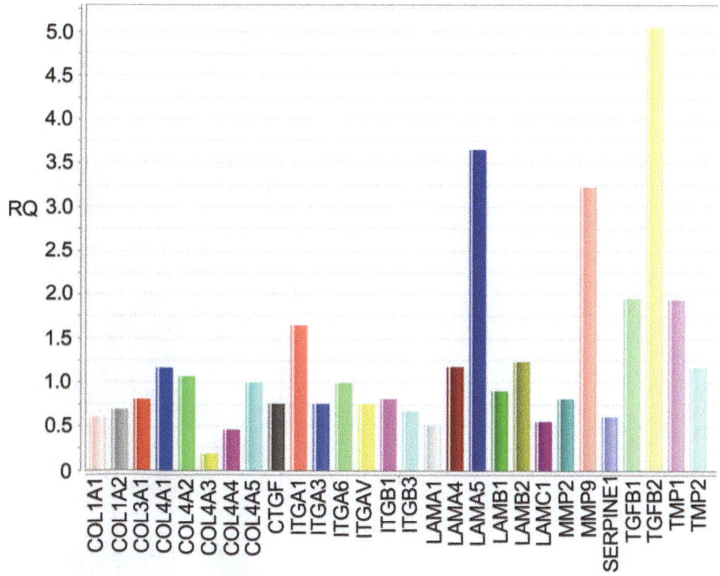

Fig. 9 Use of a custom-made TaqMan® qPCR array to determine the effect of miR-1207-5p the expression of extracellular matrix-related genes in kidney mesangial cells. Normal human mesangial cells were transfected with 30 nM miR-1207-5p mimics or negative control (NC), and total RNA was extracted 48 h after transfection. RNA was reverse transcribed and analyzed using a custom-made 96-well TaqMan® array for relative quantification of 29 extracellular matrix-related genes (*see* Fig. 2). Results were analyzed using SDS RQ Manager and DataAssist (Life Technologies) and visualized with the "*RQ Plot*" option (relative quantification or fold change) set as "*linear*" graph type. If no group is specified, the standard deviation of the ΔCt is also plotted for each sample on the "*Log2*" graph type

the fold change (RQ), RQ Min, and RQ Max are displayed. For biological groups, fold change and *p*-value will be displayed.

22. Under the panel "*Analysis Results*" you will find the "*QC Plots*," which help to visualize sample and group correlations for a quick quality check of data, and the "*Graphic results*" to represent your analyzed data (*see* **Note 53**). Figures 9 and 10 show the results obtained when we used a TaqMan® custom-made array (Fig. 2) to analyze the effect of miR-1207-5p on the expression of extracellular matrix-related genes in kidney mesangial cells. We selected "*Heat map*" (*see* **Note 54**) and "*RQ Plot*" options (*see* **Note 55**) to visualize our results. DataAssist software also provides a third option to visualize results, that is, a "*Volcano plot*," which displays *p-values* vs. fold change of groups (*see* **Note 56**).

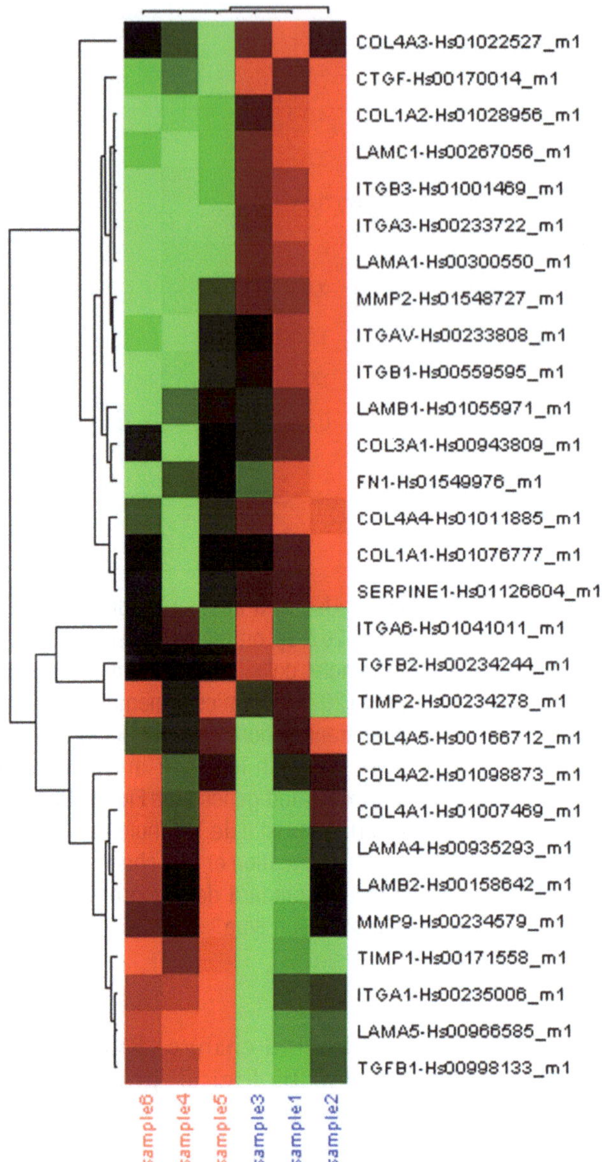

Fig. 10 Example of a cluster analysis and heat map obtained with DataAssist software. Same experiment described in the legend of Fig. 5. The results were visualized using the "*Heat Map*" option. Distances between samples and assays are calculated for hierarchical clustering based on the ΔCt values using either Pearson's correlation or Euclidean distance. The ΔCt value of the neutral expression level (mean or median) is set such that *red* indicates an increase in gene expression while *green* indicates a decrease

4 Notes

1. It is essential to use nuclease-free (RNase-free) water. The most commonly used method for eliminating RNase contamination from water, buffers, and other solutions is treatment

with diethylpyrocarbonate (DEPC), which destroys enzymatic activity by modifying -NH, -SH, and -OH groups in RNases and other proteins. However, residual DEPC can negatively affect the activity of the reverse transcriptase; therefore, DEPC-treated water should not be used.

2. We recommend using nuclease-free water that is not DEPC-treated as well as RT-PCR-grade water offered by some companies such as Life Technologies, Qiagen, and Abnova.
3. The PCR array plates only fit in 96- and 384-well real-time PCR instruments adapted with multi-well plate blocks. PCR arrays cannot be used in the Cepheid SmartCycler® or the Roche LightCycler 2.0 because they do not allow runs in multi-well plates.
4. The SYBR® Green RT² Profiler PCR Array System (Qiagen) is available in eight different plate formats, each tailored to specific real-time PCR instruments and associated blocks. Formats A, C, D, and F are 96-well plates, formats E and G are 384-well plates, format R is 100-well disc, and format H is 96×96 chip compatible (*see* Table 3). Make sure that you have the correct PCR array format for your instrument before starting the experiment (Table 3).
5. TaqMan® Gene Expression Signature Plates and Arrays come in five different formats, which are suited for the different models of thermocyclers available from the manufacturer: 96-well plate standard and FAST versions, 384-well plates and microfluidic cards, and OpenArray plates. The manufacturer provides in its home page the optimal thermocycler for each of the plate formats. Life Technologies does not provide information whether its products are suitable for thermocyclers from other manufacturers (Table 4).
6. High-quality total RNA is essential for obtaining good real-time PCR results, and it should be prepared preferably using one of the following methods, each specific for your biological sample:
   (a) *Cultured cells*: Use the miRNAeasy (if you want to isolate miRNAs) or RNeasy mini kit (Qiagen). You must perform the recommended on-column DNase treatment step to eliminate genomic DNA contamination.
   (b) *Tissue samples*: Extract total RNA from the tissue using the QIAzol/TRIzol® (Qiagen/Life Technologies) according to the manufacturer's instructions (at least ten times higher volume of reagent than tissue volume). After the ethanol precipitation step further clean up the RNA using the miRNAeasy or the RNeasy mini kit (Qiagen). You must perform the recommended on-column DNase treatment step.
   (c) *Whole blood samples*: First remove red blood cells using a density gradient centrifugation method (i.e., Lymphoprep®,

Table 3
Recommended qPCR Master Mix for each type and brand of qPCR apparatus

Format plate	Real-time qPCR instruments	Plate	Master mix
A	ABI: 5700, 7000, 7300, 7500, 7700, 7900HT, ViiA™ 7 (96-well block); Bio-Rad: iCycler®, iQ5, MyiQ, MyiQ2, Chromo4 (MJ Research); Eppendorf: MasterCycler® ep RealPlex® 2, 2 s, 4, 4S; Stratagene: Mx3005p®, Mx3000p®; Takara: TP-800	96-Well	RT^2 SYBR Green/ROX qPCR Master Mix
	Bio-Rad: iCycler®, iQ5, MyiQ, MyiQ2	96-Well	RT^2 SYBR Green/ Fluorescein qPCR Master Mix
C	ABI: 7500 Fast block, 7900HT Fast block, StepOnePlus™; ViiA™ 7 Fast block	96-Well	RT^2 SYBR Green/ROX qPCR Master Mix
D	Bio-Rad: CFX96™, Opticon® and Opticon 2 (MJ Research); Stratagene: Mx4000®	96-Well	RT^2 SYBR Green qPCR Master Mix
E	ABI: 7900HT (384-well block), ViiA™ 7 (384-well block); Bio-Rad: CFX384™	384-Well	RT^2 SYBR Green/ROX qPCR Master Mix
F	Roche: LightCycler 480 96-well block	96-Well	RT^2 SYBR Green qPCR Master Mix[a]
G	Roche: LightCycler 480 384-well block	384-Well	
H	Fluidigm BioMark	96×96 Chip	
R	QIAGEN Rotor-Gene Q	100-Well Rotor-Disc 100	

[a]Specifically designed for instrumentation that does not require a reference dye

Table 4
Instrument compatibility for TaqMan® Gene Expression and Signature arrays

Instrument/ product format	StepOne® Plus	StepOne®	7500	7500Fast	7900	ViiA7	QuantStudio
96-Well plate standard	No	No	96-Well block	No	96-Well block	96-Well block	96-Well block
96-Well plate fast	96-Well Fast block	No	No	96-Well Fast block	96-Well Fast block	96-Well Fast block	96-Well Fast block
384-Well plate	No	No	No	No	384-Well block	384-Well block	384-Well block
Microfluidic card	No	No	No	No	TaqMan® Array Card block	TaqMan® Array Card block	TaqMan® Array Card block
OpenArrays	No	No	No	No	No	No	OpenArray® block

Greiner Bio-One). Then extract RNA from the white cell fraction RNeasy mini kit (Qiagen), and perform the recommended on-column DNase treatment step. Alternatively, the PAXgene™ Blood RNA Kit can also be used to prepare total RNA from whole blood samples.

7. To obtain reliable results from the PCR arrays, all RNA samples should have high quality according to the following criteria:
   (a) *RNA concentration and purity by UV spectrophotometry*: Absorbance (A) 260:230 ratio should be greater than 1.7 (purity from organic compounds), A260:280 at least 1.8 (purity from proteins), and concentration by A260 greater than 40 μg/ml total RNA. The readings are affected by pH; therefore, be sure to perform the dilutions for spectrophotometry in RNase-free Tris–HCl pH 8.0 buffer.
   (b) *Ribosomal RNA band integrity*: A fraction of each RNA sample should be analyzed on a denaturing agarose gel or on a BioAnalyzer using an RNA 6000 Nano LabChip (Agilent). You should observe sharp bands or peaks without smearing or shoulder (signs of RNA degradation) corresponding to both the 18S and 28S ribosomal RNA. An RNA integrity number (RIN) of 7 or higher obtained using the BioAnalyzer is recommended.
8. It is essential to wear gloves throughout the whole procedure.
9. The use of RT^2 First Strand kit (Qiagen) is critical for obtaining the best results with the SYBR® Green RT^2 Profile PCR array and for detecting the RTC included in each PCR array plate (*see* **Note 17** and Subheading 3). For TaqMan® Gene Expression qPCR Arrays, the manufacturer suggests its reverse transcription kits. However, it has not been stated if kits from other manufacturers would be suitable for reverse transcription of total RNA.
10. Keep SYBR® Green RT^2 Profiler PCR array plates and system components at –20 °C for long-term storage. TaqMan® Gene Expression qPCR Arrays can be stored at room temperature until the due date determined by the manufacturer. Plates should be stored in their original box, in a clean, dry place, preferably protected from direct sunlight.
11. For a complete list of inventoried pathway- or disease-focused SYBR® Green RT^2 Profiler PCR arrays, visit the following website: http://www.sabiosciences.com/PCRArrayPlate.php. Alternatively, custom-made SYBR® Green RT^2 Profiler PCR array plates allow you to choose the genes to study and the number of sample replicates that may be needed (http://www.sabiosciences.com/custompcrplate.php). All information on TaqMan® Gene Expression Signature Plates and Arrays is listed

at http://www.invitrogen.com/site/us/en/home/Products-and-Services/Applications/PCR/real-time-pcr/real-time-pcr-assays/taqman-gene-expression/taqman-expression-arrays.html#support. For specific products, click on the manual on the left of the page. You will then find information on inventoried as well as custom-made plates and cards.

12. The 2× RT² SYBR® Green/ROX qPCR Master Mix kit was specifically designed for the following qPCR apparatus: *ABI* 5700, 7000, 7300, 7500 (Standard and FAST), 7700, 7900HT 96-well block (Standard and FAST) and 384-well block, and StepOnePlus; *Eppendorf* Mastercycler ep realplex 2/2S/4/4S; *Stratagene* Mx3000p, Mx3005p, and Mx4000; and *TaKaRa* TP-800. For the *Bio-Rad* iCycler, iQ5, MyiQ, and MyiQ2, use the RT² SYBR® Green/Fluorescein qPCR Master Mix.

13. We strongly recommend the RT² RNA QC PCR Array to test the quality of your RNA samples before proceeding with the PCR array experiment. The RT² RNA QC PCR Array is designed to assess the quality of human, mouse, or rat RNA samples before characterization with the RT² PCR Array. It contains a number of PCR controls that test for RNA integrity, inhibitors of reverse transcription and PCR amplification, and genomic and general DNA contamination. Failure of any of these controls would otherwise confound SYBR® Green-based real-time PCR results by causing false-negative or false-positive results.

14. We previously determined that the best endogenous controls for our experimental design are *18S*, *PPIA*, and *UBC* genes using a Human TaqMan® Endogenous Control Array. This array is a 384-well microfluidic card containing 16 human TaqMan® Gene Expression Assays for housekeeping genes commonly used as endogenous controls that generally exhibit minimal differential expression across different tissues.

15. If it is possible, physically separate the workspaces used for pre-PCR setup and post-PCR processing or non-PCR operations.

16. Decontaminate your PCR workspace and labware (bench, pipette barrels, tube racks, etc.) with UV light before each new use to inactivate any contaminating DNA. Alternatively, 10 % bleach can be used to chemically inactivate and degrade any DNA.

17. Close all tubes containing PCR products once you are finished adding or removing volumes. Do not leave labware (tubes and tip boxes) exposed to the air for long periods of time.

18. Do not remove the PCR array plate from its protective sealed bag until immediately ready to use.

19. Do not open any previously run and stored PCR array plate because it might release PCR product DNA into the air contaminating and confounding the results of future real-time PCR experiments.

20. We strongly recommend working with at least *three* biological replicates in order to obtain *p*-values and determine if the observed differences are significant during the final analysis of the results. Examples of biological replicates are three different batches of the same type of cells under identical treatment or three sample tissues from three different animals with the same treatment.
21. The optimal amount of total RNA depends on the relative abundance of the transcripts of interest. Lower abundance transcripts require more RNA; higher abundance transcripts require less RNA. Greater amounts of input total RNA yield a greater number of positive calls (genes expressed in the linear dynamic range of the method). Lower amounts of input total RNA not only yield a smaller number of positive calls but also increase false-negative calls. For RT^2 SYBR® Green Profiler arrays, the use of the RT^2 First Strand kit maximizes the number of positive calls at low amounts (25 ng) of total RNA over other commercial sources of reverse transcriptase and first-strand synthesis kits.
22. The RT^2 Nano PreAMP kit (Qiagen) allows for gene expression analysis from as little as 1 ng of total RNA. In case of SuperScript VILO RT reaction, the manufacturer does not specify a lower limit for RNA input. However, I would not recommend using less than 200 ng of RNA per reaction, as it may impair the detection of genes that are normally of low expression.
23. First-time users start with at least 1.0 μg of total RNA for 96-well plate format, 0.8 μg of total RNA for 100-well Rotor-Disc format, 400 ng of total RNA for 384-well (4×96) plate format, and 1.0 μg of total RNA for 384-well HT plate format PCR array.
24. To characterize and compare results between samples, it is essential to use the same amount of total RNA for all samples in a single experiment.
25. The volume of reagents used to prepare the experimental cocktail depends on the plate format (96- or 384-well PCR array) and plate format designation (A, C, D, E, F, and G). If you are not using 384-well PCR arrays, format E or G, *see* Table 5 to determine how to prepare the experimental cocktail for your experiment using the RT2 Profiler PCR array. For TaqMan® PCR Arrays, *see* Table 6.
26. Certify that all of your micropipettes are calibrated before beginning this procedure because the accuracy and precision of your pipetting determine the consistency of your results.
27. Make sure to avoid introducing air bubbles into the wells of the PCR array.

Table 5
Preparation of the experimental cocktail in a 5 ml tube according to plate format and designation

Plate format:	96-Well	384-Well (4×96)	384HT
Plate format designation:	A, C, D, and F	E and G	E and G
2× SABiosciences RT2 qPCR Master Mix	1,350 µl	550 µl	2,000 µl
Diluted first-strand cDNA synthesis reaction	102 µl	102 µl	102 µl
Nuclease-free water	1,248 µl	448 µl	1,898 µl
Total volume	2,700 µl	1,100 µl	4,000 µl

Table 6
Experimental cocktail preparation according to plate format

		Volume per well (µL)[a]							
Plates	**Components**	**1**	**8[b]**	**16**	**32**	**48**	**96**	**192**	**384**
TaqMan® 96-well fast plates	cDNA + nuclease-free water[c]	5	45	90	180	270	540	1,080	2,160
TaqMan® 384-well plates	TaqMan® Master Mix[d]	5	45	90	180	270	540	1,080	2,160
10 µl reactions	*Total volume*	10	90	180	360	540	1,080	2,160	4,320
TaqMan® 96-well Standard plates	cDNA + nuclease-free water[e]	10	90	180	360	540	1,080	N/A	N/A
20 µl reactions	TaqMan® Master Mix[f]	10	90	180	360	540	1,080	N/A	N/A
	Total volume	20	180	360	540	1,080	2,160	N/A	N/A

[a]Reactions with a 12.5 % excess in volume
[b]Number of reactions
[c]cDNA amount between 5 and 50 ng for a 10 µl reaction
[d]TaqMan® Fast Universal Master Mix (2×) or TaqMan® Gene Expression Master Mix
[e]cDNA amount between 1 and 100 ng for a 20 µl reaction
[f]TaqMan® Gene Expression Master Mix or TaqMan® Universal Master Mix

28. Carefully pipette liquids from reagent tubes, starting with the pipette tip at the top of the tube and working down slowly.
29. Make sure that you use the correct SYBR® Green Master Mix for the instrumentation in your laboratory (*see* Table 3) because each instrument uses a different reference dye to normalize its optics. As well, make sure to choose the right chemistry and block for your TaqMan® Array experiment (*see* Table 7).
30. The use of SYBR® Green RT^2 qPCR Master Mixes (Qiagen) is critical for obtaining the most accurate results from the PCR array. The chemically modified and tightly controlled HotStart enzyme in the RT^2 qPCR Master Mixes provides more accurate

Table 7
Plates and chemistry for TaqMan® Array plates

Component	96-Well standard	96-Well fast[a]	384-Well
cDNA	1–100 ng	5–50 ng	5–50 ng
Final volume	20 μl	10 μl	10 μl
TaqMan® Master Mix	TaqMan® Universal Master Mix TaqMan® Gene Expression Master Mix (2×)	TaqMan® Fast Universal Master Mix TaqMan® Gene Expression Master Mix (2×)	TaqMan® Universal Master Mix TaqMan® Gene Expression Master Mix (2×)
Cycling block	96-Well block	96-Well Fast block	384-Well block

[a]TaqMan® Fast plates can be run under standard conditions. For such, use TaqMan® Universal Master Mix or TaqMan® Gene Expression Master Mix

SYBR® Green results by preventing the amplification of primer dimers and other nonspecific products. It also helps ensure high amplification efficiencies even for those genes that are the most difficult to amplify.

31. Each 384-well RT² Profiler PCR Array kit comes with four 384EZLoad Covers in four different colors to facilitate the loading of four samples (white for sample 1, yellow for sample 2, black for sample 3, and red for sample 4). Each 384EZLoad Cover is placed on top of the 384-well plate and allows the loading of only one sample while covers the wells for the other three samples.
32. For PCR array formats A and D (*see* Table 3), seal the plate with the optical thin-wall 8-cap strips.
33. Bubbles remaining in the bottom of the wells of a PCR array plate will interfere with perfect heat exchange between PCR block and the plate generating poor PCR reaction. Therefore, inspect the plate from underneath of the plate to ensure that no bubbles are present in each well.
34. PCR array plates containing experimental cocktail and ready to be loaded in the qPCR apparatus may be stored at −20 °C wrapped in aluminum foil for up to 1 week until ready to run.
35. The PCR cycling program depends on the qPCR instrument and on the chemistry being used (*see* Table 8 for SYBR® Green PCR arrays and Table 9 for TaqMan® PCR arrays).
36. The 10-min step at 95 °C is required to activate the HotStart DNA polymerase.
37. Alternatively, a PCR protocol template file (RT² *Profiler*™ PCR Array Protocol ABI7900.sdt) can be downloaded from the SABiosciences' website (http://www.SABiosciences.com).

Table 8
qPCR cycling program recommended for each type and brand of qPCR apparatus

qPCR apparatus	Cycles	Duration	Temperatures
ABI 5700, 7000, 7300, 7500, 7700, 7900HT, step one plus, ViiA™ 7; Bio-Rad: iCycler, iQ5, MyiQ, MyiQ2, CFX96[a], CFX384[a]; Eppendorf[b]: MasterCycler ep RealPlex 2, 2S, 4, 4S; Stratagene: Mx3005p, Mx3000p, MX4000p	1 40	10 min 15 s 1 min	95 °C 95 °C 60 °C
Roche LightCycler 480[a]	1 45	10 min 15 s 1 min	95 °C 95 °C 60 °C
Bio-Rad: Opticon, Opticon 2, Chromo 4 (MJ Research); Takara: TP-800; all other instruments	1 40	10 min 15 s 30–40 s[c] 30 s	95 °C 95 °C 55 °C 72 °C

[a]Adjust the ramp rate to 1 °C/s
[b]For the Silver Thermoblock, adjust the ramp rate to 26 %; for the Aluminum Thermoblock, adjust the ramp rate to 35 %
[c]Different instruments need different lengths of time to detect the fluorescent signal. Choose the appropriate time for 55 °C annealing in your qPCR apparatus

Table 9
Cycling protocol for Fast and Standard TaqMan® Array plates

Fast plate cycling conditions on a 7900HT fast system				Standard cycling conditions on a 7900HT system			
Hold[a]	Hold[b]	PCR (40 cycles) Melt	Anneal/extend	Hold[a]	Hold[b]	PCR (40 cycles) Melt	Anneal/extend
50 °C	95 °C	95 °C	60 °C	50 °C	95 °C	95 °C	60 °C
02:00	00:20	00:03	00:30	02:00	10:00	00:15	01:00

[a]AmpErase® UNG activation step. Omit this step when using Master Mix without AmpErase UNG
[b]AmpliTaq Gold® Enzyme activation step

38. If you decide not to obtain the dissociation curve immediately, save the plates wrapped in aluminum foil at −20 °C in case you need to perform it at a later point in time for troubleshooting purposes.
39. Ensure that the threshold values are the same across all RT^2 PCR array runs in the same analysis. The absolute position of the threshold is less important than its consistent position across arrays. If the RNA sample is of sufficient quality, the cycling program has been carried out correctly, and threshold values have been defined correctly, the value of Ct PPC should be 20 ± 2 for all arrays or samples.
40. If genomic DNA contamination occurs, fold changes in gene expression may still be obtained. However, it will then be very

important to validate any results for individual genes by a separate more rigorous real-time PCR analysis that includes a "minus RT" control (sample without reverse transcriptase).

41. Apparent genomic DNA contamination comes not only from your sample but also from other reagents, tips, and tubes. The no-template control (NTC) in the RT^2 RNA QC PCR Array provides a sense of how well your technique minimizes the introduction of general DNA contamination into your assay system.
42. Double-check the A260:A280 and A260:A230 ratios of your RNA samples, and make sure to perform the dilutions for spectrophotometry in RNase-free Tris–HCl pH 8.0 buffer. If necessary, re-purify your RNA samples with the RNeasy Mini Kit (Qiagen) or any other spin column-based cleanup method.
43. Different instruments have different levels of sensitivity; therefore, an average Ct PPC value of 20 ± 2 might be difficult to obtain for some instruments. However, higher Cts are still acceptable if the observed average Ct PPC value does not vary by more than two cycles between PCR arrays being compared.
44. Make sure to save the file in ".XLS" format, not ".XLSX"; otherwise it will not work.
45. The fold change (fold difference) is calculated by the equation $2^{(-\Delta\Delta CT)}$. For the fold regulation, the software transforms fold change values less than 1 (meaning that the gene is downregulated) by returning the negative inverse. For example, if LDLR has a fold change value of 0.31, this is equivalent LDR having a fold regulation of −3.2 fold. (LDLR expression is decreased 3.2-fold.) For fold changes higher than 1, the fold regulation and fold change are equal.
46. Please ensure that you have exported or copied any figures or tables from each session because the Data Analysis Web portal does not save any results.
47. The TaqMan® array protocols recommend 5–50 ng of cDNA per reaction. The amount of cDNA input is an important step to be considered before starting the experiment. A low cDNA input will mask low-expression genes and generate many false positives, whereas an excessive input could interfere with the calculations and generate false-positive data. Therefore, one should consider the nature of the genes to be analyzed and calculate the cDNA input accordingly.
48. The plate format used in this experiment allows triplicates in each plate because each gene can be amplified three times in three different wells (Fig. 2). For a reliable qPCR experiment, three biological replicates (three independent samples with same treatment) should be assayed in triplicate (three assays from the same sample or three "technical replicates"). Ideally, three biological replicates AND three technical

replicates should be used in each experiment. If this is not possible because of the cost of PCR arrays, then work only with biological replicates. In our experiment, the same sample was analyzed three times per plate (three technical replicates) and three biological replicates were used for both treatment and control (a total of six samples); therefore, we used six TaqMan® array plates. Life Technologies offers different plate formats with different number of genes and replicates for both inventoried and custom assays. Before starting your experiment, we strongly recommend that you check which plate format adapts the best to your experimental design and how many plates you are going to need.

49. All TaqMan® assays and arrays are shipped with a CD, which contains all the information about the assay. Take good care of this CD, and keep it at a safe place. You will need this CD for your run. The CD contains the *a*ssay *i*nformation *f*iles (AIF). To view the AIF as a spreadsheet in Microsoft Excel, load the TaqMan® Assay Plate Information CD into the CD drive, navigate to the drive that contains the Information CD, right-click *ProdNum_LotNum_AIF.txt* file, and then select *Open with Excel*. You will be able to see all the information on your assay. The CD also contains the plate layout files (*ProdNum_*Platemap.html and *ProdNum_*Platemap.csv), which show the position of the assays on the TaqMan® Array Plate. Each plate layout file contains two color-coded maps. The top map shows the gene symbol representing the assay in each well. The bottom map shows the TaqMan® Gene Expression Assay ID for each well. Finally, the setup file *ProdNum_*7900_SDS.txt contains information specific to your TaqMan® Array Plate, such as the assay IDs and well locations in the file you will import to create SDS plate documents/experiments or templates.

50. In the SDS RQ Manager software (upper right section of the screen), there are three different menu bar drop-down lists:
    (a) *Table* Orientation drop-down list allows you to view the information either in "Detector Centric," "Sample Centric," or "Plate Centric" in which the detector, the samples, or the plates associated with the study appear at the top left of the screen, respectively.
    (b) Calibrator drop-down list allows you to select the tissue sample to be used as the calibrator (usually, one of the negative controls or no treatment).
    (c) Data drop-down list allows you to select the data to display (Rn vs Cycle, ΔRn vs. Cycle, CT vs. Well Position).

51. The SDS software calculates baseline and threshold values for a detector based on the assumption that the data exhibit the "typical" amplification curve" (Fig. 11a). However, experimental error such as contamination and pipetting

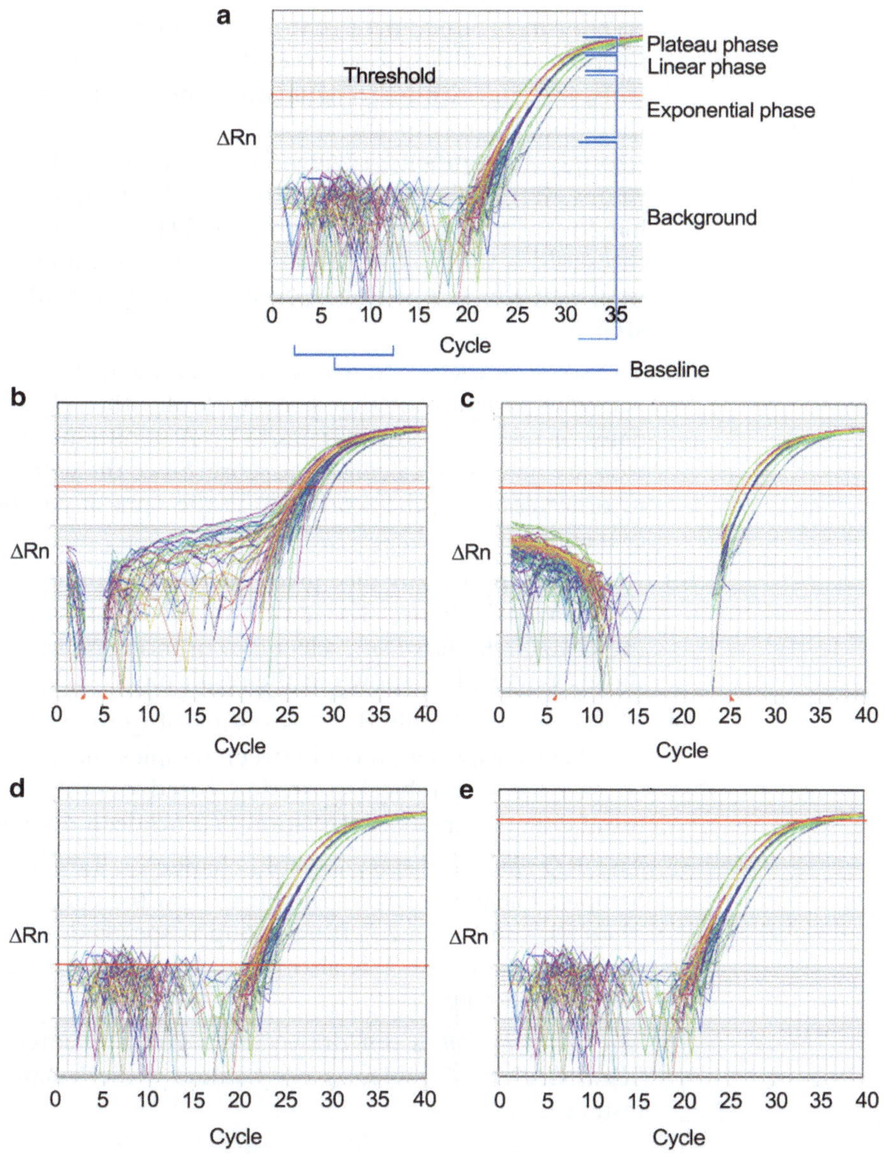

Fig. 11 Correct and incorrect setting of baselines and thresholds in the qPCR amplifications. (**a**) This is a typical amplification curve with baseline and threshold set correctly. The amplification curve begins after the maximum baseline, and the threshold is set in the exponential phase of the amplification curve; therefore, no adjustment is necessary. (**b**) Baseline set too low: The amplification curve begins too far to the right of the maximum baseline; adjust the baseline manually, and increase the end cycle value to two cycles before the amplification is detected. (**c**) Baseline set too high: The amplification curve begins before the maximum baseline; adjust the baseline manually, and decrease the end cycle value. (**d**) Threshold set too low, below the exponential phase of the amplification curve: The standard deviation is significantly higher than that for a plot where the threshold is set correctly. Drag the threshold bar up into the exponential phase of the curve. (**e**) Threshold set too high, above the exponential phase of the amplification curve: The standard deviation is significantly higher than that for a plot where the threshold is set correctly. Drag the threshold bar down into the exponential phase of the curve. Modified from Life Technologies Bulletin "Relative quantification using comparative Ct getting started guide"

errors can produce data that deviate significantly from data for typical amplification curves. Such atypical data cause the software algorithm to generate incorrect baseline and threshold values for the associated detector. Therefore, we strongly recommend reviewing all baseline and threshold values after analysis of the study data. If the automatic baseline looks like the one shown in Fig. 11b, c, or the automatic threshold is set too low or high (Fig. 11d, e), adjust the values manually as previously described in Subheading 3.2.3, **step 5**.

52. The score of each candidate or selected control is the average pair-wise variation of that gene with all other chosen candidates or selected control genes [11]. A minimum of two controls are needed to calculate the score. Since the score is relative to other controls, the score will be the same if you only have two controls.
53. Right clicking on any plot gives you the option to *Copy*, *Save as*, or *Print* the figure. To view only a subset of assays in any of the *Graphic Results Plots*, select two or more assays. Then only those assays will be shown in the result plots.
54. *Heat map* graphically displays results of unsupervised hierarchical clustering. Distances between samples and assays are calculated for hierarchical clustering based on the ΔCt values using either Pearson's correlation or Euclidean distance.
55. The *RQ plot* can display log fold change vs. Target or RQ vs. Sample. The graph types available to view the data are "linear," "Log10," and "Log2." If no group is specified, the standard deviation of the ΔCT is also plotted for each sample on the Log2 graph type.
56. The *Volcano plot* is not drawn if no group is entered in the "*Sample Design*" table, as no *p*-values are calculated in this instance.

References

1. Wong ML, Medrano JF (2005) Real-time PCR for mRNA quantitation. Biotechniques 39: 75–85
2. Higuchi R, Fockler C, Dollinger G et al (1993) Kinetic PCR analysis: real-time monitoring of DNA amplification reactions. Biotechnology (NY) 11:1026–1030
3. Heid CA, Stevens J, Livak KJ et al (1996) Real time quantitative PCR. Genome Res 6:986–994
4. Wang T, Brown MJ (1999) mRNA quantification by real time TaqMan polymerase chain reaction: validation and comparison with RNase protection. Anal Biochem 269:198–201
5. Malinen E, Kassinen A, Rinttila T et al (2003) Comparison of real-time PCR with SYBR Green I or 5′-nuclease assays and dot-blot hybridization with rDNA-targeted oligonucleotide probes in quantification of selected faecal bacteria. Microbiology 149:269–277
6. Schneeberger C, Speiser P, Kury F et al (1995) Quantitative detection of reverse transcriptase-PCR products by means of a novel and sensitive DNA stain. PCR Methods Appl 4:234–238
7. Cao H, Shockey JM (2012) Comparison of TaqMan® and SYBR Green qPCR methods

for quantitative gene expression in tung tree tissues. J Agric Food Chem 60:12296–12303

8. Holland PM, Abramson RD, Watson R et al (1991) Detection of specific polymerase chain reaction product by utilizing the 5′—3′ exonuclease activity of Thermus aquaticus DNA polymerase. Proc Natl Acad Sci U S A 88:7276–7280
9. Livak KJ, Schmittgen TD (2001) Analysis of relative gene expression data using real-time quantitative PCR and the 2–ΔΔCT method. Methods 25:402–408
10. Palmer S, Wiegand AP, Maldarelli F et al (2003) New real-time reverse transcriptase-initiated PCR assay with single-copy sensitivity for human immunodeficiency virus type 1 RNA in plasma. J Clin Microbiol 41: 4531–4536
11. Vandesompele J, De Preter K, Pattyn F et al (2002) Accurate normalization of real-time quantitative RT-PCR data by geometric averaging of multiple internal control genes. Genome Biol 3:research0034

Chapter 28

Comprehensive Meta-analysis of MicroRNA Expression Using a Robust Rank Aggregation Approach

Urmo Võsa, Raivo Kolde, Jaak Vilo, Andres Metspalu, and Tarmo Annilo

Abstract

Differential microRNA (miRNA) expression profiling by high-throughput methods has generated a vast amount of information about the complex role of these small regulatory molecules in a broad spectrum of human diseases. However, the results of such studies are often inconsistent, mostly due to the lack of cross-platform standardization, ongoing discovery of novel miRNAs, and small sample size. Therefore, a critical and systematic analysis of all available information is essential for successful identification of the most relevant miRNAs. Meta-analysis approach allows integrating the results from several independent studies in order to achieve greater statistical power and estimate the variability between the studies. Here we describe as an example the use of a robust rank aggregation (RRA) method for identification of miRNA meta-signature in lung cancer. This method analyzes prioritized gene lists and finds commonly overlapping genes, which are ranked consistently better than expected by chance. An RRA approach not only helps to prioritize the putative targets for further experimental studies but also highlights the challenges related with the development of miRNA-based tests and emphasizes the need for rigorous evaluation of the results before proceeding to clinical trials.

Key words miRNA expression profiling, Meta-analysis, Gene set enrichment analysis, Lung cancer

1 Introduction

Latest research on the field of microRNA (miRNA) biology has uncovered a startling complexity of posttranscriptional regulation at the RNA level. The expression profiling of the miRNome (entire miRNA repertoire) by high-throughput methods is becoming a routine procedure in many laboratories around the world. Until now, the most widely used methods have been microarrays, next-generation sequencing, and quantitative PCR. These techniques have revealed an aberrant regulation of a large number of miRNAs in a broad spectrum of human diseases [1]. Therefore, miRNAs have emerged as promising biomarkers for early disease detection and accurate prognosis as well as targets for more efficient treatment.

M. Lucrecia Alvarez and Mahtab Nourbakhsh (eds.), *RNA Mapping: Methods and Protocols*, Methods in Molecular Biology, vol. 1182, DOI 10.1007/978-1-4939-1062-5_28, © Springer Science+Business Media New York 2014

Unfortunately, miRNA profiling efforts have often led to inconsistent results between the studies. This is due to many factors including application of different technological platforms [2], small sample size, inconsistent annotation, ongoing discovery of novel miRNAs [3, 4], and use of different methods for data processing and analysis [5].

Here we present a meta-analysis approach to critically converge and dissect heterogeneous miRNA expression profiling datasets using lung cancer as an example. We discuss how to identify eligible studies, extract or download lists of aberrantly regulated miRNAs, standardize miRNA names across the studies, and perform actual analysis. Although the preferred method for gene expression meta-analysis involves integration of the raw expression datasets, such rigorous approach is often not possible due to the lack of cross-platform standardization of miRNA profiling technologies or the unavailability of raw data. A reasonable compromise is to aggregate published lists of aberrantly regulated miRNAs. Here we apply a recently published robust rank aggregation (RRA) method [6, 7] followed by pathway analysis. The RRA approach has been specifically designed for comparison of several ranked gene lists and identification of commonly overlapping genes (*see* **Note 1**). This method is a suitable and effective solution for identification of statistically significant miRNA meta-signature and is particularly useful when input experiments are performed by different technological platforms cover different sets of genes and full rankings of miRNAs are not available.

2 Materials

The following programs are needed to perform the analysis:

1. Statistics software R (*see* **Note 2**), freely available from the R project server at http://www.r-project.org/.
2. RStudio (http://www.rstudio.com/), a free and open-source interface, which provides more user-friendly access to different features of R. RStudio requires R 2.11.1 (or higher) and is available for Windows, Mac, and Linux.
3. For RRA analysis, RobustRankAggreg package (currently v. 1.1) is needed, freely available in the CRAN website (http://cran.r-project.org/web/packages/RobustRankAggreg/index.html).
4. To extract miRNA lists from published profiling studies, Internet connection from the academic network is needed as well as access to Scopus (http://www.scopus.com), PubMed (http://www.ncbi.nlm.nih.gov/pubmed), and Web of Knowledge (http://apps.webofknowledge.com/) databases.
5. For data formatting, use a simple text editor or a spreadsheet program such as Microsoft (MS) Excel or its free alternative, OpenOffice (OO) Calc (http://www.openoffice.org/).

3 Methods

3.1 Defining the Criteria for Study Inclusion

It is important to clearly formulate research questions and define criteria for the studies enrolled in the analysis. As an example, we will use a selection of studies that investigate the miRNA expression changes in lung cancer versus normal lung tissue. A simple PubMed search can identify hundreds of publications investigating miRNA expression changes in lung cancer. Therefore, precise and careful selection criteria have to be defined to include:

1. Original experimental articles published in English.
2. Studies comparing human lung cancer tissue versus adjacent non-cancerous lung tissue. A number of studies conducted on cell lines as well as the publications comparing different histological subtypes of lung cancer are excluded to make the study more focused and input data comparable.
3. Studies that use a hypothesis-free whole miRNome profiling approach such as microarray-based methods (regardless of the array type), qRT-PCR methods designed for parallel quantification of large number of miRNAs (96- or 384-well format), or second-generation sequencing of small RNAs. Studies that measured only a few preselected individual miRNAs or a set of preselected miRNAs are excluded.

3.2 Retrieving Publications and miRNA Expression Data from Databases

Searches in public databases such as PubMed, Web of Knowledge, or Scopus should be conducted in organized and systematic way to ensure the reproducibility of the analysis (*see* **Note 3**). Citation databases such as Web of Knowledge or Google Scholar (http://scholar.google.com/) are especially useful since they allow searches for backward (publications cited as references in the study) and forward (later works that cite the publication of interest) citations, thereby providing a broader view of the publications in the particular field.

If a comprehensive search term for publication database is used, it should be reported in the resulting publication including the date and settings of the search.

To identify studies that have investigated miRNA expression changes in lung cancer, we used the search term ("mirna*" OR "microrna*" OR "mir-*") AND "profil*" AND "lung" AND ("cancer*" OR "tumor*" OR "tumour*") from article titles, abstracts, and keywords in the Scopus database. Same search was carried out in PubMed database using "all fields" setting.

Search results should be evaluated critically, particularly the study design and the methodology used. Only datasets produced using a study design and methods in accordance with the criteria for a given meta-analysis should be included for further study.

3.3 Extraction of Ranked miRNA Lists

The results of miRNA differential expression experiments are frequently reported as a list of up- or downregulated miRNAs along with some measure of significance such as a *p*-value from statistical test and/or a logarithmic fold change, which shows the expression difference between disease and control groups. The published miRNA list is usually incomplete because it includes only a fraction of the actually tested miRNAs such as those that are most statistically significantly dysregulated or show the greatest fold change between cancer and normal tissue. Sometimes the more complete and therefore more useful miRNA lists are reported in the Supplementary Information section. Alternatively, a corresponding author of the publication or the dataset should be contacted in case that the complete list of dysregulated miRNAs is not available.

Our approach requires separate input lists for significantly up- and downregulated miRNAs, which are ranked by the magnitude of dysregulation. For example, miRNAs can be ranked according to *p*-values of differential expression. However, if the *p*-values are not reported or unusable (for example if multiple *p*-values were rounded $p<0.01$), miRNAs can be ranked by logarithmic fold changes ($\log_2$FC). Since the original datasets may have different formats, data extraction often involves laborious manual work.

3.4 Standardization of miRNA Names

For a comprehensive meta-analysis of miRNA expression, it is essential that the miRNA names are comparable across the studies and follow the same nomenclature. Because of the relative novelty of the miRNA profiling field and frequent updates in the miRBase [8], miRNA nomenclature can vary depending on when the study was conducted. The current miRBase release (v.19) has completely abandoned the older nomenclature which was using star sequences (miR*) to designate a minor product from the passenger strand. Instead, a -3p and -5p nomenclature to indicate whether a mature miRNA sequence is derived from 3′- or 5′-arm of the hairpin has been applied. On the other hand, miRBase accession numbers assigned to each stem–loop (e.g., MI0000285 for hsa-mir-205 stem–loop) and mature sequence (e.g., MIMAT0000266 for hsa-miR-205-5p with previous ID hsa-miR-205) have not been changed and thus allow tracking of annotation upgrades. So if the miRBase version is known for each technological platform and study, accession numbers can be used to convert miRNA names according to the most recent version of miRBase (*see* **Note 4**). For example, if the study uses array design based on miRBase v.14, then the miRNA name hsa-miR-205 should be converted accordingly:

miRBase v.14	miRBase accession	miRBase v.19
hsa-miR-205	MI0000285	hsa-miR-205-5p

Table 1 An example of ranked miRNA lists

Study_1	Study_2	Study_3	Study_4	Study_5
hsa-miR-9-5p	hsa-miR-205-5p	hsa-let-7a-5p	hsa-miR-21-5p	hsa-miR-144-5p
hsa-miR-205-5p	hsa-miR-155	hsa-miR-182-5p	hsa-miR-210	hsa-miR-21-5p
hsa-miR-155	hsa-miR-21-5p		hsa-miR-182-5p	hsa-miR-185-3p
hsa-miR-210			hsa-miR-31-5p	hsa-miR-182-5p
hsa-miR-9-3p			hsa-miR-9-5p	
hsa-miR-21-5p				

If a miRNA present in earlier miRBase releases is removed due to reannotation, we should omit that particular miRNA from subsequent analyses. This includes miRNAs that either cannot be mapped to the reference genome or are actually fragments of other RNAs. Several earlier studies detected not only mature miRNAs but also precursor sequences. Therefore, it is necessary to take a closer look at the array design to determine if the miRNAs reported in the lists are only mature miRNAs or the precursor miRNAs are also included. Depending on whether precursor miRNA expression is also of interest, it is then possible to rename precursors accordingly (e.g., pre-hsa-let-7a-1) or remove them from subsequent analysis.

3.5 Step-by-Step miRNA Expression Meta-analysis Using Robust Rank Aggregation Method

1. Install R and RStudio using standard installation procedures.
2. Create an analysis folder in your hard drive to store the input and output files as well as the R analysis script.
3. Create the suitably formatted analysis input file. It must be in the following format: ranked miRNA lists organized into the columns (more dysregulated miRNAs in the top) and study names or identificators in the columns. It is important to note that analyses for up- and downregulated miRNA lists are conducted separately; therefore, the input file must consist of only the lists of up- or downregulated miRNAs. In our example, we use tab-delimited sample file (*see* Table 1). The dataset consists of the ranked lists of significantly up-regulated miRNAs from five hypothetical studies (studies 1–5). This kind of table can be constructed in Excel or OpenOffice Calc and saved to specific format by selecting "File," "Save As," and then "Text" (tab delimited) in Excel menu. In this example we name the file *up_regulated.txt* and place it in the previously created analysis folder.

4. Open the RStudio R interface by double clicking on RStudio icon, and install the RobustRankAggreg package. The simplest way is to go to "Tools" and then to "Install packages" in the upper toolbar of RStudio interface. In the dialog window, leave the first drop-down menu unchanged (installation from CRAN R package repository) and write "RobustRankAggreg" to the second window. Leave other settings unchanged, and press the "Install" button. The package containing functions for RRA analysis is downloaded from CRAN website and installed into your system.
5. Create a new analysis script by going to "File," "New," and "R Script" on the upper toolbar, or just press Ctrl + Shift + N. A new script editor window opens to the left upper panel, where the command-line code for analysis is written and edited (Fig. 1). To run the written code, simply highlight the part of the code and press "Ctrl" + "Enter" or the "Run" icon in the script editor window. The code is pushed in a console window right below the editor window; the output of the script is also displayed there. Save the analysis script into your working directory for later use and editing by going to "File" and "Save As." The RStudio interface is depicted in Fig. 1.
6. Load the previously installed RobustRankAggreg package to use the functions defined by it. This is accomplished by running the command:

```
library(RobustRankAggreg)
```

7. Define the working directory for the analysis. For example:

```
setwd("C:/Your_Working_Directory/")
```

where "C:/Your_Working_Directory" must be replaced by the actual path to your previously defined analysis folder.

8. Read in the data used in the analysis. Data can be in different formats (tab delimited, comma separated, etc.), and one can use the R manuals to get a comprehensive overview of the possibilities of data import into the R. In the RStudio upper toolbar, use "Help" and "R Help." The help page now appears in the lower left window, and the relevant manuals are "An introduction to R" (paragraph 7) and "R Data Import/Export."

To read the data from previously constructed example file *up_regulated.txt* into R, use the command

```
meta<- read.delim("up_regulated.txt", na="")
```

This command reads in the aforementioned file and assigns it to the data frame named *meta*. The argument *na* = "" defines that empty cells in the table are recognized as "not available" in the data frame *meta*.

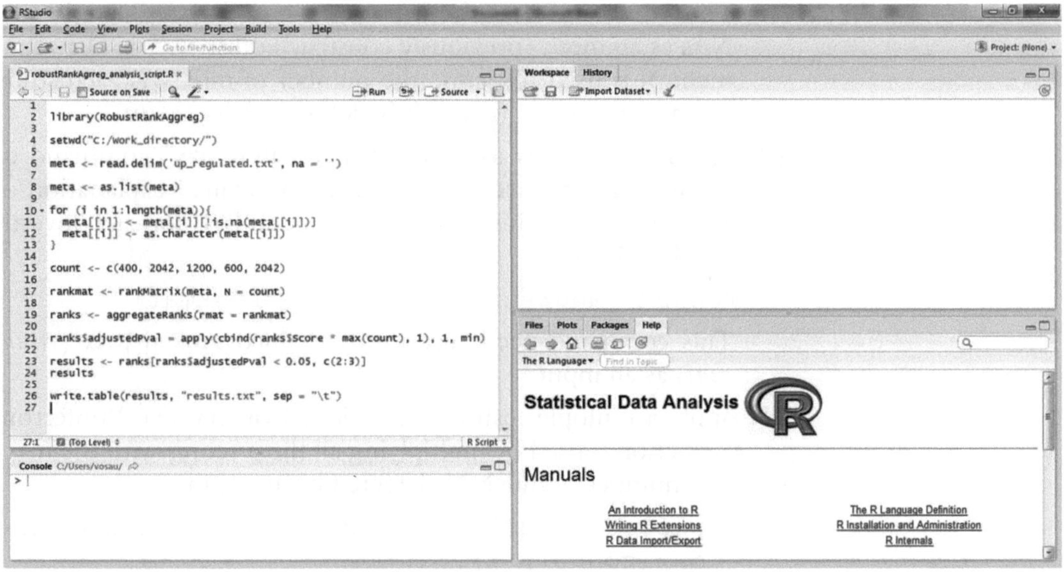

Fig. 1 Interface of RStudio v. 0.97.551. The *upper left panel* is a script editor window with the example analysis script. The *lower left panel* is an RStudio console window where the script is executed and the output displayed. The *lower right panel* shows the help page displayed after selecting "Help" and then "R Help" in the upper toolbar

9. Convert the data to a suitable format. RobustRankAggreg package needs the input to be in the format of list, which is different from data frame. To accomplish this, we convert the data frame to the list format using the command

   ```
   meta<- as.list(meta)
   ```

   To get rid of the "not available" positions in the list, we use custom helper function:

   ```
   for (i in 1: length(meta)){meta[[i]]<-meta[[i]]
   [!is.na(meta[[i]])]
   meta[[i]]<- as.character(meta[[i]])}
   ```

10. Add information about the number of miRNAs profiled in each study. We define the vector *count*, which stores the data about how many miRNAs each study was able to detect:

    ```
    count<- c(400, 2042, 1200, 600, 2042)
    ```

    In our example, study 1 used an array which was able to detect 400 different miRNAs, whereas study 2 and study 5 used small RNA sequencing which is able to detect all 2,042 currently known miRNAs (*see* **Note 5**).

11. Construct the matrix of normalized ranks using rankMatrix command:

    ```
    rankmat<- rankMatrix(meta, N=count)
    ```

This command uses the list consisting of dysregulated miRNAs as an input (previously constructed list *meta*). The argument N is used to define the number of miRNAs each study was able to profile (previously defined vector *count*).

12. Run RRA analysis using the "aggregateRanks" command to obtain a significance score (ρ-score; a detailed explanation of how to calculate this score is provided in [6]) for each investigated miRNA:

```
ranks<- aggregateRanks(rmat=rankmat)
```

This command needs previously defined rank matrix (*rankmat*) as an input.

13. Apply a multiple testing correction. Conservative Bonferroni correction is used by multiplying all the ρ-scores with the maximal number of miRNAs detected by any study:

```
ranks$adjustedPval<- apply(cbind(ranks$Score
* max(count), 1), 1, min)
```

Here the new column (adjustedPval) is created to the results table by multiplying all the ρ-scores with maximal element of "count." In this example, studies 2 and 5 using small RNA sequencing were able to profile all the 2,042 miRNAs known by today, so we multiply all the ρ-scores by 2,042.

14. Filter the results by significance. We use the command

```
results<- ranks[ranks$adjustedPval<0.05,]
```

This command allows you to construct the new data frame *results*, which consists of data about only those miRNAs with adjusted *p*-value < 0.05. Results can be displayed by running the command *results*, which displays the outcome analysis shown in Table 2.

After multiple testing corrections, only two miRNAs (hsa-miR-21-5p and hsa-miR-182-5p) are significantly up-regulated according to the results of the meta-analysis in our limited example dataset with the adjusted *p*-values 0.0005 and 0.0127.

15. Export the results table out of R with the following command:

```
write.table(results, "results.txt", sep="\t")
```

This command creates a tab-delimited results file into the analysis folder.

3.6 Enrichment Analysis of miRNA Targets

Since miRNAs regulate thousands of human genes, their aberrant expression can have a profound impact on cellular metabolism and biological processes. There are several web tools and databases available for the analysis of gene ontology (GO) and pathway enrichment for a given list of miRNAs (a detailed description of these tools is provided in the following sections). Enrichment anal-

Table 2 Results of the robust rank aggregation analysis using the list presented in Table 1 as an input

Name	Score	adjustedPval
hsa-miR-21-5p	2.309032e-07	0.0004715043
hsa-miR-182-5p	6.203219e-06	0.0126669727

ysis tests the hypothesis that a certain pathway or biological process appears in the list of specified genes more often than expected by chance using random gene list. Therefore, if a p-value for a particular pathway or GO term generated by enrichment analysis is below a threshold of significance (usually $p < 0.05$), the given pathway is considered enriched in the query gene list.

GO (http://www.geneontology.org/) provides a standardized description of a gene product in terms of its molecular function, subcellular location, and biological process [9]. Since most of these tools provide their own detailed user guides, here we present only a short overview of the most useful websites and databases.

3.6.1 Web Tools Specifically Designed for the Pathway Analyses of miRNA Targets

The tools that are designed for the direct analysis of miRNA-targeted pathways rely on different precalculated target prediction algorithms and/or databases of experimentally predicted targets. One of these tools is DIANA-miRPath (*see* **Note 6**) hosted by DIANA-Lab [10]. The targets of the input miRNAs are automatically identified from Tarbase validated target database or predicted using microT-CDS (v. 5.0) target prediction algorithm. KEGG pathways (Kyoto Encyclopedia of Genes and Genomes, http://www.genome.jp/kegg/pathway.html) significantly enriched by miRNA targets are identified and visualized. In addition, it is possible to analyze the combined effect of several dysregulated miRNAs and identify genes or KEGG pathways, which are targeted by all input miRNAs pinpointing the most interesting associations. The main advantage of this web tool is the ease of usage. There is no need for separate target identification or prediction; the necessary input consists only of miRNA names. The interactive visualizations of KEGG pathways with miRNA target sites are informative and easily interpretable.

Other tools that allow identification of pathways targeted by a set of specific miRNAs either directly or in combination with other tools include miTalos [11], starBase [12], mirTar [13], and GeneTrail [14] (*see* **Note 6**). The main limitation of these dedicated web tools is that usually enrichment analyses are supported in only one database. For example, only the enrichment of KEGG pathways can be analyzed with DIANA-miRPath. In addition, it is possible to use only the implemented target prediction algorithms such as microT-CDS in the case of DIANA-miRPath. Finally, it is usually not possible to use a custom set of miRNA sequences for target prediction. Therefore, most recently discovered miRNAs or entirely new miRNA candidate sequences cannot be used for this kind of analysis.

3.6.2 Target Prediction Followed by Enrichment Analysis

To perform more advanced and flexible miRNA-targeted pathway analysis, target prediction and overrepresentation analyses could be performed separately. Many miRNA prediction algorithms using different prediction strategies are available, and some of the most widely used are TargetScan [15], miRanda [16], PITA [17], PicTar [18], DIANA microT [19], and RNAHybrid [20] (*see* **Note 7** for server websites hosting these algorithms). In addition, a small number of databases maintain lists of experimentally verified miRNA targets such as miRTarBase [13], miRWalk [21], TarBase [22], and starBase [12] (*see* **Note 8** for websites of these databases). However, due to the limited number of validated targets available, this approach may prove to be less sensitive than using a computational target prediction.

The main advantage of performing target prediction and gene enrichment separately is the possibility of using the intersection of targets predicted by different algorithms to identify a more conservative set of potential targets, therefore adding confidence to the analyses. In the case of downloadable algorithms, it is possible to define the set of miRNAs using FASTA files downloaded from the miRBase database. In this way, it is possible to use the latest miRBase version or entirely new miRNA sequences for target prediction.

Enrichment analysis of predicted miRNA targets can be performed using any of the functional profiling tools such as DAVID [23] or g:Profiler [24]. DAVID web tool (http://david.abcc.ncifcrf.gov/) allows the identification of enriched biological processes using Gene Ontology terms, visualizes genes on BioCarta and KEGG pathway maps, and provides links to related literature as PubMed IDs. Detailed tutorials with examples are available at the DAVID website.

g:Profiler (http://biit.cs.ut.ee/gprofiler/) is useful for analyzing the enrichment in Gene Ontology terms and KEGG and REACTOME (reaction-centered manually curated, and peer-reviewed pathway database, http://www.reactome.org/ReactomeGWT/entrypoint.html) pathways as well as for getting additional information about transcription factor-binding sites from TRANSFAC database, protein–protein interactions, and associated diseases.

Depending on the research question, it is possible to choose as input either all predicted targets of dysregulated miRNAs for an estimation of the overall effect of miRNAs or each miRNA target set separately to analyze the effect of each individual miRNA.

4 Notes

1. There are other methods to perform meta-analysis using prioritized available gene lists such as a vote-counting strategy applied by Guan et al. [25]. However, the majority of them are not good at dealing with outliers (data points that are very distant from the rest of the data and may indicate either measurement error or natural variation), noise generated by high-throughput methods, or technological errors. In contrast, the RRA method presented here provides significance scores to help estimate the statistical relevance of the results.
2. Although knowledge of command line interface would greatly help, fluency in programming language is not required to start using R. Extensive documentation including installation instructions for different platforms and manual is available at the R project website (http://www.r-project.org/).
3. As with wet lab experiments, it is highly recommended to keep notes of all performed operations and generated files, either in electronic or old-fashioned pen-and-paper format, in case that you need to modify some analysis parameters. Even before starting a project, design a system for file naming and storage (for example, project name and date of analysis) so that every piece of information is found easily.
4. The sequences and IDs of hairpins and mature miRNAs for current as well as for previous releases can be obtained from miRBase download section (http://www.mirbase.org/ftp.shtml).
5. Several technologies with different technical properties and different number of detectable miRNAs have been used for miRNA profiling. For example, microarrays used until 2005 were able to detect ~400 miRNAs, while massively parallel next-generation small-RNA sequencing approach is able to detect all 2,042 mature sequences present in the miRBase v.19 as well as new previously unannotated miRNAs. To take the difference between studies into account, the number of miRNAs detectable by each study should be used when constructing rank matrix. Therefore, the examination of Materials and Methods sections of each article is highly recommended. If a commercial array is used by the study, it may be necessary to examine the manufacturer's information. If a custom array is used, the number of detectable miRNAs or miRBase version used is usually reported in the Materials section. In addition, if a study is available in Gene Expression Omnibus (GEO, http://www.ncbi.nlm.nih.gov/geo/) or ArrayExpress (http://www.ebi.ac.uk/arrayexpress/) expression databases, the number of detectable miRNAs is obtainable from raw expression data.

6. Websites for pathway enrichment servers:

   DIANA-miRPath: http://diana.imis.athena-innovation.gr/DianaTools/index.php?r=mirpath/index

   miTalos: http://mips.helmholtz-muenchen.de/mitalos/index.jsp

   starBase: http://starbase.sysu.edu.cn/

   mirTar: http://mirtar.mbc.nctu.edu.tw/human/

   GeneTrail: http://genetrail.bioinf.uni-sb.de/

7. Websites for target prediction servers:

   TargetScan: http://www.targetscan.org/

   miRanda: http://www.microrna.org/microrna/home.do

   PITA: http://genie.weizmann.ac.il/pubs/mir07/mir07_prediction.html

   PicTar: http://pictar.mdc-berlin.de/

   DIANA microT: http://www.microrna.gr/webServer

   RNAHybrid: http://bibiserv.techfak.uni-bielefeld.de/rnahybrid/

8. Websites for experimentally verified miRNA target databases:

   miRTarBase: http://mirtarbase.mbc.nctu.edu.tw/

   miRWalk: http://www.umm.uni-heidelberg.de/apps/zmf/mirwalk/mirnatargetpub.html

   TarBase: http://diana.cslab.ece.ntua.gr/tarbase/

   starBase: http://starbase.sysu.edu.cn/

Acknowledgments

This research was financed by the University of Tartu (grant "Center of Translational Genomics"), the Estonian Government (grant #SF0180142s08), the European Commission through the European Regional Development Fund in the frame of grant "Centre of Excellence in Genomics," Estonian Research Infrastructures Roadmap, and FP7 grant #313010.

References

1. Esteller M (2011) Non-coding RNAs in human disease. Nat Rev Genet 12:861–874
2. Meyer SU, Kaiser S, Wagner C et al (2012) Profound effect of profiling platform and normalization strategy on detection of differentially expressed microRNAs—a comparative study. PLoS One 7:e38946
3. Morin RD, O'Connor MD, Griffith M et al (2008) Application of massively parallel sequencing to microRNA profiling and discovery in human embryonic stem cells. Genome Res 18:610–621
4. Ryu S, Joshi N, McDonnell K et al (2011) Discovery of novel human breast cancer

microRNAs from deep sequencing data by analysis of pri-microRNA secondary structures. PLoS One 6:e16403

5. Pradervand S, Weber J, Thomas J et al (2009) Impact of normalization on miRNA microarray expression profiling. RNA 15:493–501
6. Kolde R, Laur S, Adler P et al (2012) Robust rank aggregation for gene list integration and meta-analysis. Bioinformatics 28:573–580
7. Vosa U, Vooder T, Kolde R et al (2013) Meta-analysis of microRNA expression in lung cancer. Int J Cancer 132:2884–2893
8. Kozomara A, Griffiths-Jones S (2011) miRBase: integrating microRNA annotation and deep-sequencing data. Nucleic Acids Res 39: D152–D157
9. Ashburner M, Ball CA, Blake JA et al (2000) Gene ontology: tool for the unification of biology. The Gene Ontology Consortium. Nat Genet 25:25–29
10. Vlachos IS, Kostoulas N, Vergoulis T et al (2012) DIANA miRPath v. 2.0: investigating the combinatorial effect of microRNAs in pathways. Nucleic Acids Res 40:W498–W504
11. Kowarsch A, Preusse M, Marr C et al (2011) miTALOS: analyzing the tissue-specific regulation of signaling pathways by human and mouse microRNAs. RNA 17:809–819
12. Yang JH, Li JH, Shao P et al (2011) starBase: a database for exploring microRNA-mRNA interaction maps from Argonaute CLIP-Seq and Degradome-Seq data. Nucleic Acids Res 39:D202–D209
13. Hsu SD, Lin FM, Wu WY et al (2011) miRTarBase: a database curates experimentally validated microRNA-target interactions. Nucleic Acids Res 39:D163–D169
14. Backes C, Keller A, Kuentzer J et al (2007) GeneTrail—advanced gene set enrichment analysis. Nucleic Acids Res 35:W186–W192
15. Lewis BP, Burge CB, Bartel DP (2005) Conserved seed pairing, often flanked by adenosines, indicates that thousands of human genes are microRNA targets. Cell 120:15–20
16. John B, Enright AJ, Aravin A et al (2004) Human MicroRNA targets. PLoS Biol 2: e363
17. Kertesz M, Iovino N, Unnerstall U et al (2007) The role of site accessibility in microRNA target recognition. Nat Genet 39:1278–1284
18. Krek A, Grun D, Poy MN et al (2005) Combinatorial microRNA target predictions. Nat Genet 37:495–500
19. Kiriakidou M, Nelson PT, Kouranov A et al (2004) A combined computational-experimental approach predicts human microRNA targets. Genes Dev 18:1165–1178
20. Rehmsmeier M, Steffen P, Hochsmann M et al (2004) Fast and effective prediction of microRNA/target duplexes. RNA 10:1507–1517
21. Dweep H, Sticht C, Pandey P et al (2011) miRWalk—database: prediction of possible miRNA binding sites by "walking" the genes of three genomes. J Biomed Inform 44:839–847
22. Sethupathy P, Corda B, Hatzigeorgiou AG (2006) TarBase: a comprehensive database of experimentally supported animal microRNA targets. RNA 12:192–197
23. da Huang W, Sherman BT, Lempicki RA (2009) Systematic and integrative analysis of large gene lists using DAVID bioinformatics resources. Nat Protoc 4:44–57
24. Reimand J, Arak T, Vilo J (2011) g:Profiler—a web server for functional interpretation of gene lists (2011 update). Nucleic Acids Res 39:W307–W315
25. Guan P, Yin Z, Li X et al (2012) Meta-analysis of human lung cancer microRNA expression profiling studies comparing cancer tissues with normal tissues. J Exp Clin Cancer Res 31:54

Index

C

cDNA end amplification 19–25

D

Data analysis of small RNA sequencing 93–102
Dicing assay 219–226

E

Experimental validation of predicted mammalian microRNAs of mirtron origin 245–262

G

Gene expression profile by PCR arrays 321–358
Gene regulation 4, 35, 84, 195

I

Identification of protein-coding and non-coding transcripts based on sequence 203–207
Identification of translated mRNA 173–177
In situ hybridization of microRNA 57–70
Isolation of urinary exosomes 145–169

L

Lambda cDNA library 21, 23–24
Locked nucleic acids (LNA) 57–70, 74, 124, 160, 236, 238, 261
Long non-coding RNA (lncRNA) 47, 84, 203–206, 209–217
- *in silico* function predictions 36, 279, 289, 308

M

Mapping of internal ribosome entry sites (IRES) 179–184
Mapping of RNA protein binding 187–193
Mapping of transcribed sequences
- mRNA mapping 19–25
- RNA mapping 3–18
- S1 RNA mapping 27–34
- tertiary structure mapping of primary microRNAs 43–55

Meta-analysis of microRNA expression 361–372
MicroRNA
- biogenesis 47, 219–226, 246, 247, 255, 261
- copy number variants interaction 307–319
- paraffin-embedded sample 105–118
- profiling in plasma or serum using qPCR 121–128
- profiling of exosomes from biofluids and conditioned media 131–143
- quantification by qPCR 73–80
- quantification in body fluids and formalin-fixed 105–118
- regulatory networks 279–287
- target interaction analysis using miRWalk database 289–304
- target prediction and analysis using web-based applications 265–276
- target validation using a fast luciferase reporter assay 227–243

N

Next generation sequencing (NGS) 83–94, 361
Northern blot
- non-radioactive (digoxigenin-labelling) 5, 6, 11–17
- radioactive 4–5, 9–11

P

Purification of RNA binding proteins 195–201

R

Rapid Amplification of cDNA End (RACE) 19–23
Ribonucleic acid (RNA)
- footprinting 43–55
- isolation from acellular biofluids 83–92
- labeling 37, 47–48
- secondary structure analysis 39–42
- stability 4, 45, 179, 187, 188, 196
- synthesis 3, 4, 37, 38

M. Lucrecia Alvarez and Mahtab Nourbakhsh (eds.), *RNA Mapping: Methods and Protocols*, Methods in Molecular Biology, vol. 1182, DOI 10.1007/978-1-4939-1062-5, © Springer Science+Business Media New York 2014

MIX
Papier aus verantwortungsvollen Quellen
Paper from responsible sources
FSC® C105338

If you have any concerns about our products,
you can contact us on
ProductSafety@springernature.com

In case Publisher is established outside the EU,
the EU authorized representative is:
Springer Nature Customer Service Center GmbH
Europaplatz 3, 69115 Heidelberg, Germany

Printed by Libri Plureos GmbH
in Hamburg, Germany